BIOCHEMICAL ENGINEERING

BIOCHEMICAL ENGINEERING

Second Edition

By

Shuichi AIBA
University of Tokyo

Arthur E. HUMPHREY
University of Pennsylvania

Nancy F. MILLIS
University of Melbourne

ACADEMIC PRESS, Inc., 1973
New York & London

A SUBSIDIARY OF HARCOURT BRACE JOVANOVICH, PUBLISHERS

ACADEMIC PRESS, Inc.
1973
New York and London

BIOCHEMICAL ENGINEERING, SECOND EDITION

LIBRARY OF CONGRESS CATALOG CARD NUMBER: 73-12447

ISBN: 0-12-045052-6

Originally published by
UNIVERSITY OF TOKYO PRESS

PREFACE

Biochemical processes have been carried out by man since ancient times, but it was not until a little more than 100 years ago that Louis Pasteur pointed out the role that living organisms play in these processes. In the years that followed, an increasing number of commercially important chemicals were produced by the utilization of the activities of various microorganisms.

With the discovery of the usefulness of penicillin, man began to appreciate more fully the potential of microorganisms for useful purposes. Urgent demands for penicillin throughout World War II thrust microbiologists, biochemists, and chemical engineers into a "crash" program of developing and designing processes in areas which were in many ways unfamiliar to them. As a result, most early process know-how was acquired through empirical procedures.

Following the war, industrial fermentation was rapidly developed to an advanced state. Microorganisms are now used to produce a host of complex chemicals—antibiotics, enzymes, and vitamins—and to perform highly specific changes in complicated chemical molecules. The techniques of submerged fermentation are not limited to microorganisms. They are now also being used for propagation of mammalian tissue. With these developments there has been produced a wealth of new knowledge, much of which is scattered throughout the scientific literature.

A major objective in writing this book has been to gather together the information dealing with the industrial utilization of microorganisms. This has been done not for the purpose of a literature review, but rather to develop in a single presentation an engineering approach to the subject; hence the title "Biochemical Engineering."

To achieve this objective, the authors have borrowed heavily from chemical engineering science. The microbial process is treated as a complex chemical reaction involving biological catalysts, enzymes, provided by living matter. In this way, it is possible to view a microbial process in much the same way as the chemical engineer views a chemical process and to treat the associated physical operations as "unit operations." This approach affords a rational basis on which it is possible to analyze and integrate the scattered knowledge into unit processes. The authors believe that this approach will provide the industrial worker with a useful source book that will aid his interpretation of the knowledge he has at hand, and at the same time provide the biochemical engineering student with a logical scheme of approach to the subject.

This book was written with the assumption that the background of its readers would be quite varied. It begins with an introduction to biochemical engineering. Next, the characteristics and biochemical activities of microorganisms are reviewed for the engineering reader who may be seeking a greater appreciation of the biological catalysts that he will be dealing with later on in the book. The biochemist and microbiologist may want to begin with the next chapters dealing with kinetics and continuous fermentations. The chapters dealing with aeration, agitation, and scale-up are the heart of the subject. It is in problems of scale-up where the talents of the biochemical engineer are brought to a focus. Further, scale-up is the area of greatest controversy. Much of the practice in this area is still more of an art than a science. The final chapters deal with the auxiliary operations of microbial processes and product recovery. Material in this book has been developed from first principles of physics, chemistry, and engineering. Mathematical developments require no more than an understanding of calculus.

The information presented in this book was largely gathered together as a result of a course in biochemical engineering taught jointly by all 3 of the authors at the University of Tokyo in the spring of 1963. It has been derived from the authors' experience in the pharmaceutical, chemical, and food industries; from courses taught by the authors in biology and engineering at the University of Tokyo, University of Melbourne, and University of Pennsylvania; from various researches in this field; and from numerous articles in the scientific literature. Where it has seemed judicious to do so, studies of other investigators have been reinterpreted in the light of more extensive data. Where experimental data were lacking or where phenomena were little understood, the authors have indulged in some speculation.

The application of engineering principles to microbial processes is relatively new. Recognition of biochemical engineering dates back only to July, 1947, when the Merck Chemical Company received the McGraw-Hill Award for Chemical Engineering Achievement for its process development described in the article entitled "A Case Study in Biochemical Engineering." One of the first symposia on the subject was held at the American Chemical Society meeting in Atlantic City in September, 1949. Considerable knowledge has accumulated in the last 15 years. However, much is yet to be learned. The authors hope that this book will serve to stimulate further research of a fundamental nature in the subject. The authors believe that microbial processes will have a tremendous impact, both economic and social, on the world in the next 35 years. Biochemical engineering will play a key role in developing these microbial processes to the benefit of mankind. It has been this belief which has been the primary stimulus in the writing of this book.

Finally, we should like to acknowledge with gratitude the encouragement and cooperation we have received from our friends and colleagues during the preparation of this book.

SHUICHI AIBA

December 1964   ARTHUR E. HUMPHREY

NANCY F. MILLIS

PREFACE TO SECOND EDITION

In writing *Biochemical Engineering* the authors wished to provide a bridge between the disciplines of biology and chemical engineering in the fermentation industry. We hoped that the biologist would gain an appreciation of the problems of chemical engineering and that the chemical engineer would gain insight into the behavior of biological catalysts. In this edition, we have added summaries at the end of each chapter where the essential points will be emphasized.

The second edition, like the first, has deliberately excluded certain topics such as food processing and any discussion of the selection of materials used in equipment or details of the design of vessels, since we believe these matters are adequately treated elsewhere. The second edition has extensive revisions of the sections on biological mechanisms of control, kinetics, scale-up and the methods for the measurement and control of fermentation parameters, with special emphasis on the management and control of oxygen transfer. The new chapters deal with the recovery of cells, their disruption and fractionation, and with the extraction and immobilization of macromolecules.

Finally, the encouragement, cooperation and valuable suggestions we have received from Dr. I. Chibata (Chapter 14) and many other colleagues (especially for Chapters 4, 8, and 12 to 14) during the revision are cordially appreciated.

January 1973

SHUICHI AIBA
ARTHUR E. HUMPHREY
NANCY F. MILLIS

TABLE OF CONTENTS

Chapter 3.   DIRECTING THE CHEMICAL ACTIVITIES OF MICROORGANISMS   56

BIOCHEMICAL ENGINEERING

Chapter 1

Introduction

1.1. BIOCHEMICAL ENGINEERING—THE INTERACTION OF TWO DISCIPLINES

Biochemical engineering is concerned with conducting biological processes on an industrial scale, providing the link between biology and chemical engineering.

Chemical engineers are essentially concerned with the problems of designing and operating plants to allow material to react to form new products. These operations have been developed to a high degree of sophistication using powerful, nonbiological catalysts. Biochemical engineering deals with processes where the catalysts are either living cells or extracts from them. Such processes immediately confront the chemical engineer with an unfamiliar dimension. Chemical engineering training draws heavily on the basic sciences of chemistry, mathematics and physics but generally has not provided a background in biology. Biologists, on the other hand, are quite unfamiliar with engineering concepts. Yet a process involving a biological catalyst requires a detailed understanding of the behavior of biological material and the engineering expertise to exploit the potentialities of the biological system. It is unrealistic to suggest that a biochemical engineer should undertake a full basic training in both chemical engineering and biology, including microbiology, biochemistry and genetics. The authors believe, moreover, that the heart of biochemical engineering lies in the scale-up and management of cellular processes. For this reason, the properties of living cells are discussed in this text only in so far as they seem important in telling the engineer what environmental factors must be controlled in the fermentation vessel and in making him aware of the relatively dynamic nature of biological catalysts. Attention will be focussed on the engineering principles involved in the design and operation of fermentation equipment. Although biologists may not fully comprehend this treatment of design parameters, they should become aware of the limitations of engineering design. This, we hope, will stimulate biologists to investigate the possibility of modifying the genetic make-up or the physiological environment of the cells to assist in overcoming particular design limitations. On the other hand, engineers must realize that there are limits to the modifications which can be made in biological material, and be prepared to work with the biologist in achieving maximum productivity within these limitations.

1.2. EVOLUTION OF MODERN FERMENTATION PROCESSES

Man was well aware of fermentations, even though he had little knowledge of what caused them, long before he was able to record such an awareness. The cave man discovered that meat allowed to stand a few days was more pleasing to the taste than meat eaten soon after the kill. He also was aware that intoxicating drinks could be made from grains and fruits. The aging of meat and the manufacture of alcoholic beverages were man's first uses of fermentation. In those early days man considered fermentation as some sort of mystical process. He did not know that he was profiting from the activity of invisible microorganisms.

Without even knowing that these microorganisms existed, ancient man learned to put them to work. The ancient art of cheese-making involves the fermentation of milk or cream. For thousands of years, the soy sauces of China and Japan have been made from fermented beans. For centuries, the Balkan peoples have enjoyed fermented milk, or yogurt, and Central Asian tribesmen have found equal pleasure in sour camel's milk, or kumiss. Bread, which has been known almost as long as agriculture itself, involves a yeast fermentation. Loaves of bread have been found in Egyptian pyramids built 6,000 years ago.

The discovery of fruit fermentation was made so long ago that the ancient Greeks believed wine had been invented by the god Dionysus. The manufacture of beer is only slightly less ancient than that of wine. A Mesopotamian clay tablet written in Sumerian-Akkadian about 500 years before Christ tells us that brewing was a well established profession 1,500 years earlier. An Assyrian tablet of 2000 B.C. lists beer among the commodities that Noah took aboard his ark. Egyptian documents dating back to the Fourth Dynasty, about 2500 B.C., describe the malting of barley and the fermentation of beer. Kui, a Chinese rice beer, has been traced back to 2300 B.C. When Columbus landed in America, he found that the Indians drank a beer made from corn. More than 3,000 years ago, the Chinese used moldy soybean curd to clear up skin infections, and primitive Central American Indians used fungi to treat infected wounds.

During the Middle Ages, experimenters learned how to improve the taste of wine, bread, beer, and cheese. Yet, after thousands of years of experience, men still did not realize that in fermentations he was dealing with a form of life. However, in 1857, Pasteur proved that alcoholic fermentation was brought about by yeasts and that yeasts were living cells. In addition, Pasteur showed that certain diseases were caused by microorganisms. This discovery was a turning point in medical history and the birth of microbiology.

Other observations of Pasteur's indicated that certain disease-producing microorganisms survived for only a few hours when introduced into soil. He concluded from this that certain microbes were killed by others in the soil. He also found that the bacillus which causes anthrax in cattle could thrive in the tissues of cattle but appeared to be inhibited by the presence of certain airborne microbes. This prompted Pasteur to suggest that human disease could be cured by marshaling microbe against microbe.

To avoid the risk inherent in fighting one disease with another, medical workers looked for a chemical agent elaborated from a microbe innocuous to man which would be able to destroy disease-causing microorganisms. In 1901, Rudolf Emmerich and Oscar Low, of the University of Munich, isolated "pyocyanase" from *Pseudomonas aeruginosa*, a bacterium. Several hundred patients were treated quite successfully with pyocyanase, the world's first antibiotic. But pyocyanase was ahead of its time. No techniques existed to guarantee that each batch of the substance would be equally effective. The quality controls now common to pharmaceutical manufacturing were more than 40 years in the future. Standardization was impossible, and pyocyanase was abandoned as too hazardous.

During this time, the inheritors of Pasteur's knowledge encountered better luck when they moved outside the field of medicine and sought to use microbes as production workers in industry. The production of baker's yeast in deep, aerated tanks was developed towards the end of the nineteenth century and in the early twentieth century. During World War I, Chaim Weismann almost singlehandedly rescued Britain from a serious ammunition shortage. He did it by using a bacterial cousin of the gas gangrene microbe to convert maize mash into acetone, which is essential in the manufacture of the explosive cordite. In 1923, Pfizer opened the world's first successful plant for citric acid fermentation. The process involved a fermentation utilizing the mold *Aspergillus niger* whereby ordinary sugar was transformed into citric acid.

Other industrial chemicals produced by fermentation were found subsequently, and the processes reduced to commercial practice. These included butanol, acetic acid, oxalic acid, gluconic acid, fumaric acid, and many more.

Practically nothing was done with antibiotics until 1928. It was in this year that Alexander Fleming, working with *Staphylococcus aureus*, a bacterium that causes boils, observed a strange fact. A mold of the *Penicillium* family grew as a contaminant on a Petri dish inoculated with *Staphylococcus aureus*; a clear zone was observed where the *Staphylococcus* organisms in the vicinity of the contaminating mold had been killed. Fleming nurtured the mold and then extracted a chemical from it which killed the bacteria. He named the extracted material penicillin and used most of his meager supply to clear up one infected wound.

Fleming's discovery received little notice as far as application was concerned until two Oxford University experimenters, under the stress of World War II, resolved to find an antibacterial agent of wider activity than the sulfa drugs. These two British workers, Dr. Howard Florey and Dr. Ernst Chain, were sure that earth or air could offer a yeast, mold, or fungus which, under the proper conditions, could be made to produce a therapeutic agent capable of saving the lives of war casualties. Their first candidate was the *Penicillium notatum* mold preserved from Fleming's studies. Penicillin turned out to be exactly what they were looking for; it could save thousands of lives and was needed immediately. Since all of Britain's production facilities were devoted to war work, Flory and Chain turned to the American pharmaceutical industry to help them solve their difficulties in mass-producing the antibiotic. Three American companies led the way—Merck, Pfizer, and Squibb—with the help of government laboratories.

Initially, the cultures were grown in flasks about the size of milk bottles. It was soon realized that factories larger in capacity than all the milk-bottling plants in the United States would be needed. A chance discovery in a Peoria market provided the major breakthrough. Here a government worker found a moldy cantaloupe on which was growing a new strain of penicillium, *Penicillium chrysogenum*, which would thrive when cultured in deep, aerated tanks and which gave 200 times more penicillin than did Fleming's mold.

Other antibiotics were quick to appear. From the throat of a chicken Professor Selman A. Waksman of Rutgers University isolated an actinomycete, *Streptomyces griseus*, which elaborated a new antibiotic, streptomycin. This antibiotic was particularly effective against the causative organism of tuberculosis. The search was now on. Antibiotic prospectors combed the earth for organisms that produced different and more useful antibiotics. The list of these antibiotics is long today and includes such important antibiotics as chloramphenicol, the tetracyclines, bacitracin, erythromycin, novobiocin, nystatin, kanamycin, and many others.

Progress in fermentation is continuing at an ever-increasing pace. Each year new products are added to the list of compounds derived from fermentation. Several vitamins are now produced routinely employing fermentation steps in their synthesis. Outstanding examples are B-2 (riboflavin), B-12 (cyanocobalamin), and C (ascorbic acid). Some of the more interesting fermentation processes are the specific dehydrogenations and hydroxylations of the steroid nucleus. These chemical transformations are economical short cuts used in the manufacture of the antiarthritic cortisone and its derivatives. Fermentative syntheses of the amino acids L-lysine and L-glutamic acid are also being carried out commercially. The fermentive production of nucleic acids is proving to be an important source of flavor-enhancing compounds. Important agricultural uses are being found for the new fermentation product gibberellin, a plant-growth regulator; and crystalline inclusions from a species of *Bacillus* are being used as specific insecticides in another agricultural application. Microbial attack on crude oil promises to be an important source of feed materials as well as certain highly oxidized aromatic compounds for chemical synthesis. Research is in progress on chemical transformations utilizing fermentation techniques, new fermentative biosyntheses, continuous algal culture, and submerged mammalian-tissue culture. Fermentation processes may not only be tomorrow's source of chemotherapeutic agents, but may very well be the manner in which food is produced. Many scientists have predicted that hydroponics—the submerged culture of plant cells—is the farming of the future.

1.3. ROLE OF THE BIOCHEMICAL ENGINEER IN THE DEVELOPMENT OF MODERN FERMENTATION PROCESSES

Prior to the penicillin fermentation, pure-culture requirements of fermentation processes were not strictly controlled. Earlier processes such as those used in producing yeast, citric acid, and gluconic acid were favored by pH conditions unsuitable for contaminating organisms. In other processes, such as the production of sorbose, acetone and butanol and ethanol, the concentrations of ingredients and products were sufficiently high to suppress the growth of most contaminants. The latter two fermentations are also anaerobic. The problem facing the engineer in the development of the penicillin fermentation was the design and operation of an absolutely-pure-culture fermentation in deep, aerated fermentation vessels containing an ideal environment for the growth of contaminating organisms.

The engineer thus encountered a formidable problem—the prevention of contamination. Many fermentation plants were delayed in reaching their maximum productive capacities because of contamination problems. Perhaps the most notable contribution the engineer made was in the advancement of sterile techniques, or the "contaminant-proof" philosophy, in the design and operation of the fermentation vessel and its associated maze of piping. Firstly, sterility of the equipment and the fermentation medium had to be achieved. Next, inoculum had to be passed into the fermentor without contaminating it. The process then had to be maintained in a pure state by preventing the entry of contaminating organisms during the fermentation.

To accomplish these objectives, the engineer changed methods of vessel fabrication, revised gasketing, piping, and valve design, and devised new methods of steam-sealing possible points of contaminant entry. Methods for removing samples and adding materials aseptically to the fermentation vessel were also devised. Methods for sterilizing equipment, medium, and the large quantities of compressed air required during the fermentation were developed.

In addition to designing an aseptic fermentation operation, the engineer also had to design air-compression and delivery systems and efficient methods for agitating and aerating the fermentation. Heat evolved by the metabolic reactions of the microorganism had to be removed, and methods of maintaining the temperature within a narrow range were developed. Other instrumentation and process-control problems were also encountered and solved.

The ultimate success of any chemical process depends on its successful demonstration in pilot-plant equipment and its subsequent scale-up to the production stage. The engineer has made important contributions to this facet of fermentation technology.

The process is not complete until the product is finally isolated. Numerous problems hitherto not encountered faced the engineer in product recovery. Early commercial fermentations had only been used to produce relatively simple, stable chemicals in such high concentrations that their isolation was an easy matter. Concentrations of penicillin in early fermentation media, on the other hand, were extremely low. Also penicillin could very easily be degraded to inactive material. Special isolation techniques involving filtration, extraction, adsorption, and concentration were developed to recover these small quantities in remarkably good yields. In some instances the product was contained in the cells and had to be extracted.

While the laboratory scientist—the microbiologist, the biochemist, and the microbial geneticist—continues to discover and advance desirable interactions between microorganisms and their environment, the biochemical engineer must control these interactions and translate laboratory results to production-scale operation in an economic manner. The biochemical engineer, therefore, must continue to develop, design, and scale-up new fermentation processes. Improvements in management, as well as in equipment, are needed to accomplish these aims. In addition, he must continue to operate his fermentation plant safely and efficiently, and he must see that his products meet the requirements of his customers and the standards set by his industry. Very likely, as in the past, complete engineering knowledge of a process with which he may be concerned will be unavailable at the time he needs it. Therefore, he must be prepared to offer engineering experience and judgment when needed. These latter qualities have been largely responsible for the important concepts and methods of biochemical engineering presented in this book.

1.4. Status of Biochemical Engineering in the Fermentation Industry

The engineer's contribution to the development of the penicillin fermentation has already been described as a very important one. The outgrowth of the undertaking was the pure-culture technique, carried out in aerated and agitated deep-tank fermentors. This technique, similar to its antecedent used for yeast propagation, introduced to the biochemical process industry refined fermentation equipment capable of being maintained under aseptic conditions even when vigorously aerated. The technique has now been applied widely with minor modifications to the production of other antibiotics, amino acids, steroids and enzymes.

1.4.1. Unit operations

Fermentation processes have in common many of the familiar chemical engineering unit operations. For example, aerobic fermentations involve the "mixing" of three heterogeneous phases—microorganisms, medium and air. Other unit operations include "mass transfer" of oxygen from the air to the organisms and "heat transfer" from the fermentation medium.

Analysis of fermentations by the unit-operation technique has added greatly to the understanding of their behavior. This understanding, however, is far from complete. The scale-up of fermentations for instance, is still rather empirical although the sensitive oxygen probes now available have enabled a more rational approach to scaling-up aerated, non-Newtonian fermentations.

Of the operations auxiliary to those in the fermentor, biochemical engineers have made a major contribution in establishing the theoretical bases for the design of equipment to provide large volumes of sterile medium and air. Close cooperation between biologists and engineers is needed now to devise logical methods for screening large numbers of strains, and translating the results of shaken-flask and pilot-plant experiments to production vessels.

1.4.2. Unit processes

A careful analysis of the many industrially significant fermentation processes shows that there are common reactions from a chemical as well as a physical viewpoint. Fermentation processes can be classified by the reaction mechanisms involved in converting the raw materials into products; these include reductions, simple and complex oxidations, substrate conversions, transformations, hydrolyses, polymerizations, complex biosyntheses and the formation of cells.

Unit-process classification provides a ready catalogue of the chemical activities and abilities of microorganisms for the biochemist. More importantly, for the biochemical engineer it offers a logical approach to an examination of fermentation reaction mechanisms.

1.4.3. Process design

A fermentation may be viewed as a catalyzed chemical reaction in which enzymes are the catalysts and cellular material the catalyst support. Therefore, design for fermentation processes requires an understanding of stoichiometry and reaction kinetics. In design for a batch fermentation, a consideration of the kinetics of the process often has not been necessary. However, as continuous fermentation has developed, reaction kinetics have become increasingly important. The biochemical engineer is now making important contributions to the design of practical continuous fermentation systems, based upon chemical reaction kinetics.

Successful process design depends, of course, on successful process control. Although certain process variables such as temperature, air flow and agitation were controlled in even the earliest deep-tank fermentations, other variables remained unmonitored largely because of the difficulties of obtaining sensitive measurements while maintaining aseptic conditions. Sterilizable and stable pH probes are now standard equipment and, recently, sterilizable probes for dissolved oxygen have been developed. The measurement of carbon dioxide, redox potential, specific ions, and glucose levels is currently under active investigation. Successful development of control devices for these and other variables will permit greater economy in substrate utilization, power and aeration.

Competition among different fermentation products demands increases in productivity and reductions in the costs of biological processes, and the returns from the research needed to develop new processes are being critically assessed. Advanced instrumentation (coupled to computers for data logging and analysis and linked to process controls) is now being applied to fermentations. In the fermentation industry, the diversity of operations, the ever-changing nature of fermentation products, their value, and the complexity of equipment, make costs an especially important consideration; equipment must be versatile as well as efficient and these are difficult aims to achieve simultaneously.

1.5. FUTURE DEVELOPMENTS

The products of the earliest fermentations were small molecules like ethanol, butanol, acetone, lactic, citric, gluconic, acetic, itaconic and fumaric acids. The period immediately after World War II saw a complete change in the fermentation industry; antibiotics became the major source of income, dominated by the penicillins and streptomycin. The search for new antibiotics was, and to some extent still is, a preoccupation of research and development departments and a wide array of antibiotics has been marketed. Table 1.1 lists some common antibiotics. The realization that microorganisms could be induced to form large quantities of complex molecules led to the development of processes for accumulating enzymes; some important enzymes in use at present are shown in Table 1.2. The discovery of strains of mold capable of specific modifications of the steroid nucleus opened a new area, and about this time, mutants of molds and bacteria were recognized having faulty regulation of biosynthesis; such mutants have now been exploited

TABLE 1.1

SOME IMPORTANT ANTIBIOTICS MADE BY FERMENTATION.

Amphotericin	Kanamycin	Rifamycins
Bacitracin	Natamycin	Streptomycin
Cephalosporins	Neomycin	Tetracyclines
Choloramphenicol	Novobiocin	Trichomycin
Cycloheximide	Nystatin	Tyrocidin
Erythromycin	Oleandomycin	Tyrothricin
Gentamycin	Paromomycin	Tylosin
Gramicidin	Penicillins	Vancomycin
Griseofulvin	Polymyxin	Viomycin

TABLE 1.2

ENZYMES OF INDUSTRIAL OR MEDICAL IMPORTANCE.

Amino acid acylase	Fructose-glucose isomerase
Amino acid transaminase	Invertase
Amylases	Lipases
Asparaginase	Lysozyme
Catalase	Pectinases
Cellulase	Penicillin acylase
Collagenase	Penicillinase
Dehydrogenases	Proteases
Dextranase	Rennin of bacterial origin
Glucoamylase	Streptokinase-streptodornase
Glucose oxidase	
Glutamic acid decarboxylase	
Hemicellulase	

commercially to accumulate abnormal amounts of intermediates of biosynthesis such as amino acids (glutamic acid, lysine, leucine, valine, etc.), and nucleotides (inosinic acid, xanthine monophosphate, guanosine monophosphate).

The application of new biochemical and chemical engineering knowledge can be expected to yield substantial returns in the future, both in terms of new products and of improved efficiency.

1.5.1. Exploitation of biological properties

1.5.1.1. *New sources of carbon for fermentation*
The search for new sources of food to satisfy the expanding needs of the world's population has directed attention to microorganisms as a potential source of protein. Bacteria and yeast grown on alkanes show promise. By selecting appropriate strains of yeast and controlling the environment it is possible not only to use hydrocarbons to grow yeast, but to manipulate their composition. Hydrocarbon processes thus offer a source of protein whose amino acid profile can be tailored to supplement diets deficient in particular amino acids. The French have used a mixed yeast and bacteria fermentation to produce yeast and yeast protein from the *n*-alkanes present in gas oil. This has an economic bonus, as dewaxed gas oil remaining after fermentation is more suited to further refining. In Japan and Scotland, processes have been developed using a relatively pure *n*-alkane fraction. In the alkane processes, virtually the entire substrate is usable and this greatly assists in harvesting and purification. Hydrocarbons provide a very cheap carbon substrate for the formation of metabolic products such as nucleotides and amino acids and this approach is being explored at present in Japan.

From the point of view of biochemical engineering, hydrocarbon fermentations present problems. The very reduced nature of the substrate increases the demand for oxygen—a requirement that has been estimated to be between 2.5 and 3 times greater than for the utilization of an equivalent amount of carbon in the form of carbohydrate, and growth is essentially limited to the oil/water interface. In addition, large amounts of heat must be removed from the fermentation and the very low solubility of the substrate presents a serious mass transfer problem. There is considerable interest in the possiblity of using methane or its oxidation product, methanol, to produce protein. Pure cultures of yeast cannot grow on methane, although they will grow on this substrate in association with bacteria. They will also grow on methanol in pure culture. *Methanococcus* and *Methanomonas* can oxidize both methane and methanol in pure culture and all of these organisms are being studied actively to determine the cost of protein produced in this way. Methane and methanol are attractive as substrates because of the complete absence of harmful residual substrate and the ease with which the cells can be recovered. Ethanol and acetate have been successfully used for the production of glutamate and citric acid by fermentation, and since such substrates can be made chemically, they appear to have potential in those countries which have a shortage of carbohydrate.

1.5.1.2. *Cooxidation*
In studies of substrate utilization by organisms, it has long been recognized that organisms can oxidize substrates which will not support their growth. It is thus possible to obtain partial oxidation of one substrate by cells growing at the expense of some other substrate. Such cooxidation reactions seem a profitable approach to converting relatively useless branched, cyclic or polycyclic compounds into more useful ones, for example, the formation of dimethyl-*cis*, *cis*-muconic acid from *p*-xylene.

1.5.1.3. *Production of enzymes*
One of the most exciting advances in biochemical knowledge over the past few years has been the greatly increased understanding of the regulation of protein synthesis and the control of branched metabolic pathways. Efficient processes to accumulate amino acids and nucleotides have been developed but there has been relatively little exploitation of the possibility of derepressing control of enzyme synthesis so that cells accumulate abnormally large amounts of particular enzymes. The biochemical literature cites examples where as much as 5% of cell dry weight consists of one enzyme. The recent work with extranuclear plasmids also offers a novel approach to increasing the synthesis of particular enzymes. For example, the content of β-galactosidase can be increased 15-fold in *Escherichia coli* strains containing many copies of the λ-gal episome.

With cell-free preparations of enzymes attached to solid supports and with improved stability, many highly specific reactions may become economic where previously they were unattractive because of high cost and low efficiency in recovering the enzyme. Apart from their use in detergent preparations and for hydrolytic reactions (cellulases, proteases, amylases, penicillin acylase, dextrinases, pectinases, catalase, invertase), the specificity of enzyme reactions makes them ideal analytical tools. Enzymes are sensitive, the rate of reaction is fast and specificity may extend to one particular optical isomer of the substrate. Biochemistry and medicine will find increasing use for enzymes; already glucose oxidase, urease, uricase and lactic dehydrogenase are routine diagnostic reagents. Hyaluronidase, streptodornase and proteolytic enzymes are examples of enzymes used therapeutically, and improving knowledge of the bases of physiological disorders suggests that it may be possible in the future to control some abnormal accumulations or supplement a deficient activity. For example, L-asparaginase seems promising for the treatment of neoplasia. There are clearly potential problems here since such enzymes are proteins foreign to the host.

1.5.1.4. *Polymers*
Many microbial cells excrete a little slime (often polysaccharide) outside the cell wall, but in some species, this can impart high viscosity to the culture. The glucan extracted from *Leuconostoc* cultures has been used as a plasma substitute for years and other polymers with interesting properties continue to be isolated. For example, recently, mutant strains of *Alcaligenes* and *Agrobacterium* were shown to produce β-1,3-glucan which forms thermostable gels useful in food and confectionery. Research might be expected to continue to uncover such polymers in the future.

1.5.2. Application of engineering advances

1.5.2.1. *Methods of cultivation*
For over 70 years, waste water has been treated in activated sludge plants which operate on a continuous system with recycle; for the past 20 years, it has been practical to operate pure culture fermentation with very precise controls, yet very few processes at present use this technique on an industrial scale. In research, continuous culture has provided a most powerful tool for studying the response of populations to their environment and undoubtedly will continue to yield valuable knowledge to guide the management of batch processes. Although industry has been wary of adopting continuous culture for production, there has been little appreciation of the variety of equipment which can be incorporated into continuous systems, especially of the possibility of using stirred reactors in conjunction with plug-flow reactors either as a column, a simple pipe, an Archimedes screw or with dialysis membranes. Until more real imagination and attention is given to exploiting the many possible forms of continuous culture, it seems premature to dismiss this method of cultivation as "not applicable" to the fermentation industry.

Recently, the conventional batch process has been modified to extend the period during which secondary metabolites are formed. This has been achieved by with-drawing portions of the culture during the production phase and carefully controlling the rate of regrowth to maintain maximum rates of production. This operation can be repeated 5 or 6 times with considerable economies in production.

1.5.2.2. *Automated systems and computer control*

There is probably no sphere of engineering technology which is advancing faster than systems analysis and control using computers. The fermentation industry has been slow to see the possibilities here. The lack of key analytical probes has limited the opportunities for using computers to analyze data and automatically adjust physical and chemical environmental factors including the critical nutrients and intracellular metabolites. Clearly, the usefulness of computers can only be as great as the reliability of the analytical probes providing the data and the completeness of the understanding of the biochemistry of the process, and even though there are deficiencies in both these areas, we can expect computers to play an increasingly important part in process control in the future.

1.5.2.3. *Air-lift fermentors*

The ability to aerate cultures is at the heart of modern fermentation processes and novel approaches to solving the difficulties associated with proper aeration always excite interest. Recently, data have appeared on the performance of pilot plant, air-lift fermentors. These reactors contain an open-ended cylinder suspended inside a fermentor with a sparger situated under the cylinder such that air forced through the sparger rises as an air/medium dispersion outside the cylinder. At the top, the air tends to escape and the heavier medium falls to be mixed with air and recirculated. It is claimed that the cost of aeration with this system is less than that with conventional systems and this fermentor appears to have special application for hydrocarbon substrates where the substrate itself is immiscible with water. It is doubtful, however, whether this system would be suitable for the more viscous mold fermentations.

1.5.2.4. *Membrane systems*

For many years, biologists have used membranes for dialysis and filtration. In dialysis, the concentration of solute provides the driving force, while with Millipore and ultra-filtration membranes, particles are excluded by the size of the pores in the membrane. The various Millipore membranes range from 0.2 to 0.02μ in pore diameter and are useful in retaining relatively large particles. Recently, cellulose acetate membranes with very small, uniform pores (10 to 20 Å) have been developed. These ultra-filters allow molecules of less than a certain molecular weight (M.W.) to pass, while excluding those of greater M.W. With a series of membranes of decreasing pore size, a mixture of substances of different MW can be separated provided the differences in M.W. are of the order of 5 to 10-fold. This method of concentration is very attractive for handling labile substances like enzymes since it can be carried out at low temperature with constant pH and ionic strength without change of phase and with relatively low power requirements.

Recently, technical advances have made it possible to construct membranes of such size and strength that ultra-filtration is a practical method of recovering products on an industrial scale. Ultra-filtration not only concentrates products, but solutes of small molecular weight pass through the membrane and thus a dialysis step, normally required with other methods of concentration, is avoided. There are problems associated with the use of these membranes; they are costly, have a finite life and have a tendency to clog either with material from the solution or with microbial growth.

While water normally passes across a membrane from low to high solute concentration, if sufficient pressure is applied to the concentrated solute, water can be forced in the opposite direction—this is reverse osmosis. With reverse osmosis, pressures of 200–2,000 psi (14–140 Kg/cm^2) are required in contrast to pressures of 10–100 psi (0.7–7 Kg/cm^2) for ultra-filtration. Reverse osmosis has wide application in the control of pollution. For example, a system for treating whey has been designed where casein is first removed by ultra-filtration and the permeate treated by reverse osmosis to concentrate lactose, acids, and salts. The biochemical oxygen demand (BOD) of the treated whey is thus reduced about 100-fold, and in addition, high quality protein and available carbohydrate fractions are recovered. Ultra-filtration and reverse osmosis offer the possibility of concentrating toxic substances in effluents from particular sections of a plant without the expense of treating the effluent from the entire plant.

1.5.2.5. *Immobilization of enzymes*
Most of the soluble hydrolytic enzymes in common use in industry are formed extracellularly; the wide range of intracellular microbial enzymes are virtually unexplored commercially. In order to exploit these enzymes, it is necessary to develop economic methods of purification. Reports of continuous methods for harvesting, breaking cells and fractionating their protein are major advances towards this end. Recently, enzymes have been immobilized by adsorption, encapsulation or inclusion in gels. They may also be covalently bound with a bifunctional linking agent to an insoluble polymer, covalently cross-linked with themselves or bound directly in an enzyme-polymer complex. The preparation can be packed into a column or complexed onto porous sheets and the substrate can then react with the enzyme in a batch or continuous process. Enzymes can also be contained within an ultra-filtration membrane. For instance, starch can be fed continuously into a membrane structure containing α-amylase and the products of hydrolysis collected outside the membrane. Whole cells or crude extracts can also be placed in capsules that allow substrate and products to diffuse to the enzymes.

Some properties of the enzyme, such as the pH optimum, Michaelis constant or stability may change when they are immobilized but the economies achieved by retaining the enzyme for re-use and obtaining products free of contaminating enzyme greatly outweigh any loss of stability or reduction in the rate of reaction. In many cases, indeed, the life of the immobilized enzyme is greater than that of the soluble enzyme. If substrates or products are sensitive to pH, it may be possible to select a polymer which changes the pH optimum of the immobilized enzyme to a range more favorable to the stability of the reactants.

1.6. POLLUTION CONTROL AND BIOCHEMICAL ENGINEERING

It is widely accepted in principle that industry must take responsibility for the wastes it produces, but in practice, wastes are frequently dumped into a convenient municipal system or discharged with minimum treatment (or none at all) into a nearby body of water; industry tends to adopt the cheapest expedient the law and the public will tolerate.

In the fermentation industry, the treatment of wastes should be regarded as an integral part of plant design. It is illogical and uneconomic to dilute a waste with a high available BOD to a point where the substrate concentration falls to levels that limit microbial growth, yet undoubtedly this occurs when fermentation wastes are put into a general sewage system. Such wastes should, ideally, be treated when the high BOD allows dense cell populations and rapid rates of growth. Since the effluent from a single factory is far more uniform than that reaching a sewage plant, the microbial population established on this effluent is likely to be more efficient and more readily controlled than that in a general sewage plant. Effluent treatment under these circumstances is just as amenable to optimization as any other fermentation. If basic biochemical engineering principles are applied to the treatment of wastes before they are diluted in a general waste water system, it would be expected that costs would be considerably less than those calculated from the cost of BOD removal. The treatment of wastes with high BOD may present new problems, especially with wastes developing high concentrations of ammonia, as inhibition of nitrification can occur and with wastes having very high C/N ratios, it may be necessary to supply extra nitrogen.

After treatment, wastes originally high in BOD may still constitute a problem; it may be necessary to remove inorganic nutrients to prevent excessive algal growth. Anaerobic treatment of the final effluent may be useful to decrease nitrogen by denitrification. Ions can be removed by adsorption on resins or by reverse osmosis but both of these treatments are somewhat expensive if the concentrate has no commercial value. Adsorption on charcoal is a possibility but, at the moment, this problem awaits an economic solution.

On the biological side, there is a serious lack of fundamental knowledge of the physiological behavior and of the flocculating properties of mixed microbial populations in sewage treatment plants. Design and control of plants must therefore be largely empirical and the effects of changes in the composition of the waste or the presence of particular toxic materials cannot be predicted. In trickling filters and activated sludge plants, the clarity of the effluent is greatly influenced by the protozoal population, yet we know little about the way in which protozoa detect their prey and what substances affect the efficiency of predation. This information is urgently needed.

Rivers, lakes, estuaries and beaches are properly regarded as community assets to be used for pleasure and relaxation as well as affording a final resting place for liquid effluents. Whether the cycle of changes in the various nutrients and in the microbial, protozoal and larger animal life is disturbed by the discharge of effluents is a matter of great concern. Biochemical engineering knowledge can make a contribution to the management of such large bodies of water by attempting to analyze the relationship between nutrients and the growth they support and between prey and predator along the food chain. From such analyses, it may be possible to predict when instabilities are likely to be induced in the system and to indicate what conditions lead to stability. This type of information should help to ensure that natural assets serve both recreational and industrial purposes.

Chapter 2

The Characteristics of Biological Material

2.1. TYPES OF MICROORGANISMS

Microorganisms are chemically very similar to higher animal cells, and they can perform many of the same biochemical reactions. Generally, microorganisms exist as single cells, or at most in relatively unspecialized multicellular colonies, with no capacity to control cellular temperature. While mammalian cells need to be supplied with substances like the B-group vitamins and aromatic amino acids from the diet, many microorganisms can make their requirements for growth from inorganic salts alone or from inorganic salts supplemented by a simple hexose.

The purpose of this chapter is to provide a brief introduction to the microbes which the biochemical engineer may use either to produce cell mass or some by-product of cell metabolism. Emphasis will be placed on those characteristics and properties that are significant in attaining these ends. For this discussion micro-organisms will be considered in four main groups:

1. Bacteria
2. Viruses including bacteriophages
3. Fungi including yeasts and Actinomycetes
4. Protozoa including algae

2.1.1. Bacteria

Typically, bacteria are single cells, either cocci, rods, or spirals, which are capable of independent growth. Coccal forms vary from 0.5 to 4μ in diameter, rods are from 0.5 to 20μ long and 0.5 to 4μ wide, and spirals may be greater than 10μ long and about 0.5μ wide.

Table 2.1 lists the various parts of bacteria that can be seen in electron micrographs, with some comments on their composition and functions; Fig. 2.1 shows diagramatically a transverse section of a bacterium.

Bacteria are ubiquitous in nature, in aerobic and anaerobic environments containing water. Among the genera, synthetic abilities range from those of the

TABLE 2.1

COMPOSITION OF THE VARIOUS PARTS OF BACTERIA.

Part	Size	Composition and Comments
Slime layer		
i) Microcapsule	5–10 mμ	Protein-polysaccharide-lipid complex responsible for the specific antigens of enteric bacteria and of other species.
ii) Capsule	0.5–2.0μ	Mainly polysaccharides (e.g., *Streptococcus*) sometimes polypeptides (e.g., *Bacillus anthracis*)
iii) Slime	Indefinite	Mainly polysaccharides (e.g., *Leuconostoc*) sometimes polypeptides (e.g., *Bacillus subtilis*)
Cell wall		Confers shape and rigidity to the cell.
i) Gram-positive species	10–20 mμ	20% dry weight of the cell. Consists mainly of macro-molecules of a mixed polymer of *N*-acetyl muramic-peptide, teichoic acids and polysaccharides.
ii) Gram-negative species	10–20 mμ	Consists mostly of a protein-polysaccharide-lipid complex with a small amount of the muramic polymer.
Cell membrane	5–10 mμ	Semi-permeable barrier to nutrients. 5–10% dry weight of the cell, consisting of protein 50%, lipid 28%, and carbohydrate 15–20% in a double-layered membrane.
Flagellum	10–20 mμ by 4–12μ	Protein of the myosin-keratin-fibrinogen class, M.W. 40,000. Arises from the cell membrane and is responsible for motility.
Pilus (fimbria)	5–10 mμ by 0.5–2.0μ	Rigid projections from the cell, protein. Especially long ones are formed by *Escherichia coli*.
Inclusions		
i) Spore	1.0–1.5μ by 1.6–2.0μ	One spore is formed per cell intracellularly. Spores show great resistance to heat, dryness, and antibacterial agents. Spore walls are rich in dipicolinic acid.
ii) Storage granules	0.5–2.0μ	Glycogen-like, sulphur, or lipid granules may be found in some species.
iii) Chromatophores	50–100 mμ	Organelles in photosynthetic species. *Rhodospirillum rubrum* contains about 6,000 per cell.
iv) Ribosomes	10–30 mμ	Organelles for synthesis of protein. About 1,000 ribosomes per cell. They contain 63% RNA and 37% protein.
v) Volutin	0.5–1.0μ	Inorganic polymetaphosphates which stain metachromatically.
Nuclear material	About half cell volume	Composed of DNA which functions genetically as if the genes were arranged linearly on a single endless chromosome but which appears by light microscopy as irregular patches with no nuclear membrane or distinguishable chromosomes. Autoradiography confirms the linear arrangement of DNA and suggests a M.W. of at least $1,000 \times 10^6$.

TABLE 2.2

PROPERTIES OF TYPICAL VIRUSES.

Type of Virus	Size (mμ)	Shape, Composition and Comments
Animal		
i) Cubic Symmetry		
Poliomyelitis	30	Consists of 1 molecule RNA (M.W. 2×10^6) in a spiral, surrounded by protein macromolecules 6 mμ diam. arranged as an icosahedron with no retaining membrane. Particle M.W. 10×10^6.
ii) Helical symmetry		
Influenza	100	Consists of 1 molecule RNA (M.W. 2×10^6) as a nucleo-protein macromolecule arranged in a helix, the whole coiled and enclosed in a lipo-protein sheath. Particle M.W. 100×10^6.
Plant		
i) Rods		
Tobacco mosaic	300×15	The whole virus particle is rod-shaped and consists of 1 molecule RNA (M.W. 2×10^6) associated with protein macromolecules arranged in a helix. Particle M.W. 39×10^6.
ii) Sphere		
Tomato bushy stunt	30	An icosahedron consisting of 16% RNA (M.W. 1.6×10^6) and protein. Particle M.W. 9×10^6.
Insect		
Silkworm	280×40	The actual virus is rod-shaped. DNA constitutes about 8% of dry weight but *in vivo* the virus rods are embedded in polyhedral crystalline aggregates of protein 0.5-15μ diameter.
Bacteriophages		
i) Double-stranded DNA		
T-even of *E. coli*	Head: 90×60 Tail: 100×25 Tail fibrils: 130×2.5	A tadpole-shaped phage with DNA (M.W. 130×10^6) confined to the head. The tail is protein some of which is contractile; long tail fibrils are involved in attachment to the host cell. Particle M.W. 250×10^6.
ii) Single-stranded DNA		
ϕX174 of *E. coli*	22	A dodecahedron with 12 subunits. DNA (M.W. 1.6×10^6) 25% dry weight. Particle M.W. 6.2×10^6.
iii) RNA f2 of *E. coli*	20	A polyhedron containing RNA (3×10^{-12} μg/virus) and protein. Nucleic acid content is probably similar to that of ϕX174.

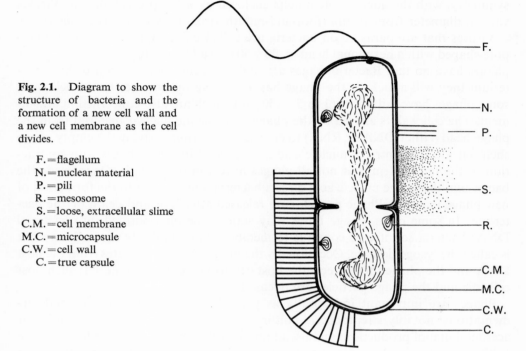

Fig. 2.1. Diagram to show the structure of bacteria and the formation of a new cell wall and a new cell membrane as the cell divides.

    F.=flagellum
    N.=nuclear material
    P.=pili
    R.=mesosome
    S.=loose, extracellular slime
C.M.=cell membrane
M.C.=microcapsule
C.W.=cell wall
    C.=true capsule

autotrophic species, which require only inorganic compounds for growth, to those of heterotrophic species, which may have considerable synthetic ability or so little that they must be grown in tissue culture. Similarly, there is an enormous range of degradative ability, from species which may use native protein and complex chitins, to those that cannot degrade large molecules and so require amino acids and hexoses supplied as such. Because of these diverse capabilities, bacteria have been exploited industrially to accumulate both intermediate and end products of metabolism. They are also a rich source of both synthetic and degradative enzymes; but the exploitation of bacteria in this field is still at an early phase of development.

2.1.2. Viruses

Viruses are the smallest microbes; they are obligate intracellular parasites of animals, plants, insects, fungi, algae, or bacteria. Some of their properties have been listed in Table 2.2. They contain no water and have little or no synthetic or metabolic activity in themselves. Growth and multiplication take place intracellularly, where the virus directs the host cell to synthesize a new virus; this frequently results in damage and death. The genetic material of viruses may be either ribonucleic acid (RNA) or deoxyribonucleic acid (DNA). Many plant and animal viruses consist of a single molecule of RNA, associated with protein subunits arranged in a polyhedron showing cubic symmetry. The myxoviruses show helical symmetry, with the nucleoprotein helix enclosed by a lipoprotein sheath. Viruses vary in diameter from 10 mμ (foot-and-mouth virus) to 300 mμ (vaccinia virus).

Viruses that are parasitic on bacteria are called bacteriophages; some are tadpole-shaped with a hexagonal head (c. 90×60 mμ) and a tail (c. 100×25 mμ); a few phages have no tail. Bacteriophages are highly specific to the strain of host bacterium they will attack. If the phage has a tail, the tip attaches to the cell wall; some phages have tail fibrils (3 to 5×150 mμ) which are also concerned in attachment. The cell wall is altered by the phage to allow the nuclear material from the phage head (either DNA or RNA) to enter the bacterium, leaving the empty phage shell outside. The entry of nucleic acid is associated in some phages with contraction of the tail protein, but not all phages have contractile tails. Once inside the bacterium, the phage nucleic acid directs bacterial metabolism to the formation of new phage material. Phage particles are released after lysis and death of the bacterium. In some cases, phage DNA may actually be incorporated into bacterial DNA; bacteria so infected continue to metabolize and multiply normally in what is called the lysogenic state. Occasionally the phage DNA dissociates from the host DNA and the phage then directs the host cell to make phage material; such host cells die and the phage particles are released.

Phages are important industrially as possible contaminants of bacterial fermentations, notably streptococcal cultures used in cheese-making, clostridia for acetone-butanol production, and pseudomonads for organic-acid production. The cultivation of animal viruses to test antiviral drugs and for vaccine production is also an important undertaking, although on a relatively small scale.

2.1.3. Fungi

Fungi are widely spread in nature in environments of lower relative humidity than those which favor bacteria. The metabolism of fungi is essentially aerobic; they form long filamentous, nucleated cells (hyphae) 4 to 20μ wide, which are much branched and which may or may not have cross walls. In Table 2.3 is a summary of some of the properties of fungal hyphae, while Fig. 2.2 shows a diagrammatic cross section of a hypha.

Many species of fungi have complicated life cycles during which both sexual and asexual spores are formed, these spores being often carried on special fruiting bodies. The hyphal mass (mycelium) on a solid surface may be only 2 mm wide, or as much as 2 m, depending on the species and the growth medium.

Generally, fungi are free-living saprophytes, but a few are parasitic on animals and many are serious pathogens of plants. They have very wide degradative and synthetic capabilities and have proved a fruitful source of industrially important organic acids (e.g., citric, gluconic, gibberellic), numerous antibiotics (e.g., penicillin, griseofulvin) and enzymes (e.g., cellulase, protease, amylase). In addition, they may cause spoilage of paper, fabrics, and food, particularly in high humidity.

Some yeasts form elliptical cells 8 to 15μ by 3 to 5μ, others are almost spherical.

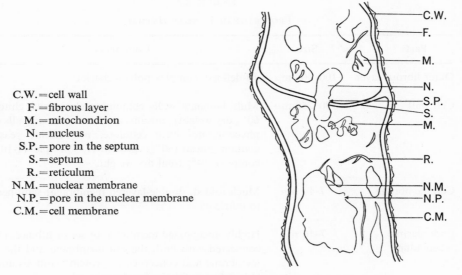

C.W. = cell wall
F. = fibrous layer
M. = mitochondrion
N. = nucleus
S.P. = pore in the septum
S. = septum
R. = reticulum
N.M. = nuclear membrane
N.P. = pore in the nuclear membrane
C.M. = cell membrane

Fig. 2.2. Diagram to show the structure of septate fungal
hyphae with a nucleus migrating from one cell to the next.

Asexual spores are occasionally observed, but they form sexual spores in the same way as *Ascomycetes*, and are therefore classified with them. Yeasts differ from most fungi, however, in that vegetative growth is by budding so that long hyphae are rarely formed, their cell wall contains mannans and glucans, not chitin or cellulose as is common in fungi, and they are capable of both aerobic and anaerobic growth. Growth of yeast on a solid medium resembles a bacterial colony. They are industrially important in making beverage alcohol and baker's yeast and in producing protein for stock feed or human use.

Actinomycetes are a group of organisms intermediate in properties between bacteria and true fungi. They form long, much-branched hyphae with no cross walls, spores are budded from the tip of aerial hyphae, and they can form hetero-karyons (cells with discrete nuclei derived from different parents, coexisting in a common cytoplasm). In these characteristics they are similar to true fungi. On the other hand, the cells are smaller, being only 0.5 to 1.4μ wide; the cell wall polymer contains N-acetyl muramic-peptide, including diaminopimelic acid; the nuclear material is not enclosed in a nuclear membrane; and the cells are lysed by lysozyme and attacked by specific phages. In these properties and in the small cell width, Actinomycetes resemble bacteria.

Industrially, the group is extremely important as a source of powerful antibiotics for the control of microbial infections of man, animals, and plants. Their growth in fermentation tanks is very similar to that of fungal fermentations; for this reason they have been grouped with the fungi in this discussion.

TABLE 2.3

PROPERTIES OF FUNGAL HYPHAE.

Part	Size	Comments
Outer fibrous layer	100–500 mμ	Ill-defined, complex polysaccharide.
Cell wall	100–250 mμ	Multi-laminate walls containing chitin and chitosan (2-60% dry weight), glucans, and/or mannans; walls of some phycomycetes have cellulose, not chitin; yeast walls contain glucan (30%), protein (13%), lipid (1-10%) and comprise 30% total dry weight.
Cell membrane	7–10 mμ	Much-folded, double-layered membrane; semi-permeable to nutrients.
Endoplasmic reticulum	7–10 mμ	Highly invaginated membrane or set of tubules, probably connected with both the cell membrane and the nuclear membrane and concerned in protein synthesis and probably other metabolic functions.
Nucleus	0.7–3μ	Surrounded by a double membrane (10 mμ), containing pores 40 to 70 mμ wide. Nucleus is flexible and contains cytologically distinguishable chromosomes. Nucleolus about 3 mμ. In Actinomycetes there is no nuclear membrane. The nucleus is capable of migration.
Mitochondria	0.5–1.2μ by 0.7–2μ	Analogous to those in animal and plant cells, containing the electron transport enzymes and bounded by an outer membrane and an inner one forming cristae. They probably develop by division from pre-existing mitochondria.
Inclusions		Lipid and glycogen-like granules are found in some fungi and ribosomes have been seen in all fungi examined.

2.1.4. Protozoa

Protozoa are widely distributed in fresh and salt water, in soil, and in animals. They may be unicellular or multicellular and exhibit a wide variety of morphological forms; it is convenient to divide them into two subgroups: the algae, which are largely photosynthetic and resemble primitive plants, and the protozoa which are not photosynthetic and resemble primitive animals.

The algae may be either uni- or multicellular. Some species have complex life cycles; all are capable of photosynthesis and contain chlorophyll. Associated with the photosynthetic pigments are many other pigments such as carotenoids, xanthophylls, and phycocyanins, which give different colors to the algae. The diatoms are algae with siliceous exoskeletons; diatomaceous earth, a useful filter aid, is composed of diatom bodies. The largest of the algae are the multicellular sea-weeds; kelp (a brown alga) has a large, highly differentiated thallus. Some species of algae are capable of heterotrophic as well as photosynthetic growth; the photo-synthetic blue-green algae are also able to fix gaseous nitrogen.

The protozoa are a very diverse group. Taxonomy is based on their means of motility; the *Sarcodina* have amoeboid movement (*Amoeba, Foraminifera*); the *Ciliophora* move by means of cilia (*Paramecium, Chlamydomonas*); the *Mastigo-phora* have compound flagella. Some flagellates cause serious diseases in man, while others like *Euglena* are free-living and harmless. There is also a nonmotile group, the *Sporozoa*, to which the important malaria parasite (*Plasmodium vivax*) belongs.

Protozoa are known to have a very important role in removing bacteria from waste water in trickling filters and activated sludge plants but little is known at present of the kinetics of their growth in these systems. Protozoa have not been grown to any extent on an industrial scale, although *Chlorella* (a unicellular alga) was grown as a protein supplement in Japan after World War II. Seaweeds, of course, are gathered in large quantities from natural sources and constitute an important dietary item in Asia. They are a commercial source of agar, carrageens and alginates, but so far seaweeds have not been cultivated on an industrial scale.

2.2. CHEMICAL COMPOSITION

Complete detailed analyses of microorganisms are not often made and would be of little use. There is abundant evidence that the amount of any particular com-ponent found depends greatly on the composition of the growth medium, the age of the culture, and its rate of growth. The data in Table 2.4 merely indicate gross differences in the composition of the different groups of organisms and give some idea of the numbers and the dry weight of the population likely to be achieved in culture.

Table 2.4 shows that while some viruses consist entirely of nucleoprotein, others contain some lipid. Viruses have virtually no enzymic activity, though viruses infecting animal cells contain a small amount of an enzyme involved in the polymerization of nucleic acids. Bacteriophages contain an enzyme which attacks the bacterial cell wall, permitting entry of the phage nucleic acid. Adenosine triphosphate (ATP), which is important for contraction of muscle in higher ani-mals, has been detected in those phages with contractile tail protein.

All microorganisms, except viruses, contain about 80% water. Bacteria, yeast, and unicellular algae are also very similar in protein content—about 50% of the dry weight; the protein is largely enzymatic. It is estimated that *Escherichia coli*

TABLE 2.4

CHEMICAL ANALYSES, DRY WEIGHTS, AND THE POPULATIONS OF DIFFERENT MICROORGANISMS
OBTAINED IN CULTURE.

Organism	Composition (% dry weight)			Population in culture (numbers/ml)	Dry weight of this culture (g/100 ml)	Comments
	Protein	Nucleic acid	Lipid			
Viruses	50–90	5–50	<1	10^8–10^9	0.0005*	Viruses with a lipoprotein sheath may contain 25% lipid.
Bacteria	40–70	13–34	10–15	2×10^8– 2×10^{11}	0.02–2.9	*Mycobacterium* may contain 30% lipid.
Filamentous fungi	10–25	1–3	2–7		3–5	Some *Aspergillus* and *Penicillium* sp. contain 50% lipid.
Yeast	40–50	4–10	1–6	1–4×10^8	1–5	Some *Rhodotorula* and *Candida* sp. contain 50% lipid.
Small unicellular algae	10–60 (50)	1–5 (3)	4–80 (10)	4–8×10^7	0.4–0.9	Figure ( ) is a commonly found value but the composition varies with the growth conditions.

* For a virus 200 mμ diam.

has about 2,000 different enzymes, as well as the genetic potential to form many
more should suitable environmental conditions be supplied.

Structurally more complex organisms, like fungi and seaweeds, contain less of
the metabolically active proteins and nucleic acids, while the inert polysaccharide
components of the cell walls constitute a larger proportion of the total dry weight.
As an approximation, about 10% of the dry weight of yeasts, bacteria, and unicellular algae is nitrogen, but nitrogen comprises only 5% of the dry weight of fungi.

Lipids are essential components of all organisms, except for some viruses, and,
although they are vital to the structure of semipermeable membranes and cell
walls, they are not usually a major component of the dry weight. There are a few
exceptions however; among bacteria, *Mycobacterium* species may contain 30%
mycolic acid; *Rhodospirillum rubrum*, during growth on acetate, may synthesize
a polymer of β-hydroxybutyric acid constituting about 20% of its dry weight.
Lipid inclusions are common among the fungi and, in certain yeasts and *Ascomycetes*, may constitute 50% of dry weight under special conditions. The fatty acid
content of the lipid is variable, but it is rich in C_{16} and C_{18} fatty acids, particularly
palmitic, oleic, and stearic acids.

2.3. REQUIREMENTS FOR GROWTH AND FORMULATION OF MEDIA

2.3.1. Requirements for growth

It is important to appreciate that the cultural conditions that achieve maximum cell mass may not necessarily be those that give maximum yield of some product of metabolism. For example, *Aspergillus niger* gives optimum yields of citric acid when growth is restricted by semi-starvation concentrations of nitrogen, phosphorus, and trace metals, but a high concentration of sugar in media.

The best temperature for cultivation varies with the species, but organisms naturally occurring in soil, air, or water usually grow best at 25° to 30°C, while those isolated from animals grow best at 37°C. Some organisms are in fact thermophilic; some of industrial importance, like lactobacilli, cellulose digesters, and methane producers, grow best at 40° to 45°C. Organisms with high temperature optima offer the technical advantage that contaminants usually grow better at lower temperatures, and thus the growth of contaminants is likely to be inhibited at the temperature of the fermentation. In fermentations where growth and product formation are not concurrent, it might well be profitable to test different temperatures for the growth and product-forming stages.

The products of microbial metabolism often cause major shifts in pH; hence to maintain rapid growth, the pH must be kept close to the optimum for the particular organism. For many organisms this is close to pH 7. As with the thermophilic organisms, however, it is useful to take advantage of any tolerance of extremes of pH which an organism may happen to possess. For example, yeast and lactobacilli grow well at pH 4.5, while many fungi and the thiobacilli readily tolerate pH 2. Fermentations maintained at these pH levels are relatively free from contamination by other organisms. The pH for optimum product formation may be different from that for optimum growth, as is the case with *Clostridium acetobutylicum*, which gives very low yields if the pH is maintained near neutrality, although at this pH growth is rapid.

When acid by-products accumulate in the medium, causing an unwanted fall in pH, it is sometimes possible to feed ammonia slowly to the culture, so supplying nitrogen for growth while controlling pH. Calcium carbonate may also be used for pH control, but this is only practicable if the required product is water soluble. When cell mass or some insoluble metabolite is required, acidic products are most conveniently neutralized by adding sodium hydroxide as required.

Microorganisms vary in their need for oxygen. Fungi, algae, and a few bacteria are obligate aerobes, a few bacteria are strict anaerobes, while yeasts and many bacteria can grow in both situations (facultative aerobes). If organisms are capable of facultative growth, a substrate that is metabolized aerobically gives a much larger cell yield than the same weight of substrate metabolized anaerobically (see

Section 2.3.2.2, Source of carbon). When a product of anaerobic metabolism is required, it may be profitable to grow a dense population of cells aerobically and then allow them to metabolize the remainder of the substrate anaerobically. The conversion of substrate to product is thus achieved much faster than when the cells are grown slowly under completely anaerobic conditions. This principle has been successfully exploited in the production of ethanol by yeast.

The obligate anaerobes and facultative aerobes, grown for some product of anaerobic metabolism, can be cultivated in very large tanks provided with adequate facilities for mixing the nutrients with the organisms. Obligate aerobes or facultative aerobes, grown aerobically, require a much more complicated plant. The details of air sterilization, and the problems of design and aeration of cultures, are fully discussed in Chapters 6 and 10.

Heterotrophs often fix carbon dioxide during growth, but normal air usually supplies sufficient carbon dioxide for this purpose. Autotrophs depend on carbon dioxide for their main source of carbon; improved growth of autotrophs has been found in an atmosphere enriched with carbon dioxide.

2.3.2. Formulation of media

Detailed investigation will be required in order to establish the most economic medium for any particular fermentation, but certain basic requirements must be met by any growth medium; these will be discussed below.

2.3.2.1. *Source of energy*

Growth processes are endergonic, so that energy for growth must come from either the oxidation of medium components or from light. ATP is the most important compound in energy transformations in cells. The coupling of ATP to thermodynamically unfavorable reactions enables them to proceed at a useful rate. Photosynthetic bacteria and algae can utilize the energy from light for ATP formation; autotrophic bacteria can generate ATP from the oxidation of inorganic compounds; while heterotrophic bacteria, yeasts, and fungi form ATP while oxidizing organic compounds.

In present-day industrial fermentations, the commonest source of energy is starch or molasses, but the growing world population and the increasing cost of plant products may force a change in this pattern; autotrophic microorganisms and those which can oxidize cheap hydrocarbon substrates like alkanes for energy are therefore of potential industrial interest.

2.3.2.2. *Source of carbon*

Frequently, the carbon needs of a cell are supplied with the energy source, but the autotrophic and photosynthetic bacteria use carbon dioxide. The pathway by which heterotrophs metabolize substrate carbon is important in determining the amount of carbon converted to cell material. It is found that facultative organisms incorporate about 10% of substrate carbon in cell material when metabolizing anaerobically, but 50–55% of substrate carbon is converted in cells with fully aerobic metabolism. Since 50% of the dry weight of cells is carbon, it is possible to calculate how much carbon must be supplied in the medium in order to give any particular weight of cells. For example, if 40 g/l dry weight of cells are required, and metabolism is aerobic, then the carbon required in the medium will be $40/2 \times 100/50 = 40$ g. If this is supplied as a hexose, then $40 \times 180/72 = 100$ g/l of hexose are required.

2.3.2.3. *Source of nitrogen*
This can be supplied for most industrially important organisms by ammonia or ammonium salts, although growth may be faster when using organic nitrogen. Certain organisms, however, have absolute requirements for organic nitrogen. These requirements vary from single amino acids to the entire range of amino acids, as well as purines, pyrimidines, and vitamins. Some organisms have complex, unknown requirements, which can be met only by crude animal or plant extracts, and it is impossible to supply the needs of viruses in cell-free systems. When products with a high nitrogen content are required, the greater the amount of nitrogen that can be assimilated by the organism, the greater the chance of a high yield of product, provided the organism has the genetic potential to produce the product.

Organic nitrogen compounds suitable for media are relatively expensive. The cheapest forms are soy bean, peanut, fish, or meat meals, malt combings, yeast extract, whey, casein, and various enzyme digests of protein-rich materials. Plant by-products, like molasses and corn steep liquor, can often provide the small

TABLE 2.5

INORGANIC CONSTITUENTS OF DIFFERENT MICROORGANISMS.

Element	Bacteria	Fungi (g/100 g dry weight)	Yeast
Phosphorus	2.0 – 3.0	0.4 – 4.5	0.8 – 2.6
Sulphur	0.2 – 1.0	0.1 – 0.5	0.01 – 0.24
Potassium	1.0 – 4.5	0.2 – 2.5	1.0 – 4.0
Magnesium	0.1 – 0.5	0.1 – 0.3	0.1 – 0.5
Sodium	0.5 – 1.0	0.02 – 0.5	0.01 – 0.1
Calcium	0.01 – 1.1	0.1 – 1.4	0.1 – 0.3
Iron	0.02 – 0.2	0.1 – 0.2	0.01 – 0.5
Copper	0.01 – 0.02		0.002 – 0.01
Manganese	0.001– 0.01		0.0005 – 0.007
Molybdenum			0.0001 – 0.0002
Total ash	7 – 12	2 – 8	5 – 10

amounts of growth factors that are either essential for growth or greatly accelerate the rate of growth.

Nitrogen constitutes 10% of the dry weight of most organisms, so that the minimum nitrogen content of the medium can be calculated for a desired cell yield; but the form in which it is supplied will depend on the organism, the cost and the purpose of the fermentation.

2.3.2.4. *Source of minerals*

Table 2.5 shows the range of concentrations of some elements found in bacteria, fungi and yeast. Clearly, phosphorus, potassium, sulphur and magnesium are major components and these and other elements present in significant amounts must be supplied in the medium. Trace metals like iron, copper, cobalt, manganese, zinc and molybdenum are essential but are usually present as impurities in other ingredients of the medium.

2.4. Reproduction in Microorganisms

Different microorganisms multiply in different ways; it is therefore important to know how this is achieved, in order to understand variability in fermentations, to maintain stock cultures without deterioration, and to breed strains for particular purposes.

2.4.1. Reproductive cycles in microorganisms

2.4.1.1. *Bacteria*

Bacteria divide vegetatively by fission. Growing cells increase in mass in a continuous process, the cell wall and cell membrane extending as the protoplasm increases. As the cell mass increases, the membrane, followed by the cell wall, begins to grow inwards, until the membrane and cell wall from the circumference meet in the center. At the same time the nuclear material elongates, and forms dumbbell-shaped masses with the narrow part at the site of the future wall. The cytoplasm and nuclear material separate equally between the two cells, with no indication that one cell is the mother and the other the daughter. After the cell wall and membrane have been laid down between the two cells, in some species they detach immediately, but often a bacterium which appears by simple stain to be one cell has several cross walls and several sets of nuclear material.

Some bacterial species (e.g., *Escherichia coli* and *Pseudomonas aeruginosa*), show a primitive system of sexual reproduction between positive and negative strains. The nuclear material of a bacterium contains only one chromosome with one set of genes (the units of genetic information which fix its characteristics) which are contained in one double-stranded DNA polymer, the whole polymer constituting one chromosome. The nucleus of other microbes may contain more than one chromosome; when a nucleus has only one set of chromosomes it is haploid, when it has pairs of similar chromosomes it is diploid. In sexual reproduction in higher plants and animals, haploid nuclei derived from different diploid parents combine to form a zygote, which develops into a diploid offspring. In conjugation between positive and negative strains of bacteria, only some of the genes from the positive strain enter the negative strain; this results in the formation of a partial diploid. The characteristics of the offspring of the cross show, however, that there has been a mixing and recombining of the genes from both parents, in the same way as occurs in the nuclei from higher cells, but there is only one offspring from the cross, as an incomplete chromosome cannot give rise to a viable cell.

2.4.1.2. *Viruses*

Animal viruses are adsorbed onto the host cell, penetrate the wall, and shed their protein subunits, thus exposing the viral nucleic acid. The virus suppresses in the host cell processes that are directed to host requirements, and directs metabolism to the formation of viral constituents. New viral nucleic acid and protein subunits are assembled into macromolecules, which, in viruses with cubic symmetry, are cytologically identifiable intracellularly as complete virus particles. Viruses having a lipoprotein sheath acquire this as they leave the cell.

For phages, the process of multiplication is very similar to that of animal viruses, except that only the phage nucleic acid enters the bacterium; the phage nucleic acid then directs the formation and assembly of new phage material. The bacterium then lyses, so that phage particles are released. This cycle of multiplication is known as a lytic cycle, but phage and bacteria can form a different type of association. The so-called temperate phages adsorb onto the host cell, and their DNA penetrates the host cytoplasm just as with lytic phages, but instead of directing the cell to form new phage particles, the viral DNA becomes attached to the bacterial DNA. The mixed DNA complex replicates as a unit, and divides with the host bacterium. This cycle is called a lysogenic cycle and can be repeated through hundreds of cell divisions.

The lysogenic cycle may change to a lytic cycle spontaneously at a low rate of frequency. When this occurs, the viral DNA becomes detached from the bacterial DNA and directs the host metabolism to the formation of phage particles. During this cycle, sometimes a phage head contains bacterial, instead of phage, DNA. If such a phage attacks another bacterium, the bacterial DNA from the first bacterium may enter and become incorporated into the DNA of the second bacterium, with the possibility that the second bacterium may acquire new characteristics. This process is known as transduction.

In bacteria that are capable of entering the lysogenic state, but without sexual reproduction, transduction could provide a rational approach to the problem of breeding strains of bacteria with desirable fermentation characters. So far, however, this technique has mainly been used in academic research.

2.4.1.3. *Fungi*

Vegetative growth of filamentous fungi and Actinomycetes is by simple extension of the hyphae and asexual division of the nuclei. As the hyphae elongate, the older parts of the mycelium become extensively vacuolated and inactive in nutrient absorption. Cross walls are formed in all fungi except members of the *Oomycetes* and *Zygomycetes*, but the cross wall has a pore through which cytoplasm and nuclei can pass from cell to cell. This situation is quite different from that in plants and animals, where nuclei are confined to their respective cells. Asexual spores are budded from the tip of hyphae and may be either uni- or multinucleate.

Sexual spores are produced following conjugation of two haploid nuclei. In the *Ascomycetes*, the zygote divides immediately to form four or eight haploid ascospores enclosed in a special sack or ascus. In the *Basidiomycetes*, the zygote is formed inside a special cell or basidium and divides to form four haploid nuclei which migrate to outgrowths of the basidium. The mature basidiospores are discharged from these outgrowths.

Yeasts grow vegetatively by budding from the mother cell, or by fission. During bud formation, the cytoplasm of the bud is continuous with that of the mother cell and is bounded by a typical cell membrane and cell wall. Most workers consider that nuclear division does not involve cytologically recognizable chromosomes. The mother nucleus enlarges and migrates in the direction of the bud; the nuclear membrane constricts the nucleus, so that one half remains in the bud and the other in the mother cell; the cell membrane is then completed between the bud and the mother cell, and the cell wall is laid down. The readiness with which the bud is detached from the mother cell is a characteristic of the species, and the site of the former bud can be seen as a bud scar. One cell may have as many as 100 daughter cells. When sexual spores are produced in yeast, two haploid nuclei fuse, and the diploid zygote divides to give four haploid ascospores inside the original vegetative yeast cell.

This very brief outline of the ways in which microorganisms divide is to serve as an introduction to a discussion of variability encountered in fermentations and of the rationale behind strain-breeding programs.

2.4.2.   Replication of nuclear material

As has been stated in previous sections, the genes determining the properties of cells (with a few exceptions) are found in the DNA of the nucleus. The DNA molecule has been shown to be a double-stranded helical polymer of nucleotides; the deoxyribose of the nucleotide has either a purine (adenine or guanine) or a pyrimidine (cytosine or thymine) base attached at the $1'$ position. The bases are so arranged that adenine in one strand is held by hydrogen bonding to thymine in the other, while cytosine is bonded to guanine. It has also been established that the two strands are antiparallel; the deoxyribose of each nucleotide has a phosphate at the $5'$ position and a hydroxyl at the $3'$ position with the diester linkages between the nucleotides 5′→3′ in one strand, and 3′→5′ in the other. When the DNA replicates, the strands (with bases attached) separate and nucleotides complementary to the bases already present in the old strands are sequentially added to both strands of the DNA. These features of DNA replication are illustrated in Fig. 2.3.

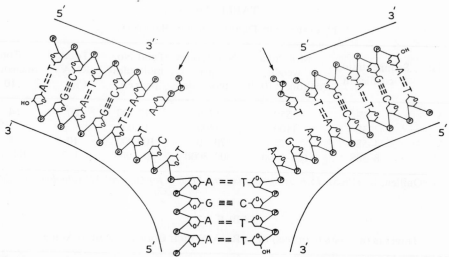

Fig. 2.3. Replication of deoxyribonucleic acid (DNA). When DNA replicates, the hydrogen bonds between the pairs of bases are cleaved and the two component strands of nucleotides separate. Two new strands of nucleotides are polymerized according to the templates provided by each of the original strands —adenine is paired with thymine, and guanine with cytosine, and hydrogen bonds form between the complementary bases. On the right of the figure, thymidine triphosphate is in position to be polymerized to the 5′ end of the new strand, opposite adenine in the old strand; as the thymine nucleotide is added to the polymer, inorganic pyrophosphate will be released. The two double-stranded molecules formed in this way thus contain one strand from the original DNA molecule and one newly synthesized strand. A=adenine; C=cytosine; G=guanine; T=thymine; ℗=phosphate.

2.4.3. Transfer of information from the nucleus

Although genes are the repository of the heritable properties of a cell, the characteristics actually observed are a reflection of the enzymes it contains. In order to control fermentations intelligently, it is important to know how genetic information is translated into enzyme protein and how this can be manipulated. The salient features of the mechanism of transfer of information from DNA to cellular protein and the genetic code are summarized in the following sections.

2.4.3.1. *Code for amino acids*
It has been established that amino acids are specified by triplets of bases (termed codons) in one strand of DNA and that the order of these triplets in DNA determines the order of the amino acids polymerized into protein. The code is redundant since several different triplets may code for a single amino acid.

2.4.3.2. *Ribonucleic acids and protein synthesis*

In the cell, there are three major classes of ribonucleic acids: messenger RNA (m-RNA), transfer RNA (t-RNA) and ribosomal RNA (r-RNA). Table 2.6 shows the amounts of these classes of RNA in bacteria.

TABLE 2.6

NUCLEIC ACID DISTRIBUTION IN BACTERIA.

Material	Sedimentation coefficient	Molecular weight (daltons $\times 10^{-3}$)	Nucleotides per molecule	Total No. molecules per cell	Total RNA (%)	Total nucleotides $\times 10^{-6}$
r-RNA	16S	550	1700	12,000	80	64
	23S	1100	3400	12,000		
t-RNA	4–5S	25	70–80	100,000	10	8
m-RNA	8–30S	100–1500	600–5000	500–1000	1–2	0.8–1.6

From McQuillen, K. (1965). *15th Symposium of the Society for General Microbiology*, p. 134.

TABLE 2.7

TRIPLETS OF BASES (CODONS) IN m-RNA WHICH CODE FOR AMINO ACIDS.

UUU	Phenylalanine	CUU	Leucine	GUU	Valine	AUU	Isoleucine
UUC	"	CUC	"	GUC	"	AUC	"
UUG	Leucine	CUG	"	GUG*	"	AUG	Methionine
UUA	"	CUA	"	GUA	"	AUA	Isoleucine
UCU	Serine	CCU	Proline	GCU	Alanine	ACU	Threonine
UCC	"	CCC	"	GCC	"	ACC	"
UCG	"	CCG	"	GCG	"	ACG	"
UCA	"	CCA	"	GCA	"	ACA	"
UGU	Cysteine	CGU	Arginine	GGU	Glycine	AGU	Serine
UGC	"	CGC	"	GGC	"	AGC	"
UGG	Tryptophan	CGG	"	GGG	"	AGG	Arginine
UGA	None	CGA	"	GGA	"	AGA	"
UAU	Tyrosine	CAU	Histidine	GAU	Aspartic	AAU	Asparagine
UAC	"	CAC	"	GAC	"	AAC	"
UAG	None	CAG	Glutamine	GAG	Glutamic	AAG	Lysine
UAA	None	CAA	"	GAA	"	AAA	"

* GUG at the beginning of m-RNA codes for *N*-formyl methionine.

About 2% of the total RNA is m-RNA. Messenger RNA is a single-stranded polymer of nucleotides with one of the four bases uracil (U), adenine (A), cytosine (C), or guanine (G) attached at the 1′ position of each ribose. When m-RNA is synthesized, the strands of the appropriate portion of DNA separate and one strand forms the template upon which m-RNA is synthesized; adenine, cytosine, guanine, and thymine in DNA are complementary to uracil, guanine, cytosine, and adenine, respectively in m-RNA. Thus the order of bases specified in DNA (the genetic message) is preserved in m-RNA (see Fig. 2.4). Table 2.7 shows the triplets of bases in m-RNA (codons) which specify the various amino acids. It can be seen that while the first two bases may be sufficient to specify some amino acids, other have unique codons.

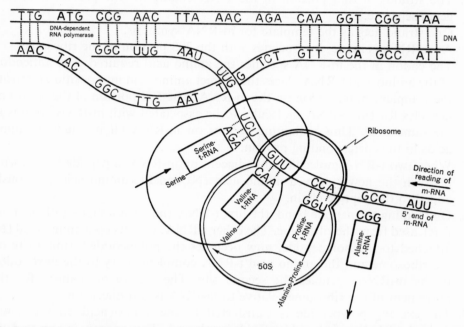

Fig. 2.4. Protein synthesis. The strands of DNA are shown separating temporarily while m-RNA is transcribed, in this case, from the upper strand. The 30S portion of a ribosome combines with the newly synthesized m-RNA. Two t-RNA-amino acid molecules associate with the 50S portion of the ribosome; via their complementary anticodons, they align amino acids as directed by the codons on the m-RNA. The newly coded amino acid forms a peptide bond with the amino acid most recently added to the peptide chain. When the peptide bond is formed, the first t-RNA is released from the ribosome and this allows the next t-RNA-amino acid to combine with the ribosome. The ribosome thus "moves" along the m-RNA molecule until all the codons are read, then the completed polypeptide is released from the last t-RNA and the ribosome is free to combine with a new m-RNA. The bases in the nucleic acids are indicated by A=adenine; C=cytosine; G=guanine; T=thymine; U=uracil; t-RNA=transfer RNA; The ribosome with a single outline indicates the position of the m-RNA relative to the ribosome when the next codon, UCU, is read.

Transfer RNA constitutes about 10% of total RNA. There is a family of t-RNA molecules, each of different composition and having the ability to combine with a specific amino acid. The t-RNA molecules are so folded that one end can combine reversibly with the amino acid and another part of the molecule is arranged so that one triplet of bases (anticodon) can combine with a complementary codon on m-RNA. For instance, serine t-RNA has the anticodon UCG and thus can combine with the m-RNA codon, AGC, which specifies serine.

About 80% of total RNA is r-RNA. This RNA is found combined with protein in organelles called ribosomes and these are the site of protein synthesis. The organelle can be fractionated *in vitro* into two subunits (30S and 50S) that have been shown to have different functions in protein synthesis.

When a molecule of protein is synthesized, the following events occur:

1. The strands of that portion of the DNA coding for the protein (i.e., that gene) separate.
2. One strand acts as the template for m-RNA synthesis.
3. As m-RNA is made, it associates with the 30S portion of the ribosome.
4. The codon in the m-RNA specifies the amino acid required. The anticodon of the amino acid t-RNA places the correct amino acid into position opposite the complementary codon on the m-RNA. The 50S portion of the ribosome has sites for two t-RNA molecules to be associated with m-RNA codons at the same time. Thus, adjoining codons on m-RNA align neighbor amino acids in the corresponding protein.
5. When two t-RNA molecules are adjacent on m-RNA, a peptide bond forms between the newly arrived amino acid (N) and the amino acid previously added to the peptide chain.
6. On forming the peptide bond, the first t-RNA (now without its polypeptide) is released from the ribosome. The other t-RNA, with its own amino acid (N) attached to the polypeptide, now occupies the polypeptide-forming site on the ribosome and the amino acid t-RNA complementary to the next codon in the m-RNA occupies the second site. The precise mechanism for the movement of the ribosome relative to m-RNA is still uncertain.
7. The existing polypeptide is transferred to the amino acid on the newly arrived t-RNA and the old t-RNA is released. This process is repeated till the ribosome reaches the end of the message.
8. One m-RNA may be translated by a number of ribosomes simultaneously, and thus many copies of the protein may be made from the one m-RNA molecule. Figure 2.4 illustrates the steps just described.

2.4.3.3. *Control of protein synthesis*

Although the nucleus of an organism may contain the genetic information to produce a wide variety of enzymes, only some enzymes (constitutive) are produced at all times irrespective of the presence of their substrate. The concentration of others (inducible) is greatly influenced by the presence of their respective substrates. In the absence of the inducing substrate, the enzyme-synthesizing system is inhibited but this can be reversed by the substrate. In biosynthetic systems another type of control is found. Here, an intermediate metabolite may control the amount of enzyme produced. This system can best be illustrated by reference to Fig. 2.5, in which a series of biosynthetic reactions, A to D, each controlled by a different enzyme, E_a, E_b, etc., is shown leading to an amino acid end product. If the amino acid is formed faster than needed for protein synthesis, its concentration

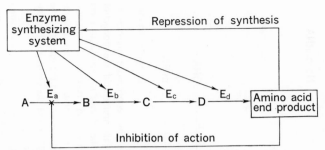

Fig. 2.5. System for the control of the formation and action of repressible enzymes involved in the biosynthesis of an amino acid.

rises; this inhibits the first enzyme (E_a) involved in the formation of the amino acid. This effect is immediate and tends to decrease the formation of end product. It is freely reversible should the concentration of amino acid fall. Further, the amino acid also represses the formation of all the enzymes of the series. This effect will be slow, since the existing enzymes will remain in the cell until they are either diluted out by cell division or destroyed in normal protein turnover.

These processes allow the cell to regulate its enzyme content in direct response to the composition of the environment. They also permit the most economic use of nutrients, as they prevent formation both of end products already present in excess of the needs of synthesis and of superfluous enzyme protein. Genetic studies have established that the regulatory mechanism itself is under genetic control. The pioneering work in this field was done by Jacob and Monod, who studied the control of the inducible enzymes concerned in lactose metabolism by *Escherichia coli*. They proposed the *operon* model to explain this control system. The model has been modified and invoked to explain control of both inducible and repressible enzymes in a wide variety of cells. It should be pointed out that experimental proof of the existence of this type of control is available for only a limited number of systems and organisms, though it is assumed to operate widely. The model is summarized below and in Figs. 2.6 (a) and (b). As an example of the control of an inducible enzyme system, the features of the lactose operon will be outlined.

1. The genes coding for the production of the enzymes of the operon are contiguous on the DNA molecule and are called structural genes.
2. The operator (*o*) gene is contiguous with the first structural gene and is involved in the control of the transcription of all of the structural genes.
3. A small gene, the promoter (*p*), is also necessary for transcription of the structural genes. In the lactose operon, there is one promoter situated between the *i* and *o* genes, but there may be more than one promoter per operon.
4. The regulator gene (*i* in the case of the lactose operon, but more generally designated by *r*) is located some distance from the genes of the operon. The regulator gene makes repressor (R), an allosteric protein with affinity for both the *o* gene and for the inducing substrate, lactose.

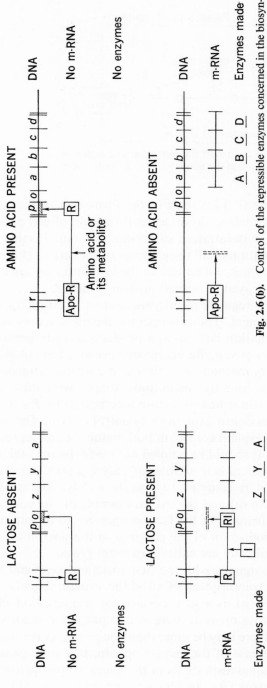

Fig. 2.6 (a). Control of the inducible enzymes of the lactose operon. In the absence of the inducer (lactose), the product of the *i* gene (repressor) combines with the operator gene and prevents transcription of the genes of the lactose operon. When lactose is present, it combines with the repressor and prevents the repressor combining with the operator. The genes of the operon can then be transcribed to m-RNA and the enzymes for lactose metabolism synthesized at the ribosomes.

Genes: *o* = operator; *p* = promoter; *z* = β-galactosidase; *y* = permease; *a* = transacetylase; *i* = regulator (located outside the operon). R = repressor; I = inducer; Z, Y, and A = enzymes from *z*, *y*, and *a*, respectively.

Fig. 2.6 (b). Control of the repressible enzymes concerned in the biosynthesis of an amino acid. In the presence of the amino acid, the product of the *r* gene (apo-repressor) combines with the effector metabolite to form repressor. The repressor binds to the operator and prevents transcription of the genes of the operon. In the absence of the amino acid, the apo-repressor continues to be made but as it does not bind to the operator, the genes of the operon are transcribed to m-RNA and the corresponding enzymes made.

Genes: *o* = operator; *p* = promoter; *a*, *b*, *c* and *d* = genes for the enzymes of the operon; *r* = regulator (located outside the operon). Apo-R = apo-repressor; R = repressor; A, B, C, and D = enzymes from *a*, *b*, *c*, and *d*, respectively.

5. In the absence of lactose, the repressor binds with *o*. This prevents transcription of m-RNA and hence enzymes are not made. In the presence of lactose, the repressor combines preferentially with lactose. The complex now cannot bind to *o*, so the structural genes are transcribed and the enzymes are synthesized.

The model just described is modified to account for control in biosynthetic pathways.

1. The arrangement of genes in the operon is similar but the product of the *r* gene is a protein which is nonfunctional as a repressor, called apo-repressor.
2. Apo-repressor cannot bind to *o* but combines readily with the end product (or an end-product complex) of the biosynthetic pathway to make functional repressor.
3. In the presence of the end product, repressor is formed and combines with *o* to prevent transcription of the structural genes.
4. In the absence of the end product, apo-repressor is made, but as it does not bind to *o*, the structural genes are transcribed and the enzymes for biosynthesis made.

While the operon model of control of protein synthesis by repression of the transcription of DNA to m-RNA is well established for many systems, there is evidence that a further control may also operate. Studies of the production of tryptophan synthetase and of histidine biosynthesis in bacteria suggest that the rate of translation of existing m-RNA to protein may be regulated by the concentration of an intracellular metabolite; this controls the translation process by feedback. The types of compound which it is speculated might have this controlling function are an enzyme or enzymes of the particular biosynthetic pathway, a particular amino acid-t-RNA complex or its synthetase, or the end product metabolite itself.

2.4.4. Mutations

The previous sections have shown how the characteristics of cells are preserved from one generation to the next in the DNA and how the structure of proteins is related to the structure of DNA. When haploid cells like bacteria divide, the daughter cells exhibit the same properties as the parent cells. However, very occasionally, cells arise which have different properties from their parents and breed true to the changed properties. Such cells are mutants. It has been established that mutants arise spontaneously from alterations to the DNA of the normal or wild-type cell.

An incorrect base may be inserted during replication of DNA. For instance, guanine might be inserted opposite thymine instead of adenine, and when this altered DNA is copied, the strand carrying guanine will be paired with cytosine. Thus the A-T pair in the wild-type organism will be replaced in the mutant by a C-G pair, as in Fig. 2.7. If the base change occurs in either of the first two positions of a codon, a different amino acid will usually be inserted into the protein but this may not occur if the change is in the third base. For instance, alanine is coded in m-RNA by all four of the codons GCU, GCC, GCG, and GCA. Thus a base change in the third position is not expressed while a change in either of the first two bases changes the amino acid.

Mutations can also arise if one or more bases are deleted from DNA. If three bases or multiples of three bases are deleted, the remaining codons will be unchanged and the mutant will have a protein (which may or may not be functional) that lacks the amino acids corresponding to the deletion (see Fig. 2.8). If a number of bases not divisible by three is deleted, the reading frame for the m-RNA codons

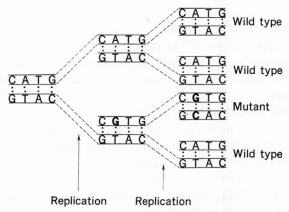

Fig. 2.7.   Error in replication of DNA. Guanine is incorrectly paired with thymine and, after two nuclear divisions, a mutant arises in which guanine-cytosine has replaced adenine-thymine.
A=adenine; C=cytosine; G=guanine; T=thymine.

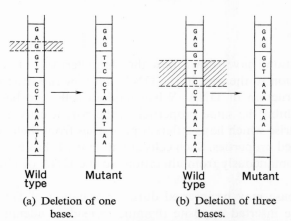

Fig. 2.8.   Deletion of bases from DNA.   (a) Deletion of a single base causing a shift in reading frame. (b) Deletion of three bases causing minimal perturbation of the reading of the code but a change in the amino acid composition of the protein.
A=adenine; C=cytosine; G=guanine; T=thymine.

will be shifted and it is highly probable that no protein at all or a nonfunctional polypeptide will result. If deletions are large, even those which cause no change in the reading frame, activity is usually lost. Whether this loss is lethal depends on the gene involved.

Similarly, mutants arising from an inversion of part of the DNA molecule can cause a small or a very extensive change in the structure of the resulting protein.

2.4.5. Practical implications of mutations

For the fermentation industry, mutations are of great practical importance; selection of desirable mutants can improve the performance of strains but, at the same time, processes must be managed so as to minimize the emergence of unwanted mutants.

Mutations to a faster growth rate occurring while the inoculum is building up are especially troublesome. If the mutation occurs late in the growth cycle of the main fermentation, the yield may not be seriously affected; provided the finished fermentation is not used as the inoculum for the next batch, no great harm results. The practice of inoculating a new fermentation from the previous one is to be avoided, as it tends to perpetuate mutations. The inoculum for each fermentation should be built up from a stock culture; a possible protocol is set out below:

Stage 1.   Stock or master culture.

Stage 2.   Numerous sub-master cultures made at one time
            from the master culture.

Stage 3.   Each sub-master culture used for a new production
            run.

Cultures at Stages 1 and 2 should be held so as to minimize the opportunities for nuclear variation (see Section 2.4.5.2.).

Growth in continuous culture presents a special problem. Theoretically, it offers infinite time for the selection of mutants arising spontaneously during growth, and the very nature of the apparatus automatically selects the fastest-growing individuals. Hence marginal advantages in growth rate, which would never be significant in batch browth, could be significant in prolonged continuous culture.

Relatively little work has so far been carried out on the genetic stability of industrially important organisms in continuous culture. The production of fodder yeasts from sulphite-waste liquors has been highly successful in continuous cultivation, but the selection of the faster growers imposed by the method of cultivation is also in the best interests of the process.

In pilot plant investigations, *Aerobacter aerogenes* gave high yields of 2,3-

butylene glycol after two months of continuous growth, while *Salmonella typhi-murium* showed no change in antigenic potency after 14 days. *Pasteurella pestis* degenerated in continuous culture, but full antigenic potency could be retained by using a two-stage system with the first stage at 28°C and the second at 37°C. Studies of antibiotic production have been discouraging; novobiocin yields from *Streptomyces niveus* declined in 28 days, and those of penicillin from *Penicillium chrysogenum* in 25 days. With *Streptomyces venezuelae*, chloramphenicol yields fell to zero in synthetic media in 6 to 40 days depending on growth rate. With complex media, yields were maintained at fast growth rates for 58 days but excessive wall growth complicated the interpretation of results. It is possible to overcome the problem of strain degeneration to some extent by replacing the degenerate culture with a new inoculum at appropriate intervals.

2.4.5.1. *Strain breeding*
Although in a normal bacterial population about one mutant arises in 10^6 cells, the rate of mutation is far higher than this since many mutations are lethal and so are not detected, while others involve base changes that do not change the amino acid read from the m-RNA codon. This low rate of spontaneous mutation is unsatisfactory when attempting strain improvement. Mutagens are available, however, that markedly increase the rate of mutation.

Some mutagens act as homologues of the natural bases of DNA. For example, 5-bromouracil (5-BU) added to a growing culture will replace some of the thymine in DNA. Adenine usually pairs with 5-BU (just as it would have done with thymine) but once in several thousand replications, 5-BU pairs with guanine. The further replication of this 5-BU-G pair results in a mutant in which a C-G pair replaces the original T-A pair. Another mutagen, nitrous acid, reacts with adenine and causes a G-C pair to replace an A-T pair. Other mutagens cause purines to be deleted from one strand without otherwise changing the DNA; then any base may fill the space left by the purine. Nitroso-guanidine, an extremely potent mutagen whose mode of action is complex, produces up to 50% mutants among the surviving population. The acridine dyes, being planar molecules, can fit between base pairs in the DNA helix thus causing frame-shift mutations. Irradiation by X-rays and ultraviolet light can also increase the rate of mutation either by altering bases or by inducing breaks and deletions in the DNA. Table 2.8 summarizes the effects of some mutagens.

It is thus relatively easy to produce mutants. Strain improvement requires painstaking effort, ingenuity in devising screening tests, and luck. The difficult problem is to select desirable mutants and this is accentuated when the parent strain is already a high producer. In this situation, any improvement is likely to be minor and quantitative rather than major and qualitative; major changes are most useful in the very first stages of a development program and in basic research.

Improvement in product formation can be achieved in a number of ways. Mutants can be selected which have the following properties:

TABLE 2.8

SOME MUTAGENS AND THEIR EFFECT ON DNA.

Mutagen	Mode of action	Effect on DNA
5-Bromouracil	Analogue of thymine	A-T → G-C* or G-C → A-T
Aminopurine	Analogue of adenine	A-T → G-C or G-C → A-T
Nitrous acid	Deaminates (a) adenine, (b) guanine or cytosine	(a) A-T → G-C (b) G-C → A-T
Ethyl ethanesulfonate and Ethyl methanesulfonate	Alkylating agents	G-C → A-T or alkylated bases deleted and random bases inserted
Acridines	Inserted between bases in DNA helix	Reading frame shift
X-rays		Causes breaks in DNA and alters bases
Ultraviolet light	(a) Hydrates pyrimidines (especially cytosine) (b) Forms thymine dimers	(a) G-C → A-T or addition or deletion of a base (b) May prevent DNA strands from separating for replication

* A=adenine; C=cytosine; G=guanine; T=thymine

1. Defective enzymes in a biosynthetic pathway, so that useful intermediates accumulate.
2. Lack of feedback control by the end product itself or by a key intermediate on the pathway to the end product.
3. Defective cell membranes, so that intracellular metabolites are secreted rapidly.
4. Lack of some undesirable by-product, such as a pigment.
5. Reduced toxicity to a precursor molecule such as phenyl acetate for penicillin production.

In the transcription of m-RNA from DNA, the DNA-dependent RNA polymerase reacts first with the promoter, and then transcription of the structural genes of the operon begins (see Section 2.4.3.3). Studies of the rate of transcription of different operons indicate that some promoters permit faster transcription of m-RNA than others. Therefore, it should be possible to improve the rate of formation of product in strains with a slow promoter by introducing a fast promoter from another strain. This has been achieved by transduction and conjugation in strains of *Escherichia coli*, where 2.5% of cell protein is in the form of *lac* repressor protein.

In eucaryotic organisms, it is possible to cross strains with different genetic properties and look among the progeny of the cross for those which have acquired, by recombination during DNA replication, the desirable properties from both parents. Figure 2.9(a) shows how this reassortment of genes occurs in a cross between haploid organisms. Figure 2.9(b) shows how haploid gametes are produced in a diploid organism and how recombination may occur. In a cross between diploid organisms, both parents produce four haploid gametes which may fuse in any combination to form diploid offspring in which considerable rearrangement of genes from the parents is possible.

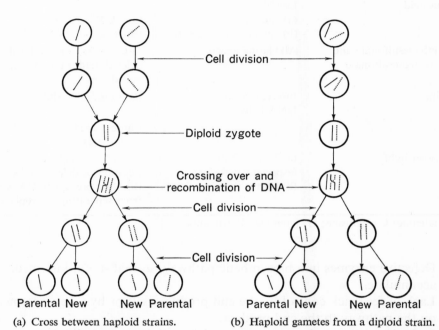

(a) Cross between haploid strains.   (b) Haploid gametes from a diploid strain.

Fig. 2.9. Formation of recombinants during sexual reproduction in eucaryotic organisms. The solid line represents a chromosome with the genes in one form and the dotted line represents the chromosome with the same genes in a different form. For simplicity, only one chromosome is shown, though the nucleus normally contains more than one chromosome.

(a) Cross between haploid parents. The nuclei fuse to form a diploid zygote; the DNA of the two chromosomes replicates, but during replication the molecules break and recombine though not necessarily with their original parts. Thus, chromosomes arise that are different from those of either parent.

(b) Formation of haploid gametes from a diploid organism. The nucleus is shown with two representatives of the same chromosome. The chromosomes pair and the DNA replicates. Crossing over and recombination of the DNA molecules occurs as described in (a) and after two further cell divisions, four haploid gametes (two with chromosomes like the parents, and two with new types of chromosome) are formed.

2.4.5.2. *Stock cultures*

In all the industrially important bacteria and fungi, except some yeasts, the main vegetative phase of growth is haploid, so that variation in genetic material arises only from mutations, provided transduction does not occur in bacteria, and provided sexual spores are not formed in fungi. In maintaining stock cultures, genetic change must be minimized; this is best achieved by preventing nuclear divisions, since most mutations occur as errors in DNA replication.

The method of choice is to store cells or spores (if these are produced) in sealed ampoules at very low temperatures($-200°C$) in liquid nitrogen. This method has the great advantage that the culture can be stored almost indefinitely, thawed and used immediately as an inoculum without loss of viability or diminution in metabolic rate. Cultures kept at $-20°$ to $-60°C$ are satisfactory but less active than those kept in liquid nitrogen. Although storage at $0°$ to $4°C$ allows some growth, this is better than storage at room temperature. Lyophilization (freeze-drying) is widely used and is very convenient since freeze-dried cultures retain viability without genetic change for years when stored at room temperature. It may be noted that all of these methods are, in effect, techniques to immobilize intracellular water and yet retain viability.

2.5. CHANGES DUE TO ALTERATION IN ENVIRONMENT

Extremes of pH, temperature, and osmotic pressure or the presence of sublethal concentrations of inhibitory substance, can cause aberrant morphology, metabolic malfunction, and slow growth rates. Such gross effects can be readily avoided in industrial processes, but even in favorable conditions, the composition and behavior of cells change in response to the environment. Some factors affecting the composition and efficiency of cells will now be discussed.

2.5.1. Changes in the composition of cells with age and with growth rate

In batch culture, cells multiply in a closed system until some nutrient is exhausted or some by-product accumulates to toxic levels. The developing population passes through a number of phases:

1. Lag phase, in which cell mass increases but no division occurs.
2. Logarithmic (log) phase, in which cell numbers increase at a constant rate.
3. Stationary phase, when the rates of death and multiplication are equal.
4. Decline phase, when the rate of death is faster than the rate of multiplication.

The time scale for this cycle is very variable for different organisms; coliforms may complete their cycle in 3–5 hr with optimal conditions, while the tubercle bacillus requires two weeks. If the nutrient concentration is very low, as in sewage treatment, the system may never attain the maximum rate of growth. In addition, phases of increased and decreased rates of growth occur for a brief time at the beginning and end of the log phase.

It is important to appreciate that, in a batch system, the environmental conditions are not constant, even during the phase of constant growth rate. When the composition of bacteria from different stages of batch cultivation was examined, some very significant differences were found. These observations have been collated by Herbert from his own data and from those of other workers and are presented in slightly idealized form in Fig. 2.10. Ordinates on the left of the figure represent (on a logarithmic scale) the numbers and the dry weight of cells per ml, while the ordinates on the right represent (on an arithmetic scale) the weight of individual cells and their content of RNA and of DNA. All scales have arbitrary units chosen so that each initial value is 1.0.

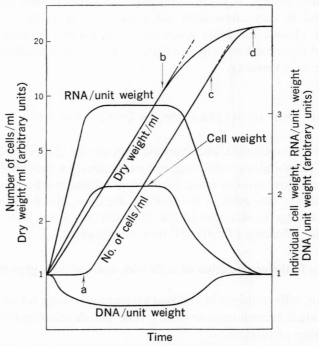

Fig. 2.10. Schematic representation of changes in cell size and chemical composition during growth of bacteria in batch culture. (from Herbert, D. (1961). *11th Symposium of the Society for General Microbiology*, p. 395.)

When resting cells are inoculated into a fresh culture medium with complex nutrients, the cell weight increases logarithmically immediately and is accompanied by a sharp increase in the RNA content per cell. The individual cell weight and the RNA content rise to a certain level, indicated at time *a* in the figure, then both cell numbers and cell mass increase logarithmically. The period from inoculation to time *a* is the time required for each individual cell to initiate cell division, that is, the lag phase.

During the exponential growth phase, the values of RNA per unit cell weight and the individual cell mass remain relatively constant. When some nutrient in the medium is exhausted, the individual cell weight falls, at time b in the figure. The RNA content per cell and the individual cell weight also begin to fall at b, since exponential cell division continues until time c, while the rate of synthesis of RNA decreases during this period. At time c, the rate of cell division declines, eventually entering a resting stage, shown at d in the figure. The amount of DNA is small compared with that of RNA but varies with growth rate; cytological evidence shows that rapidly growing cells contain two or three nuclei per cell while declining cells have only one.

Table 2.9 shows examples of variation in the composition of cells grown in batch culture at different rates or on different substrates. It also indicates that widely different microbes exhibit similar variation.

TABLE 2.9

VARIATION IN COMPOSITION OF MICROORGANISMS GROWN
IN BATCH CULTURE.

Organism	Medium and Conditions	Doubling time (hr^{-1})	$\dfrac{\text{RNA}}{\text{DNA}}$ ratio	Protein (% dry weight)	Reference
Aerobacter aerogenes	G* + casamino acids + yeast extract	0.52	9.6		Rosset, R. *et al.* (1966). *J. Mol. Biol.* **18**, 308
	G + casamino acids	0.67	8.6		
	G + NH$_4$	0.83	5.7		
	Rh† + NH$_4$	5.0	3.3		
Escherichia coli	G + casamino acids + yeast extract	0.55	9.9		
	G + casamino acids	0.71	8.0		
	G + NH$_4$	1.0	6.1		
	Rh + NH$_4$	3.3	3.1		
Escherichia coli	G + NH$_4$ (at 10°C)	10.6	1.7	72	Shaw, M.K., and Ingraham, J.L. (1967). *J. Bacteriol.* **94**, 157
Saccharomyces cerevisiae	G + casamino acids + yeast extract	3.0 (log)	94	33	Polakis, E.S., and Bartley, W. (1966). *Biochem. J.* **98**, 883
		Stationary	32	28	
		Early decline	21	24	
Penicillium atrovenetum	G + asparagine (still growth)	13.0 (log)	7.5	42	Gottlieb, D., and Van Etten, J.L. (1964). *J. Bacteriol.* **88**, 114
		Stationary	5.8	10	

G* = Glucose; Rh† = Rhamnose.

Analyses of cells are of little value unless the conditions of growth are precisely defined, but, strictly speaking, it is impossible to do this for a batch culture since conditions change from minute to minute. However, conditions of growth can be maintained constant indefinitely in a continuous system maintained at steady state.

With such a system, it is possible to select one nutrient to limit growth—or, indeed to ensure that no nutrient limits growth—and in addition, it is possible to select the rate at which the cells will grow under these conditions. It then becomes possible to correlate cell composition and performance with conditions of growth.

Herbert observed *Aerobacter aerogenes* growing at different rates in continuous culture in a mineral-salts medium with glycerol as the growth-limiting substrate; the specific rate of growth (μ) was controlled by changing the dilution rate. The mean cell mass and the percentages of RNA, DNA, and protein per unit dry weight were measured and these are shown in Fig. 2.11. The percentages of protein and DNA changed very little with specific growth rate but both mean cell mass and the percentage of RNA increased markedly with increase in growth rate. Similar changes in RNA, DNA, protein, and cell mass were observed if the substrate limiting growth was changed from glycerol to ammonium ion; this implies that the

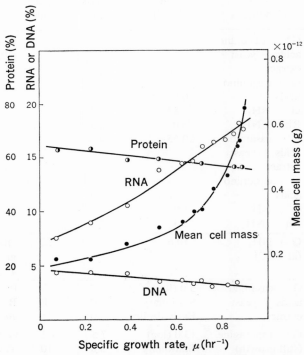

Fig. 2.11. Effect of different rates of growth on the cell mass and chemical composition of *Aerobacter aerogenes* maintained in continuous culture. (from Herbert, D. (1961). *11th Symposium of the Society for General Microbiology*, p. 401.)

specific growth rate was important in determining the composition and mass of cells and that this was independent of the nutrient limiting growth. Essentially similar effects were observed following changes in the rate of growth of *Bacillus megaterium* in continuous culture.

2.5.2. Effect of the substrate limiting growth on the composition of cells

Using continuous cultivation techniques, it is possible to investigate the effects of limiting growth with different substrates while maintaining any desired growth rate. Table 2.10 presents data from studies of eucaryotic and procaryotic cells where growth has been limited by either carbon or nitrogen deprivation, with all other nutrients in excess. A number of important facts are illustrated here.

1. The amounts of DNA and protein remain relatively constant for any one organism, irrespective of the rate of growth or the substrate limiting growth.
2. The amount of RNA increases with increase in growth rate irrespective of the substrate limiting growth.
3. However, cells grown at different temperatures (even though the rate of growth and the medium are identical), show an increase in RNA content at lower temperatures. For instance, at a dilution rate $D=0.05$ with nitrogen limiting, *Candida* contained 3.2, 4.6, and 9.8 g RNA/100g dry weight at 30°, 20° and 15°C, respectively.
4. When nitrogen limits growth and carbon is in excess, cells contain greater amounts of carbon-rich storage materials, either polysaccharides or lipids, or both.
5. More storage materials are made at slow growth rates than at fast ones.
6. Growth rate and nitrogen limitation may not be the only factors influencing the amount of storage material; 31 g of storage material were made by *Azotobacter* growing at a dilution rate of 0.05 when ammonia limited growth, whereas 48 g were made when nitrogen fixation limited growth.

The elegant studies of Tempest and others at Porton show continuous culture to be a most powerful tool for investigating cell physiology. Table 2.11 summarizes some of their studies of *Aerobacter aerogenes*. They limited growth by deprivation of magnesium, carbon, potassium, ammonia or phosphate and the cells were analyzed for their contents of magnesium and RNA at different rates of growth. In addition, the RNA content of the cells was varied by growing the cells at different temperatures, and similar analyses were carried out. Their results indicated that a large amount of the magnesium required by bacteria is needed to make functional ribosomes. Other experiments with potassium-limited and phosphorus-limited cells showed a similar correlation between the concentrations of these elements in the cells and their contents of RNA. Considered together, these data were interpreted as indicating that potassium, as well as magnesium, was required to maintain functional ribosomes. The molar ratios they observed under a wide variety of conditions and rates of growth for magnesium: potassium: RNA (as

TABLE. 2.10

Contents of DNA, RNA, Protein (Prot), Polysaccharide (Poly), and Poly-β-hydroxybutyrate (PHB) When Different Substrates Limit Growth.

Organism and Limiting substrate	Dilution rate (hr⁻¹)	Temperature (°C)	Carbon-limitation					Nitrogen-limitation					Reference
			DNA	RNA (g/100 g dry weight)	Prot	Poly	PHB	DNA	RNA (g/100 g dry weight)	Prot	Poly	PHB	
Torula utilis Glucose or Ammonia	.32					22					21		Herbert, D. (1961). *11th Symp. Soc. Gen. Microbiol.* p. 408
	.22					21					28		
	.07					18					35		
Candida utilis Glucose or Ammonia	.35	30	0.52	9.5	31	23		0.58	11.9	48	24		Brown, C. M., and Rose, A. H. (1969). *J. Bacteriol.* **97**, 261
	.20		0.30	7.6	30	23		0.44	4.3	24	32		
	.10		0.30	6.8	34	23		0.34	4.3	19	34		
	.05		0.35	4.2	35	20		0.36	3.2	21	32		
	.20	20	0.32	14.0	29	25		0.37	10.4	45	27		
	.10		0.39	7.9	33	24		0.34	4.3	26	30		
	.05		0.39	6.9	32	22		0.30	4.6	22	28		
	.10	15	0.34	8.6	30	29		0.36	9.2	44	26		
	.05		0.35	8.5	35	23		0.39	9.8	36	24		
Azotobacter chroococcum Mannitol or Ammonia	.27	30	1.8	13.8	63	5		*	23.4	65	5	0	Dalton, H., and Postgate, J.R. (1969). *J. Gen. Microbiol.* **56**, 307
	.15		2.0	10.2	57	9	3		12.8	69	8	8	
	.05			5.4	57	7	4		5.3	63	19	13	
Mannitol or N₂ fixation	.20	30	1.8	16.8	64	5	3	1.9†	21.0	62	8	8	
	.15		2.4	16.8	58	6		2.7	6.9	40	10	27	
	.05			7.1	55	3	<1	2.3	0.7	39	20	28	

* Dilution rate = .20 .15 .04   † Dilution rate = .22 .10 .05

Where no figure is given the data were not presented.

nucleotide): phosphorus, were 1:4:5:8. Similar ratios were found for *Aerobacter aerogenes* by other workers.

TABLE 2. 11

MAGNESIUM AND RNA CONTENT IN *Aerobacter aerogenes* GROWN IN
A CHEMOSTAT WITH VARIOUS SUBSTRATES LIMITING GROWTH.*

Dilution rate (hr^{-1})	Magnesium content (g/100 g dry weight)					RNA content (g/100 g dry weight)				
	Limiting substrate					Limiting substrate				
	Mg^{2+}	C	K$^+$	NH$_4^+$	PO$_4^{3+}$	Mg^{2+}	C	K$^+$	NH$_4^+$	PO$_4^{3+}$
0.1	0.11	0.13	0.12	0.18	0.10	8.4	9.4	7.3	7.5	5.7
0.2	0.17	0.18	0.16	0.22	0.13	11.0	10.5	11.7	10.0	8.0
0.4	0.22	0.20	0.20	0.26	0.18	15.1	14.1	15.0	13.6	12.5
0.8	0.27	0.30	0.24	0.30	0.31	16.4	18.3	17.6	18.3	20.0

* Tempest, D.W. (1969). *19th Symposium of the Society for General Microbiology*, p. 92.

When these types of experiment were extended to gram-positive organisms, important differences were seen. Magnesium-limited *Bacillus subtilis* contained 300% more potassium and 50% more phosphorus than *Aerobacter aerogenes* grown similarly, and even when potassium was used to limit growth, *Bacillus* cells contained magnesium: potassium: RNA (as nucleotide): phosphorus, in the ratios of 1:13:5:13. Clearly these cells contained phosphorus-rich material not found in gram-negative cells. Why this should affect the potassium concentration was unknown. Further experiments with cells grown with phosphate limitation showed that the amount of intracellular phosphorus was now lower and that the ratio of RNA: phosphorus was the same as that for gram-negative cells. It was then found that the cell walls of magnesium-limited cells had the normal amount of teichoic acid (a phosphorus-rich polymer), while the walls of phosphate-limited cells contained little teichoic acid but large amounts of teichuronic acid (a phosphorus-free uronic acid polymer) instead. The relatively high proportion of phosphorus in magnesium-limited cells was due to teichoic acids in the cell walls and this is virtually absent from both gram-negative cells and from *Bacillus* cells grown with phosphate limitation.

The potassium content of both phosphate-limited and potassium-limited *Bacillus subtilis* remained higher, however, than that found for gram-negative cells. In experiments with *Bacillus megaterium*, it was possible to grow cells lacking both teichoic and teichuronic acids. Both of these polymers are strong anions and it is possible that they bind potassium preferentially within the cell wall, thus raising the content of potassium in the cells. In support of this hypothesis, *Bacillus megaterium* cells lacking both polymers were found to have ratios for magnesium: potassium: RNA (as nucleotide): phosphorus that were similar to those of gram-negative cells.

These studies illustrate that great flexibility is inherent in microbial cells, enabling them to adapt to changes in environment. Increasing knowledge of these effects will enable the biochemical engineer to adjust the environment to produce the type of cell best suited to his purpose. Knowledge of factors influencing the composition of cell walls, for example, seems to have special application for vaccine production and for those interested in cell fragility and the extraction of intracellular products.

2.6. Summary

1. Bacteria and molds, including yeasts and streptomyces, are the microbes of industrial importance. Bacteriophages (viruses) may be important as unwanted contaminants lysing bacterial cultures.

2. Bacteria contain 50–60% protein, yeasts, 40–50% and molds, 20%, most of which is enzymic. Some species have a high lipid (or carbohydrate) content but, generally, this is less than 10%. Viruses contain up to 50% nucleic acids, bacteria c. 20%, yeast c. 10% and molds c. 3%.

3. Viruses are obligate parasites and require a specific host cell for their multiplication, often killing the host in the process. Bacteria are largely unicellular and haploid (one set of genes) with the nuclear material lying free in the cytoplasm. They multiply by simple fission, though a primitive mating system has been found in the enteric bacteria and the pseudomonads. Fungi are multicellular with many migrating nuclei. They multiply by apical extension of hyphal filaments and may produce both sexual and asexual spores. Yeasts, however, grow by budding from single, ovoid cells. Depending on the fungal species, the main growth phase may be either haploid or diploid (two sets of genes).

4. The genetic properties of microbial cells are preserved in their DNA. Four nucleotides (AMP, TMP, CMP, and GMP) are polymerized in two strands bonded to each other to form a double helix. The bases are arranged such that adenine in one strand is complementary to thymine in the other, while guanine is complementary to cytosine. When the DNA replicates, the order of bases in DNA is conserved by the two strands separating, each acting as a template, while nucleotides complementary to those of the existing strands are polymerized to form two double-stranded molecules identical with the original one. Errors may occur in this copying process; these errors may give rise to mutants.

5. Viable mutants are found to the extent of about 1 in 10^6 cells; a tank of 8,000 l with a population of 10^8/ml could have about 10^9 mutants at the end of one batch run. It is thus dangerous to inoculate new fermentations from previous ones. For the purpose of strain development, the general rate of mutation can be raised, but particular genes cannot be selected for mutation.

6. When a protein is required by a cell, the information for its synthesis is transcribed from one strand of the DNA into m-RNA (a labile species of RNA); the order of bases in DNA codes for the order of bases in m-RNA. The m-RNA

associates with ribosomes (nonspecific, relatively stable organelles) where the protein is synthesized. The amino acids are activated and associate with a species of t-RNA (each amino acid has a specific t-RNA). The t-RNA has a triplet of bases (anticodon) which is complementary to the triplet of bases (codon) in m-RNA specifying that particular amino acid.

7. Each amino acid is specified by at least one codon but may have as many as six. At the ribosome, amino acids are polymerized into protein in the order specified by m-RNA; the growing polypeptide is attached to the ribosome until the last amino acid is added, when the ribosome is free to combine with a new m-RNA. Free of the ribosome, m-RNA is hydrolyzed and its component nucleotides enter the nucleotide pool for resynthesis.

8. The fact that a cell contains the gene for a particular property does not mean that it will necessarily always exhibit that property. The ability to control the expression of genetic material is itself a genetic property. The expression of the gene may be inhibited or induced by metabolites or substrates.

9. Changes in the environment can also affect the cell. Differences in cell composition are found when changes are made in temperature, rate of growth, composition of medium or in the particular substrate limiting growth.

NOMENCLATURE

a = Transacetylase gene of the lactose operon
A = Adenine
ATP = Adenosine triphosphate
C = Cytosine
DNA = Deoxyribonucleic acid
G = Guanine
i = Regulator gene in the lactose operon
m-RNA = Messenger RNA
o = Operator gene
p = Promoter gene
P = Phosphate
r = Regulator gene
RNA = Ribonucleic acid
r-RNA = Ribosomal RNA
T = Thymine
t-RNA = Transfer RNA
U = Uracil
μ = Specific growth rate
y = Permease gene for the lactose operon
z = β-galactosidase gene for the lactose operon

TEXTBOOKS AND ARTICLES FOR FURTHER READING AND REFERENCE

General Texts
Abercrombie, M., Hickman, C. J., and Johnson, M. L. (1966). *A Dictionary of Biology* 5th ed. Penguin, England.
Brock, T. D. (1970). *Biology of Microorganisms* Prentice-Hall Inc., New Jersey.
Hawker, L. E., and Linton, A. H. (1971). (Eds.) *Micro-organisms: Function, Form and Environment* Arnold Ltd., London.
Haynes, R. H., and Hanawalt, P. C. (1968). (Eds.) *The Molecular Basis of Life* Freeman & Co., London.
Norris, J. R., and Ribbons, D. W. (1969, 1970, 1971, 1972). (Eds.) *Methods in Microbiology* Vols. 1, 2, 3a, 3b, 4, 5a, 5b, 6a, 6b, 7a, 7b. Academic Press, London.
Stanier, R. Y., Doudoroff, M., and Adelberg, E. A. (1970). *General Microbiology* 3rd ed. Macmillan, London.

Cytology and Structure
Ainsworth, G. C., and Sussman, A. S. (1965). (Eds.) *The Fungi* Vol. I: *The Fungal Cell* Academic Press, London.
Gould, G. W., and Hurst, A. (1969). (Eds.) *The Bacterial Spore* Academic Press, London.
Gunsalus, I. C., and Stanier, R. Y. (1960). (Eds.) *The Bacteria* Vol. I: *Structure* Academic Press, London.
Rose, A. H., and Harrison, J. S. (1969). (Eds.) *The Yeasts* Vol. 1: *Biology of Yeasts* Academic Press, London.
Valentine, R. C. (1962). (Ed.) *Electron Microscopy* Brit. Med. Bull. **18**, 179, British Council, London.

Classification and Taxonomy
Ainsworth, G. C. (1961). *Dictionary of the Fungi* 5th ed. Commonwealth Mycol. Inst., Kew, Surrey.
Alexopoulos, C. J. (1962). *Introductory Mycology* 2nd ed. Wiley, London.
Breed, S. R., Murray, E. G. D., and Smith, N. R. (1957). *Bergey's Manual of Determinative Bacteriology* 7th ed. Williams & Wilkins, Baltimore.
Skerman, V. B. D. (1967). *A Guide to the Identification of the Genera of Bacteria* 2nd ed. Williams & Wilkins, Baltimore.
Smith, G. (1969). *An Introduction to Industrial Mycology* 6th ed. Arnold, London.
Smith, G. M. (1955). *Cryptogamic Botany* 2nd ed. Vol. I: *Algae and Fungi* McGraw-Hill, New York.
Waksman, S. A. (1961). *The Actinomycetes* Vol. II: *Classification, Identification and Description of Genera and Species* Williams & Wilkins, Baltimore.

Genetics and Reproduction
Beckwith, J. R., and Zipser, D. (1970). (Eds.) *The Lactose Operon* Cold Spring Harbor, New York.
Fincham, J. R. S., and Day, P. R. (1971). *Fungal Genetics* 3rd ed. Blackwell, Oxford.
Frisch, L. (1966). (Ed.) In *Cold Spring Harbor Symposia on Quantitative Biology* **31**, 311–408 "Control of Gene Expression." Cold Spring Harbor, New York.
Hartman, P. E., and Suskind, S. R. (1969). *Gene Action* 2nd ed. Prentice-Hall Inc., New Jersey.
Hayes, W. (1968). *The Genetics of Bacteria and their Viruses* 2nd ed. Blackwell, Oxford.

Stahl, F. W. (1969). *The Mechanics of Inheritance* 2nd ed. Prentice-Hall Inc., New Jersey.
Stent, G. S. (1971). *Molecular Genetics* Freeman & Co., San Francisco.

Growth Requirements and Composition of Cells

Ainsworth, G. C., and Sussman, A. S. (1965). *The Fungi* Vol. I: *The Fungal Cell* Academic Press, London.

Dean, A. C. R., Pirt, S. J., and Tempest, D. W. (1972). (Eds.) *Environmental Control of Cell Synthesis and Function* Academic Press, London.

Foster, J. W. (1949). *Chemical Activities of Fungi* Academic Press, New York.

Gunsalus, I. C., and Stanier, R. Y. (1960, 1962). (Eds.) *The Bacteria* Vol. I: *Structure* Vol. IV: *The Physiology of Growth* Academic Press, New York.

Herbert, D. W. (1961). "The Chemical Composition of Microorganisms as a Function of their Environment" *Microbial Reaction to Environment Symp. Soc. Gen. Microbiol.* **11**, 391, Camb. Univ. Press.

Lewin, R. A. (1962). (Ed,) *Physiology and Biochemistry of Algae* Academic Press, New York.

Rose, A. G., and Harrison, J. S. (1971). (Eds.) *The Yeasts* Vol. 2: *Physiology and Biochemistry of Yeasts* Academic Press, London.

Chapter 3

Directing the Chemical Activities of Microorganisms

It would be completely out of place in this text to attempt to cover what is known of the very diverse metabolic activities of microorganisms. References at the end of this chapter will provide a starting point for those who wish for detailed discussions on this subject.

The purpose of this chapter is to provide a basic outline of the ways in which organisms obtain energy and building units for biosynthesis, how these processes are controlled and how they can be manipulated to improve the yield of useful products. It is important for biochemical engineers to be aware of these activities in order that they may know what questions can profitably be asked of expert biochemists and to appreciate which factors in the environment are likely to influence the yield of product and, therefore, must be controlled with precision.

3.1. BIOLOGICAL OXIDATIONS AND THE TRANSFER OF ENERGY

Energy generated in biological oxidations or from light must be conserved in a chemical form before it can be used for endergonic reactions. In the cell, energy is trapped in organic compounds with so-called high energy bonds. These compounds contain sulfur or phosphate in one of the configurations shown below.

$$
\begin{array}{cccc}
& \overset{\displaystyle O}{\overset{\displaystyle \|}{}} & & \overset{\displaystyle O}{\overset{\displaystyle \|}{}} \\
-C-O\sim P-OH & & -N-C-N\sim P-OH & \\
\overset{\|}{R} \quad \overset{|}{OH} & & \overset{\|}{N} \quad \overset{|}{OH} &
\end{array}
$$

$$
\begin{array}{ccc}
\overset{\displaystyle O}{\overset{\displaystyle \|}{}} \quad \overset{\displaystyle O}{\overset{\displaystyle \|}{}} \quad \overset{\displaystyle O}{\overset{\displaystyle \|}{}} & & -C\sim S- \\
-P-O\sim P-O\sim P-OH & & \overset{\|}{O} \\
\overset{|}{OH} \quad \overset{|}{OH} \quad \overset{|}{OH} &
\end{array}
$$

$\sim$ = bond which releases a large amount of energy on hydrolysis

Adenosine triphosphate (ATP) is one of the most important high energy compounds in cellular metabolism—it is the energy currency of the cell. When the two terminal phosphate groups of ATP are hydrolysed, 12,000 cal are released per phosphate, but hydrolysis of adenosine monophosphate yields only 1,500 cal.* Although ATP contains two "high energy" phosphate bonds, it is common (though not always so) for the terminal phosphate only to be involved in reactions.

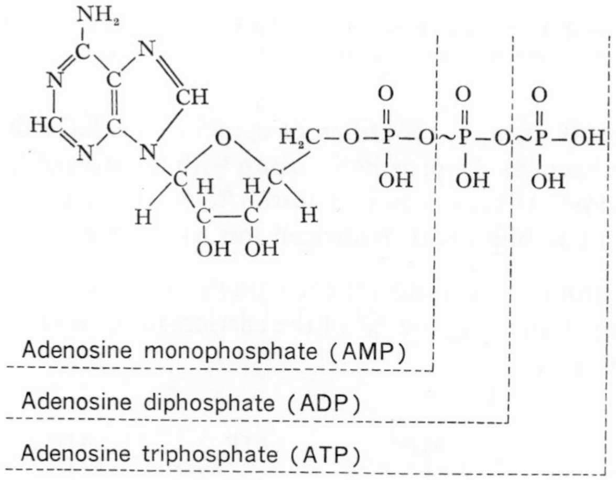

Adenosine monophosphate (AMP)

Adenosine diphosphate (ADP)

Adenosine triphosphate (ATP)

Cellular oxidations are mediated by enzymes which have a cofactor or prosthetic group that accepts protons and electrons from the substrate and passes these via a series of electron-accepting compounds, each of higher oxidation/reduction potential (E_0') than the donor compound, to oxygen or some other acceptor. The most important carriers of this electron transport pathway are listed below.

1. Nicotinamide adenine dinucleotide (NAD). The nicotinamide portion of this nucleotide is capable of accepting hydrogen from a reduced substrate as shown below.

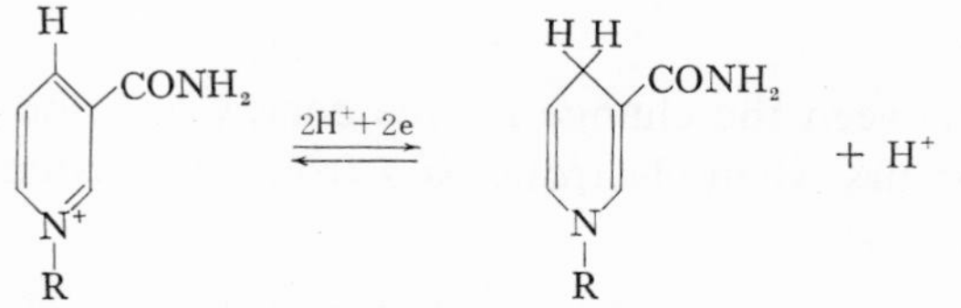

When R = ribose-phosphate-phosphate-ribose-adenine, the prosthetic group is nicotinamide adenine dinucleotide (NAD)

When R = ribose-phosphate-phosphate-ribose, 2-phosphate-adenine, the prosthetic group is nicotinamide adenine dinucleotide phosphate (NADP)

2. Flavin adenine mononucleotide (FMN) and flavin adenine dinucleotide (FAD). Riboflavin is the portion of these two nucleotides that accepts hydrogen from $HADH_2$ or directly from the substrate.

---

* The amounts of energy released on hydrolysis of ATP are dependent upon the conditions of hydrolysis, particularly on the pH and the concentration of reactants. Under physiological conditions, it is thought that the number of calories released on hydrolysis of the phosphate groups is of the order given in the text.

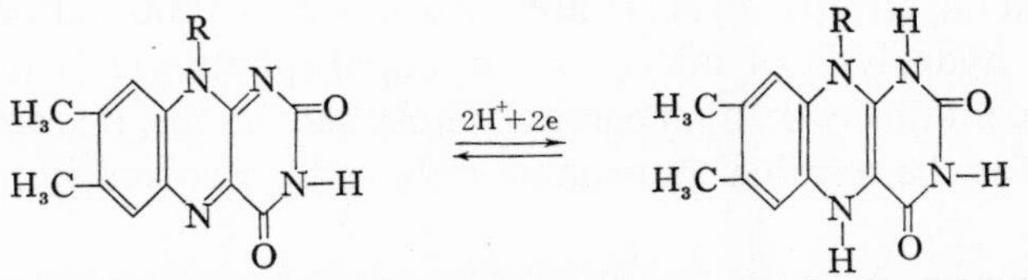

When R = ribose-phosphate, the prosthetic group is flavin mononucleotide (FMN)
When R = ribose-phosphate-phosphate-adenine, the prosthetic group is flavin adenine dinucleotide (FAD)

3. Cytochromes (Cyt. a, b, c). Different cytochromes have different E_0' values but all have haem as the prosthetic group with the iron of the haem accepting electrons from the flavin carriers of lower E_0' and passing them to oxygen. The oxygen then combines with hydrogen ion to form water.

Figure 3.1 sets out the reactions for the oxidation of a substrate via the electron transport pathway, indicating the E_0' of the carriers in the series, and the reactions where ATP is formed.

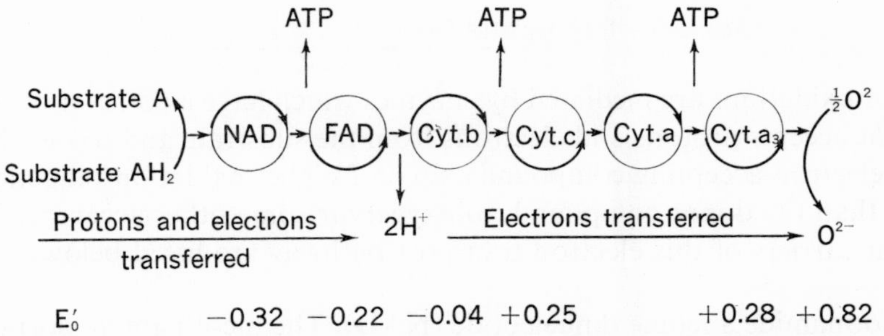

Fig. 3.1. Electron transport pathway showing the oxidation/reduction potential (E_0') of the reactants; protons and electrons pass via a dehydrogenase, a flavoprotein, and cytochromes b, c, a, and a_3 to oxygen as the final electron acceptor. The sites of ATP formation are also indicated.

The relationship between the change in free energy and the change in potential difference which occurs when electrons pass from one system to another can be expressed as follows:

$$\Delta F° = -nF\Delta E_0$$

where

$\Delta F°$=free energy change (cal/mole) at standard state
n=number of electrons transferred
ΔE_0=potential difference (volts)
F=faraday (23,063 cal/volt equivalent)

For biological systems, n is generally 2 and $F°$ is replaced by F' as conditions are not at the standard state. Then,

$$\Delta F' = -46,126 \times \Delta E_0'$$

TABLE 3.1

THE OXIDATION/REDUCTION POTENTIALS OF SOME BIOLOGICALLY IMPORTANT COMPOUNDS ARE SHOWN ON THE LEFT. THE CALCULATED CHANGES IN POTENTIAL AND FREE ENERGY WHICH OCCUR WHEN A COMPOUND AT THE E_0' OF X IS OXIDIZED TO THE E_0' OF OXYGEN VIA 2 INTERMEDIATES, Y AND Z, IS SHOWN ON THE RIGHT. THE NUMBER OF HIGH ENERGY PHOSPHATE BONDS OBSERVED TO BE FORMED IN MITOCHONDRIA IN SOME OF THE OXIDATION STEPS IS ALSO INDICATED.

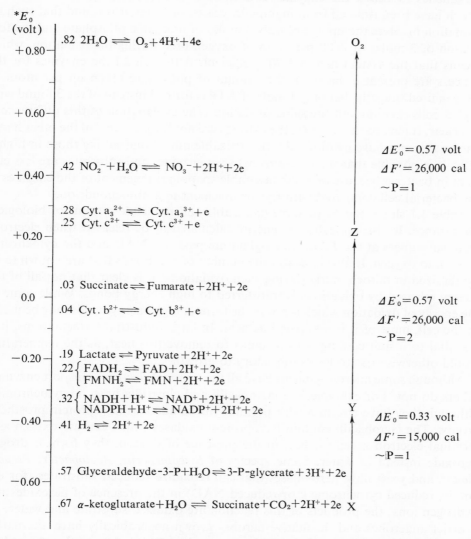

$*E_0'$
(volt)

+0.80 — .82 $2H_2O \rightleftharpoons O_2 + 4H^+ + 4e$ O_2

+0.60 —

$\Delta E_0' = 0.57$ volt
$\Delta F' = 26{,}000$ cal
$\sim P = 1$

+0.40 — .42 $NO_2^- + H_2O \rightleftharpoons NO_3^- + 2H^+ + 2e$

.28 $Cyt.\ a_3^{2+} \rightleftharpoons Cyt.\ a_3^{3+} + e$
.25 $Cyt.\ c^{2+} \rightleftharpoons Cyt.\ c^{3+} + e$ Z

+0.20 —

.03 $Succinate \rightleftharpoons Fumarate + 2H^+ + 2e$

0.0 — .04 $Cyt.\ b^{2+} \rightleftharpoons Cyt.\ b^{3+} + e$

$\Delta E_0' = 0.57$ volt
$\Delta F' = 26{,}000$ cal
$\sim P = 2$

−0.20 — .19 $Lactate \rightleftharpoons Pyruvate + 2H^+ + 2e$
.22 $\begin{cases} FADH_2 \rightleftharpoons FAD + 2H^+ + 2e \\ FMNH_2 \rightleftharpoons FMN + 2H^+ + 2e \end{cases}$

.32 $\begin{cases} NADH + H^+ \rightleftharpoons NAD^+ + 2H^+ + 2e \\ NADPH + H^+ \rightleftharpoons NADP^+ + 2H^+ + 2e \end{cases}$ Y

−0.40 — .41 $H_2 \rightleftharpoons 2H^+ + 2e$

$\Delta E_0' = 0.33$ volt
$\Delta F' = 15{,}000$ cal
$\sim P = 1$

.57 $Glyceraldehyde-3-P + H_2O \rightleftharpoons 3-P-glycerate + 3H^+ + 2e$

−0.60 —

.67 α-ketoglutarate $+ H_2O \rightleftharpoons Succinate + CO_2 + 2H^+ + 2e$ X

* These E_0' values may not apply to the electron carriers of all species but their order relative to each other is the same.

Table 3.1 was compiled from data presented by DOLIN, M.I. (1961). *The Bacteria* (Eds.). GUNSALUS, I.C., and STANIER, R.Y. Vol. II: *Metabolism* p. 319, "Microbial electron transport mechanisms." Academic Press, London.

If the potential difference of a system before and after oxidation is known, from the above equation it is possible to calculate the amount of energy released during an oxidation. For example, if electrons pass from $NADH+H^+$ to oxygen, $\Delta E_0' = 1.14$ volt and so $\Delta F' = -52,000$ cal. Theoretically, this would allow the synthesis of 4 high energy phosphate bonds. When these reactions are studied in mitochondria (organelles in which the enzymes concerned in electron transport are situated) which have been isolated from mammals, yeasts, or fungi, it is found that such an oxidation involves the uptake of only 3 moles of inorganic phosphate and the formation of 3 moles of ATP per atom of oxygen used during the oxidation, which means that the system is about 70% efficient. Although all the enzymes for the process are present in bacteria, the amount of phosphate taken up per atom of oxygen used suggests that only 1 mole of ATP is formed instead of the 3 found with higher cells carrying out the same oxidation. The explanation of this difference is not clear. It may be that *in vitro*, the extracts are not fully active. On the other hand, bacteria do not have cytologically recognizable mitochondria like those in higher cells. It is possible that the reactions between enzymes and substrates are less efficient in bacteria because of a less favorable spatial arrangement of the enzymes in the bacterial cell compared with the arrangement in a mitochondrion.

Table 3.1 shows the E_0 of some oxidizable and reducible systems of biological importance. It also indicates the energy calculated to be released when electrons from substances at the E_0' of α-ketoglutarate pass, via NAD and the cytochrome system, to oxygen. It also indicates the number of $\sim$P bonds that are known to be synthesized in mitochondria during such oxidations; it is clear that not all of the energy released by oxidations is transferred to high energy bonds. While some of the energy of oxidation which is not in the form of high energy bonds may be useful to the cell, most of it is dissipated as heat. In large industrial fermentations, it is essential to install cooling coils in tanks to remove this heat, as the temperature would otherwise rise to levels inhibitory to growth.

Although some microorganisms have all the terminal electron transport enzymes, others do not. For example, the lactobacilli and clostridia have no cytochromes, although they have enzymes with pyridine nucleotide and flavoprotein prosthetic groups. The lactobacilli contain flavoprotein oxidases which can use oxygen as a terminal electron acceptor, but, in the presence of oxygen, they form hydrogen peroxide instead of water. Some species of *Streptococcus, Acetobacter, Pseudomonas,* and yeast have peroxidases that can reoxidize reduced substrates (for example, reduced cytochrome c or reduced NAD) in the presence of peroxides and hydrogen ions; the products of this reaction are oxidized substrate and water.

Strict anaerobes and facultative aerobes grown anaerobically have alternative ways of reoxidizing their reduced hydrogen carriers. This is frequently achieved by coupling the oxidation of one substrate with the reduction of another, and often the reduced compound accumulates as an end product. For example, under anaerobic conditions, the lactobacilli reoxidize dehydrogenases reduced during glycolysis, by passing the hydrogen to pyruvate to form lactate.

$$CH_3 \cdot CO \cdot COOH + NADH + H^+ \rightleftharpoons CH_3 \cdot CHOH \cdot COOH + NAD^+$$

The species of clostridia that ferment hexoses anaerobically reoxidize their reduced-hydrogen carriers by reducing acetoacetate (as the coenzyme A complex) to butyric acid or butanol; yeast reduces acetaldehyde to ethanol; some enteric bacteria reduce the dicarboxylic acids oxalacetic, malic, and fumaric to succinic acid; and acetoin is reduced to 2,3-butylene glycol in order to oxidize reduced flavin and nicotinamide carriers. A few species of *Pseudomonas* and a *Thiobacillus* can use nitrate as an electron acceptor, reducing the nitrate to gaseous nitrogen in the process; many species of bacteria and fungi can reduce nitrate to nitrite, and a few species of bacteria reduce sulphate to sulphide as a means of oxidizing hydrogen carriers.

Clearly, microorganisms and particularly bacteria, can reoxidize hydrogen carriers in the absence of oxygen in a number of ways and sometimes commercially useful products accumulate in the medium, for instance, lactic acid and ethanol.

3.2.   Systems to Provide Energy and Metabolites for Growth

As Kluyver observed in the 1930's, there are remarkable similarities in the ways in which cells of widely different types obtain energy and form the intermediate compounds they need for biosynthesis. Microbes of many types attack carbohydrates using the glycolytic or Embden-Meyerhof-Parnas (EMP) and the hexose monophosphate (HMP) pathways and, under aerobic conditions, oxidize the substrate further via the tricarboxylic acid (TCA) or Krebs cycle. Not all microorganisms, however, attack hexoses using these routes. Although the chemical transformations achieved in the reactions of these pathways by different organisms are the same, each organism catalyzes the reactions with enzymes which are somewhat different from the enzymes carrying out the same reaction in other cells. In the next sections, the pathways common to many cells will be outlined, followed by a discussion of some pathways peculiar to particular groups of microbes. Biochemical texts cited at the end of the chapter should be consulted for chemical formulae, the enzymes involved, their cofactors and equilibrium constants.

3.2.1.   Glycolysis or Embden-Meyerhof-Parnas (EMP) pathway

The intermediates of the glycolytic pathway are shown in Fig. 3.2. Points of importance in the sequence are listed below.

1. Activation of glucose with ATP; isomerization and a second phosphorylation to give fructose-1,6-diphosphate and 2 ADP.
2. Cleavage of fructose-1,6-diphosphate to give 2 molecules of triosephosphate.
3. Oxidation of 3-phospho-glyceraldehyde with reduction of NAD and uptake

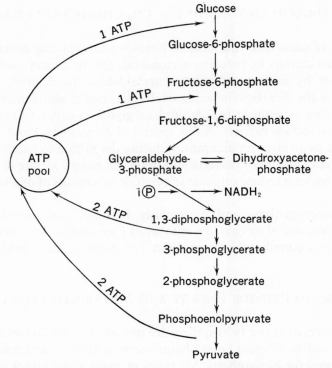

Summary: Glucose+2 ATP+2 NAD⁺ ⟶ 2 Pyruvate+4 ATP+2 NADH₂

Fig. 3.2. The glycolytic pathway.

i$\circled{P}$=H_2PO_3; ATP=adenosine triphosphate; NAD=nicotinamide adenine dinucleotide.

of inorganic phosphate (i $\circled{P}$), resulting in the formation of a high energy bond in 1,3-diphospho-glycerate.

4. Transfer of the ~P from 1,3-diphospho-glycerate to ADP.
5. Isomerization of 3-phospho-glycerate, followed by dehydration to form a ~P in phospho-enol-pyruvate.
6. Transfer of the ~P of phospho-enol-pyruvate to ADP and the formation of pyruvate and ATP.

There is thus a net gain of 2 moles of ATP per mole of glucose oxidized, and 2 moles of NAD are reduced. The glycolytic pathway to pyruvate is widely distributed among all types of cells. Oxygen is not required for any of its reactions, but of course the reduced NAD must be regenerated by some mechanism. In the absence of oxygen, lactobacilli reduce pyruvate to lactate and yeast reduces acetaldehyde to ethanol.

3.2.2. Hexose monophosphate pathway (HMP)

The hexose monophosphate pathway, in a much abbreviated form, is shown in Fig. 3.3. The HMP is very important in providing the pentoses needed for the synthesis of nucleic acids and nucleotide-containing prosthetic groups, in providing the starting materials for the synthesis of aromatic amino acids and vitamins, and in supplying reduced pyridine nucleotides needed in many biosynthetic reactions. In addition, "active" (C_2) and (C_3) fragments needed for the biosynthesis of such intermediates as leucine, isoleucine and valine can be derived from this pathway. The HMP does not provide energy directly but, potentially, $NADPH_2$ produced in the pathway could be a source of ATP if the electrons from it were passed

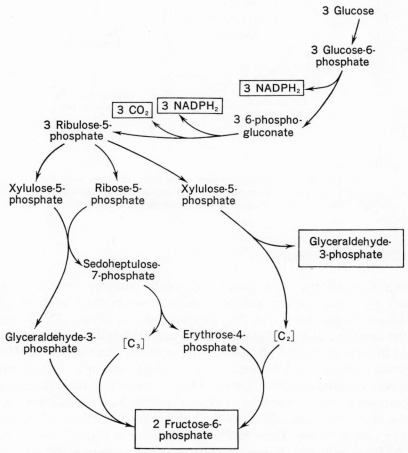

Summary: 3 Glucose-6-phosphate + 6 NADP⁺ ⟶ 2 Fructose-6-phosphate + Glyceraldehyde-3-phosphate + 3 CO_2 + 6 $NADPH_2$

Fig. 3.3.   Oxidation of glucose via the hexose monophosphate pathway (HMP).

by the electron transport chain to oxygen. Thus, the HMP is largely concerned in biosynthetic metabolism and it is important to see its links with glycolysis and to appreciate that the cell must regulate the amount of glucose which is metabolized via glycolysis and via the HMP. Figure 3.4 shows in schematic form the interconnections between glycolysis and the HMP and indicates the compounds withdrawn from the pathway for biosynthetic purposes.

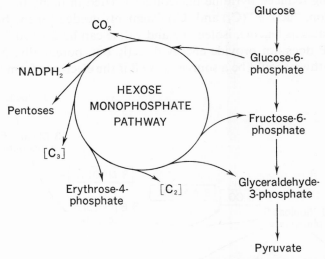

Fig. 3.4. Schematic representation of the links between glycolysis and the hexose monophosphate pathway and the intermediates provided for biosynthesis from the monophosphate pathway.

3.2.3. Heterolactic lactobacilli

This group of organisms has a modified glycolytic pathway, since the organisms lack a key glycolytic enzyme, aldolase, which converts fructose diphosphate to glyceraldehyde-3-phosphate and dihydroxyacetone-phosphate. These organisms oxidize the number 1 carbon of glucose and decarboxylate the resulting 6-phosphogluconate to form ribulose-5-phosphate, just as in the first steps of the HMP.

The pentose phosphate is isomerized and then cleaved in a reaction which involves uptake of inorganic phosphate and the formation of the high energy phosphate compound, acetyl~phosphate; glyceraldehyde-3-phosphate from the other half of the pentose is metabolized by the conventional reactions of the glycolytic pathway to yield pyruvate. These reactions are set out in Fig. 3.5; the reactions involving ATP are indicated. If acetyl~phosphate gives rise to acetic acid, the high energy bond is conserved and the net gain of ATP is 2, as in glycolysis. If, however, acetyl~phosphate is reduced to ethanol, the high energy bond is lost and the net gain of ATP is only 1.

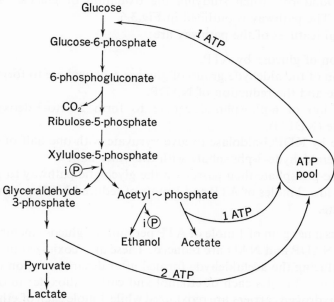

Fig. 3.5.   Metabolism of glucose by heterolactic lactobacilli; the net gain of ATP will be 1 mole if ethanol is produced from acetyl~phosphate, but 2 moles if acetic acid is formed. Glyceraldehyde-3-phosphate is metabolized by the glycolytic pathway. i℗=H₂PO₃.

3.2.4. Entner-Doudoroff pathway

This pathway, which has some reactions in common with glycolysis, was found by

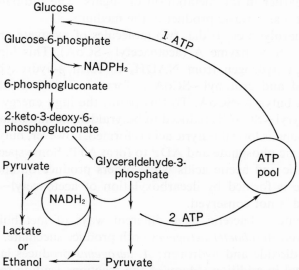

Fig. 3.6.   Entner-Doudoroff pathway for the oxidation of glucose; there is a net gain of 1 ATP. The reactions from glyceraldehyde-3-phosphate to pyruvate are those of glycolysis.

Entner and Doudoroff when studying the oxidation of glucose by species of *Pseudomonas*. The pathway is outlined in Fig.3.6.

The essential features of the pathway are:

1. Activation of glucose by ATP.
2. Oxidation of the aldehyde group of glucose-6-phosphate to form 6-phospho-gluconate and the reduction of NADP.
3. Dehydration of 6-phospho-gluconate to form 2-keto-3-deoxy-6-phospho-gluconate (KDPG).
4. Cleavage by KDPG-aldolase to give pyruvate with one half of the molecule and glyceraldehyde-3-phosphate with the other half.
5. The triose-phosphate then passes via the glycolytic pathway to pyruvate and this provides 2 moles of ATP and 1 mole of reduced NAD per mole of triose-phosphate.

There is thus a net gain of 1 mole of ATP per mole of glucose metabolized, and 1 mole each of NADP and NAD are reduced. These are reoxidized in *Pseudomonas lindneri* by reducing the acetaldehyde formed after decarboxylation of 2 moles of pyruvate, to form 2 moles each of ethanol and carbon dioxide; in other pseudomonads, the hydrogen carriers are reoxidized while 1 mole each of ethanol, lactate, and carbon dioxide are formed.

3.2.5. Anaerobic metabolism of pyruvate

Different organisms metabolize pyruvate under anaerobic conditions by different pathways. These reactions function mainly to reoxidize the reduced hydrogen carriers formed earlier in the metabolism of sugars. The ultimate hydrogen acceptor accumulates as a waste product in the medium.

With the saccharolytic clostridia, two molecules of pyruvate are condensed to form acetoacetyl~S coenzyme A (acetoacetyl~SCoA). This high energy compound can accept hydrogen from NADH$_2$ to form β-hydroxybutyryl~SCoA, water is removed and crotonyl~SCoA is formed. This is further reduced by NADH$_2$ to form butyryl~SCoA. To this point, the high energy bond has been conserved. If butyryl~SCoA is reduced to butyraldehyde and butanol by NADH$_2$ the high energy bond is lost; if butyric acid is formed, however, butyryl~SCoA can react with inorganic phosphate and ADP to form ATP. Some species of clostridia accumulate butyric and acetic acids while others produce butanol, ethanol and acetone. Acetone is formed by decarboxylation of acetoacetyl~SCoA and the high energy bond is not conserved.

The various enteric bacteria have different ways of metabolizing pyruvate. *Escherichia coli* and *Aerobacter aerogenes* both produce succinate, acetate, lactate, ethanol, carbon dioxide and hydrogen; *A. aerogenes* produces large amounts of 2,3-butylene glycol in addition. Many of the reactions leading to these products oxidize reduced hydrogen carriers, but the only reaction to yield energy to the cell is that forming acetic acid and ATP from acetyl~SCoA. Most species of *Salmonella* metabolize pyruvate in much the same way as *E. coli*. *Salmonella typhi*, however, lacks the enzyme formic hydrogen lyase and therefore accumulates formic acid instead of hydrogen and carbon dioxide as *E. coli* does.

Some lactobacilli (homofermentative) oxidize virtually all the $NADH_2$ produced from glycolysis by reducing pyruvate to lactate. Similarly, yeast regenerates $NADH_2$ by reducing acetaldehyde (formed by decarboxylation of pyruvate) to ethanol. Neither of these reactions yields ATP.

These different pathways of pyruvate metabolism have been summarized in Fig. 3.7.

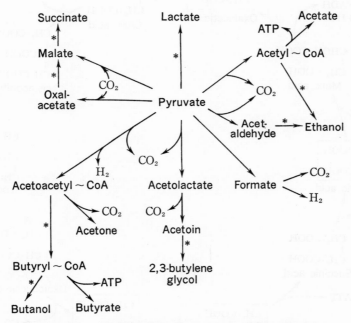

Fig. 3.7.   Anaerobic metabolism of pyruvate by different organisms.

$* =$ Reactions that oxidize $NADH_2 \longrightarrow NAD^+$

Organism	Major Products
Clostridia	Butyric, acetic acids, butanol, ethanol, acetone, CO_2, H_2.
Enteric bacteria	Acetic acid, ethanol, CO_2, H_2 (or formic acid), lactic acid, succinic acid, 2,3-butylene glycol.
Yeast	Ethanol, CO_2.
Homofermentative lactobacilli	Lactic acid.

3.2.6. Tricarboxylic acid (TCA) cycle

When oxygen is available, many microbes can oxidize pyruvate by a series of reactions known as the tricarboxylic acid (TCA) or Krebs cycle. These reactions are shown in Fig. 3.8. This cycle is of the greatest importance in biosynthesis, as it

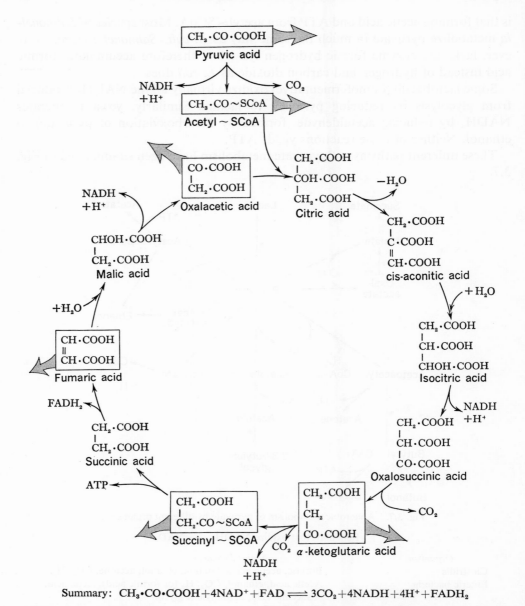

Summary: $CH_3 \cdot CO \cdot COOH + 4NAD^+ + FAD \rightleftharpoons 3CO_2 + 4NADH + 4H^+ + FADH_2$

Fig. 3.8. Tricarboxylic acid cycle. The compounds put inside squares are the starting compounds for synthetic reactions; α-ketoglutaric, fumaric, oxalacetic, and pyruvic acids are all concerned in amino acid formation, succinyl~SCoA is used for porphyrin synthesis, and acetyl~SCoA is needed for acetylation reactions.

provides both the carbon skeletons needed as starting materials and the energy needed for the reactions.

For most microorganisms, glutamate is the key amino acid synthesized *de novo* from ammonia and carbohydrate; it is formed by combining ammonia with α-ketoglutarate. Many bacteria, however, can combine ammonia with fumarate to form aspartate; by transamination, these amino acids act as amino-donors to α-keto acids such as pyruvic, oxalacetic, and α-ketoisovaleric to form other amino acids. Another important biosynthetic pathway begins when succinyl$\sim$SCoA from the TCA cycle combines with glycine to form the pyrroles which are needed for the formation of porphyrins; the haem prosthetic group of the cytochromes is an iron-porphyrin and the prosthetic group of chlorophyll is a magnesium-porphyrin.

Since intermediates of the TCA cycle are used extensively for biosynthesis, these must be replenished from some other source to keep the cycle functioning. This is achieved by fixing carbon dioxide to C_3 intermediates from glycolysis. A number of such reactions have been found in microorganisms:

1. *PEP carboxylase*
   Phosphoenolpyruvate $+$ CO_2 $\longrightarrow$ Oxalacetate $+$ iP

2. *PEP carboxykinase*
   Phosphoenolpyruvate $+$ CO_2 $+$ ADP $\rightleftharpoons$ Oxalacetate $+$ ATP

3. *PEP carboxytransphosphorylase*
   Phosphoenolpyruvate $+$ CO_2 $+$ iP $\rightleftharpoons$ Oxalacetate $+$ iPP

4. *Pyruvate carboxylase*
   Pyruvate $+$ CO_2 $+$ ATP $\longrightarrow$ Oxalacetate $+$ ADP $+$ iP

5. *Malic enzyme*
   Pyruvate $+$ $NADPH_2$ $+$ CO_2 $\rightleftharpoons$ Malate $+$ $NADP^+$

Reaction 1 seems to be the main replenishing mechanism for enteric bacteria and reaction 4 for yeasts, but regulation of the activities of this group of enzymes is very complex and incompletely studied and may well vary with different organisms and substrates.

Energy is derived from the reoxidation of hydrogen carriers reduced by the reactions of the TCA cycle. Four reactions are catalyzed by dehydrogenases with nicotinamide prosthetic groups and one by a dehydrogenase with a flavin prosthetic group. In the presence of oxygen, these carriers are reoxidized by the reactions of the electron transport pathway shown in Fig. 3.1, and for eucaryotic cells, three high energy phosphate bonds are generated. Experimentally, the number of sites of ATP formation can be determined by measuring the number of moles of phosphate esterified per atom of oxygen consumed (termed the P/O ratio). The number of moles of ATP formed during the oxidation of glucose to carbon dioxide and water via glycolysis and the TCA cycle is set out in Table 3.2. With bacteria, there are difficulties in the interpretation of data. Direct measurements of P/O ratios generally give values less than 1, but indirect estimates suggest that there may be more than 1 site of ATP formation in bacteria. Hence, the estimated number of moles of ATP produced by bacteria shown in Table 3.2 may be too conservative.

TABLE 3.2

AMOUNT OF ENERGY DERIVED BY MICROORGANISMS DURING THE OXIDATION OF GLUCOSE VIA
GLYCOLYSIS AND THE TRICARBOXYLIC ACID CYCLE.

Pathway	Number of moles of ATP formed	
	Algae, Yeast and Fungi	Estimated for bacteria
Glycolysis		
Glucose $\longrightarrow$ 2 Pyruvate	2	2
Oxidation of: 2 NADH + 2 H$^+$	$2 \times 3 =$ 6	$2 \times 1 =$ 2
TCA Cycle		
Pyruvate $\longrightarrow$ 3 CO$_2$ + 3 H$_2$O		
Oxidation of: 4 NADH + 4 H$^+$	$4 \times 3 = 12$	$4 \times 1 = 4$
1 FADH$_2$	$1 \times 2 = 2$	1
Succinyl$\sim$SCoA $\longrightarrow$ 1 ATP	$\dfrac{1}{15}$	$\dfrac{1}{6}$
2 Pyruvate $\longrightarrow$ 6 CO$_2$ + 6 H$_2$O	$2 \times 15 =$ 30	$2 \times 6 =$ 12
Glucose + 3 O$_2$ $\longrightarrow$ 6 CO$_2$ + 6 H$_2$O	38	16

3.2.7. Yield of cells

It was recognized very early that the crop of cells from microbial cultures was closely related to the way in which the energy substrate was metabolized and that pathways that gave high yields of ATP also gave large amounts of cells.

Bauchop and Elsden (for ref. see p. 71) studied the yields from cultures grown anaerobically in a complex medium, where the source of energy limited growth. They expressed the yield (Y_s) as g dry weight of cells/mole of energy substrate used, or as Y_{ATP} = g dry weight of cells/mole of ATP synthesized during the metabolism of a mole of the energy substrate. They found that, while Y_s varied with different substrates and organisms, the mass of cells produced per mole of ATP was almost constant, averaging 10 ± 2 g, irrespective of the organism or the substrate. Since then, a number of studies by other workers has confirmed these findings (see Table 3.3). Hernandez and Johnson (for ref. see p. 71) emphasized, however, that the value of Y_{ATP} is a constant only if the energy substrate is indeed the one factor limiting growth. If inhibitory substances are present, Y_{ATP} values are likely to be less than 10, and when cells have energy substrate in excess, yields may be greater than 10 because of polymerization of the substrate into storage material. Provided the energy substrate does limit growth, the Y_s of cultures can be useful in studies of unknown metabolic pathways. For example, if anaerobic metabolism of glucose yields 40 g of cells per mole of glucose, it is likely that there are 4 sites of ATP synthesis. If there is evidence for glycolysis, 2 ATP will be produced by this route,

TABLE 3.3

YIELDS OF DIFFERENT ORGANISMS GROWN ANAEROBICALLY WITH THE
ENERGY SUBSTRATE LIMITING GROWTH.

Organism	Energy substrate	Calculated ATP (mole)	Y_S (g/mole substrate)	Y_{ATP} (g/mole ATP)	Reference
Streptococcus					
faecalis	Glucose	2.0	22	11	Bauchop, T., and
	Glucose	2.0	23	11.5	Elsden, S.R. (1960).
	Glucose	2.0	18.5	9.3	*J. Gen. Microbiol.* **23,**
	Ribose	1.67	21	12.6	457
	Arginine	1.0	10	10	
Saccharomyces					
cerevisiae	Glucose	2.0	21	10.5	
Pseudomonas					
lindneri	Glucose	1.0	8.3	8.3	
Propionibacterium					
pentosaceum	Glucose		37.5		
Lactobacillus					
plantarum	Glucose	2.0	18.8	9.4	Oxenburg, M.S., and Snoswell, A.M. (1965). *J. Bacteriol.* **89,** 913
Aerobacter					
aerogenes	Glucose	2.55	26.1	10.2	Hadjipetrou, L.P. *et*
	Fructose	2.50	26.7	10.7	*al.* (1964). *J. Gen. Microbiol.* **36,** 139
Clostridium					
tetanomorphum	Glutamate	0.62	6.8	10.9	Twarog,R.,and Wolfe, R.S. (1963). *J. Bacteriol.* **86,** 112
Aerobacter					
cloacae	Glucose	2.15	27.1	12.6	Hernandez, E., and
	Glucose	2.43	21.8	9.0	Johnson, M.J. (1967). *J. Bacteriol.* **94,** 991
Escherichia					
coli	Glucose	2.55	24.0	9.4	
Desulfovibrio					
desulfuricans	Pyruvate	1.0	9.4	9.4	Senez, J.C. (1962).
	Lactate	1.0	9.9	9.9	*Bact. Rev.* **26,** 95

leaving 2 other sites to be recognized. Conversely, if the energy yield is known, it is possible to predict the yield of cells likely to be achieved for a given weight of energy substrate. Some caution is needed, however, before applying the value $Y_{ATP}=10$ as a universal, biological constant; some bacteria give other values for Y_{ATP} and there is no explanation at present for this deviation.

The determination of Y_{ATP} for aerobically grown bacteria is troublesome. Some of the energy substrate is used for synthesis, media components other than the supposed energy substrate may be metabolized for energy, and the extent and control of uncoupling of oxidation and phosphorylation is uncertain. In addition, organisms vary in the amount of energy used for maintenance. Since direct measurements of P/O ratios give values less than 1, indirect estimates of the efficiency of ATP production have been proposed based on the yield of cells either per mole of oxygen used for energy production or per "available" electron in the energy substrate. Using the data from oxygen yields and assuming $Y_{ATP}=10$, different workers using different organisms and substrates have calculated P/O ratios from 0.5. to 3.0. It seems that the question of the efficiency of ATP production in microorganisms during electron transport cannot yet be resolved.

3.2.8.   Glyoxylic acid cycle and metabolism of fats

Certain bacteria and fungi are able to grow when C_2 substrates are supplied as the sole source of carbon. These organisms have all the enzymes for the operation of the TCA cycle but have two enzymes in addition—one which splits isocitrate to succinate and glyoxylate, and a second which condenses glyoxylate and acetyl$\sim$ SCoA to form malate. The operation of these two enzymes enables the cell to synthesize dicarboxylic acids which will then permit the complete TCA cycle to operate for energy and to provide intermediates for biosynthesis. Phosphoenolpyruvate formed by decarboxylation of the oxalacetate is a key compound for the biosynthesis of pentoses and hexoses from C_2 substrates, since it is the starting point for the reactions which reverse glycolysis. Figure 3.9 shows the TCA and glyoxylic acid cycles, the links between these cycles, and the synthesis of pentoses and hexoses.

When organisms grow with a fat as the sole source of carbon, the triglyceride is first cleaved into glycerol and its three constituent fatty acids. The glycerol enters the glycolytic pathway via dihydroxyacetone-phosphate; the fatty acids are "activated" by ATP in the presence of coenzyme A to form an acyl$\sim$SCoA ester. This is oxidized at the β-carbon atom and the resulting β-keto-acyl$\sim$SCoA is hydrolyzed to give acetyl$\sim$SCoA and an acyl$\sim$SCoA ester with two less carbon atoms. The oxidative reactions are then repeated with the removal of further acetyl$\sim$SCoA units till either a $C_3\sim$ or a $C_4\sim$CoA ester remains. The glyoxylate cycle is vital to such cells since they are growing essentially on acetate.

The reactions taking place during the oxidation of a fat are set out in an abridged form in Fig. 3.10. Notice that in addition to providing acetate, the β-oxidation of fatty acids yields two hydrogen pairs per acetate. These hydrogen pairs, passing by the electron transport pathway to oxygen, are an important source of ATP.

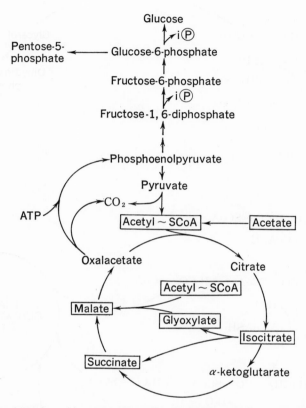

Fig. 3.9. Glyoxylic acid cycle and its connections with the TCA cycle; the compounds concerned in the glyoxylate cycle are placed in squares. When growing on acetate, cells decarboxylate oxalacetate to provide phosphoenolpyruvate which begins the biosynthetic pathway to pentoses and hexoses. i℗=H_2PO_3

3.2.9. Metabolism of hydrocarbons

Many species of bacteria and mold can grow with hydrocarbons as the sole source of carbon. Straight-chain *n*-alkanes are the most readily attacked hydrocarbons, and substituted or branched alkanes are more resistant to attack; cyclic and polycyclic hydrocarbons can also be oxidized but by a more restricted group of organisms and, again, substitution in the molecule or the addition of branched chains usually makes attack more difficult.

Cells growing on *n*-alkanes oxidize a terminal methyl group and the first stable intermediate is the corresponding primary alcohol. Experiments with $^{18}O_2$ show that oxygen itself, rather than the oxygen from water, is incorporated into the molecule. The alcohol is then oxidized via the aldehyde to a fatty acid. This is then

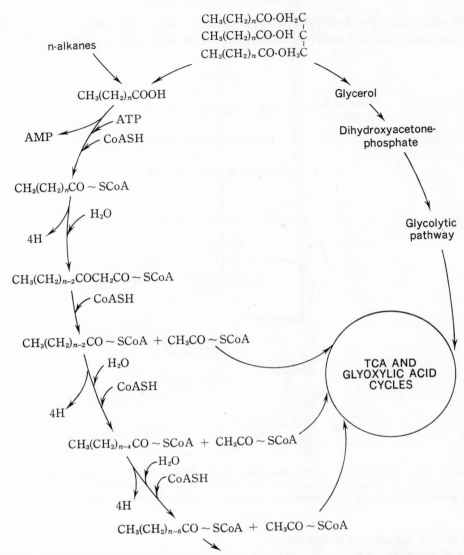

Fig. 3.10. Oxidation of fat. The fatty acid derived from hydrolysis of the triglyceride is activated and oxidized at the β-carbon atom; the resulting β-keto acid is hydrolyzed to yield acetyl~SCoA units which are further metabolized via the glyoxylic acid and TCA cycles. n-Alkanes are first oxidized to the corresponding fatty acid, then activated by ATP and CoA; further oxidation occurs by β-oxidation.

metabolized by β-oxidation as described in Section 3.2.8. While this is the major route of metabolism, both terminal methyl groups of the alkane may be oxidized, yielding dicarboxylic acids which may be further metabolized by β-oxidation. The unusual feature of metabolism of alkane-utilizing organisms is their ability to carry out the first oxygenation of the hydrocarbon molecule—after this, metabolism is by the usual enzymes found in organisms growing on conventional substrates. In fact, most alkane-utilizing organisms prefer such substrates to hydrocarbons.

Organisms growing on aromatic compounds incorporate molecular oxygen into the molecule at adjacent carbon atoms. Depending on the organism, the ring is cleaved either between adjacent hydroxyls or between a hydroxylated and a non-hydroxylated carbon atom. Ring cleavage is accompanied by incorporation of more molecular oxygen. Figure 3.11 sets out two possible routes for the metabolism of benzoate via catechol. Here, as with the metabolism of alkanes, the initial reactions incorporating oxygen and cleaving the hydroxylated ring compound are the ones peculiar to those organisms which metabolize ring compounds. The products of ring fission are directly metabolized via the TCA cycle.

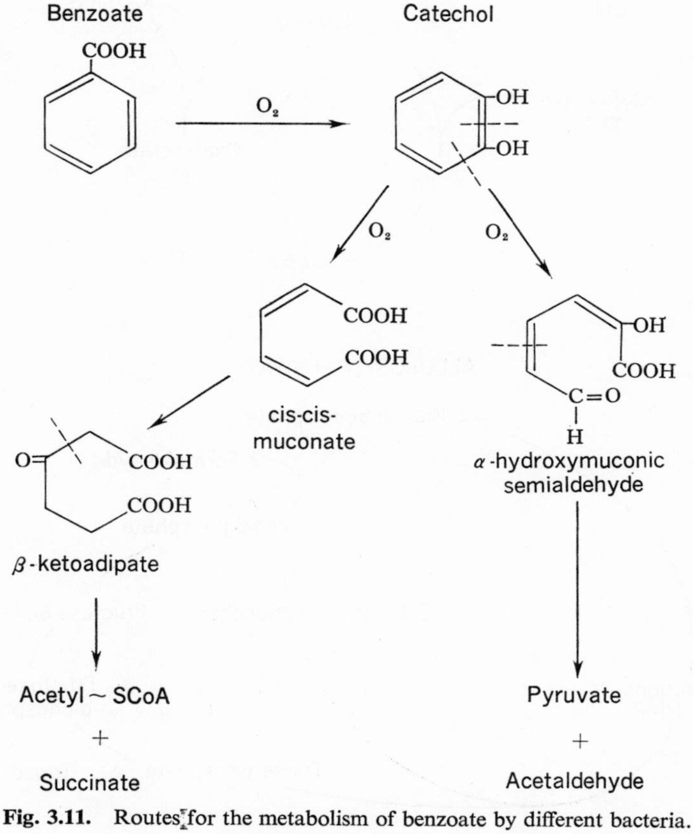

Fig. 3.11.   Routes for the metabolism of benzoate by different bacteria.

Quayle and his collaborators have greatly contributed to our understanding of the metabolism of the pseudomonads growing on methane or methanol as the sole source of carbon. These organisms oxidize methane as follows:

$$CH_4 \longrightarrow CH_3OH \longrightarrow HCHO \longrightarrow HCOOH \longrightarrow CO_2$$

SERINE PATHWAY

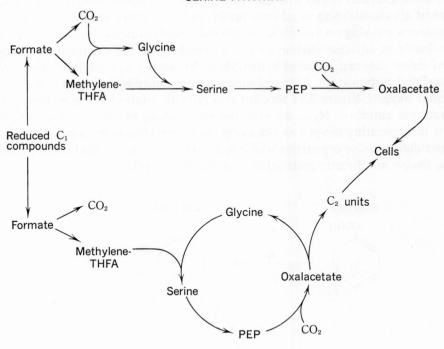

ALLULOSE PATHWAY

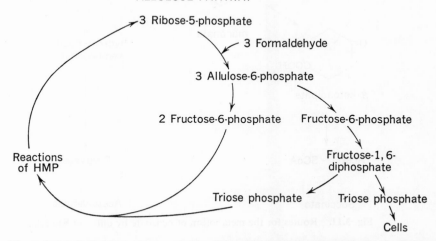

Fig. 3.12. The serine and allulose pathways used by different bacteria for the synthesis of cellular material from reduced C_1 compounds. The serine pathway is shown with two alternative schemes for glycine production—the upper indicates *de novo* synthesis of glycine and the lower a cyclic regeneration of glycine. At present, it is not certain which of these two schemes is correct. Methylene-THFA = $N^{5,10}$-methylene-tetrahydrofolic acid; PEP = phosphoenolpyruvic acid. Diagrams modified from those of Ribbons, D.W., Harrison, J.E., and Wadziaski, A.M. (1970). *Ann. Rev. Microbiol.* **24**, 135.

Two major pathways for the incorporation of C_1 substrates have been found: one involves the amino acid serine, and the other, the keto hexose allulose. The serine pathway occurs in many organisms which are obligate users of methane or methanol, while the allulose pathway has been found only in *Pseudomonas methanica* and *Methylococcus capsulatus*. The two pathways are set out in Fig. 3.12. In the serine pathway, the C_1 substrate is first oxidized to formate, then converted to methylene-tetrahydrofolic acid and condensed with glycine to form serine. It is not known whether the glycine is synthesized *de novo* or whether it arises by a cyclic process from oxalacetate. For the allulose pathway, the C_1 substrate is oxidized to formaldehyde when it is condensed with ribose-5-phosphate to form allulose-6-phosphate. This is then isomerized to fructose-6-phosphate which can enter the HMP and glycolytic pathways. Thus, ribose-5-phosphate is regenerated to condense more formaldehyde, and cell material is synthesized.

3.3.   ACCUMULATION OF METABOLITES

The previous sections of this chapter have suggested something of the range of catabolic activities found in microorganisms and the ways in which they obtain small molecules and energy for growth. Modern industrial fermentations are, however, primarily concerned with producing large amounts of the metabolites of growth, though it is often desirable to have a large cell mass as well. Success in achieving these aims has largely been empirical and reflects our ignorance of the routes of biosynthesis and their control. Biochemical knowledge is rapidly extending our understanding of the principles of cellular control and this will allow a more rational approach in the future.

Feedback regulation is achieved either by repression of enzyme synthesis (see Section 2.4.3.3) or by inhibition of enzyme action. Mutants can be selected which lack these controls. They may have a mutation or deletion of the operator or regulator genes, or the genetic change may code for an enzyme altered in such a way that it is no longer sensitive to its controlling metabolite. In industrial processes, strains with faulty regulation, altered permeability or metabolic deficiencies may be exploited to accumulate products. Sometimes the precise reason for the accumulation is not known; it may be lack of repression or inhibition, or both.

3.3.1.   Production of amino acids

Studies of the control of biosynthesis of amino acids have greatly contributed to the understanding of the control of cellular processes in general. Although the work has been done with a fairly restricted range of organisms dominated by studies of *Escherichia coli*, it is already clear that organisms with identical pathways of biosynthesis have different methods of control. Persons managing industrial fermentations must appreciate that it is not safe to assume that an organism used in production will have a system of regulation identical with that established for another organism. If precise control of the biosynthesis of a metabolite is vital to the success of an industrial process, then it will be necessary to establish the system of regulation prevailing in the production strain.

In pathways leading to a single product, this product often inhibits the activity of the first enzyme (E_A) of the series, while intermediates of a related pathway may activate E_A (see Fig. 3.13(a)). In addition, intermediates early in the pathway may activate enzymes occurring later in the sequence (see Fig. 3.20). With branched pathways leading to several products, total inhibition of E_A by a high concen-

(a) Unbranched pathway to one product. P inhibits the first enzyme.

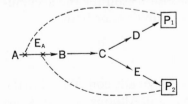

Activating metabolite

(b) Branched pathways to several products.
I. *Multivalent feedback.*
P_1 or P_2 alone has no effect, or very little effect, but together (concerted feedback inhibition) they inhibit E_A completely.

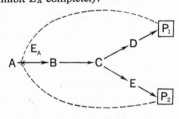

II. *Cooperative feedback.*
P_1 or P_2 alone inhibits E_A partially, together the inhibition is more than simply additive.

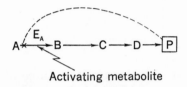

III. *Reversal of inhibition.*
E_A is inhibited by P_1 but this can be partly reversed by the activating metabolite, F.

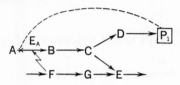

IV. *Inhibition by branch-point compounds.*
Intermediates D and F inhibit E_A while the concentrations of D and F are themselves controlled by the end products P_1, P_2, and P_3.

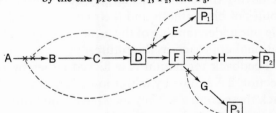

V. *Iso-enzymes with specific effects.*
Iso-enzymes catalyzing the first reaction are inhibited by P_1 or P_2, respectively.

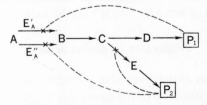

Fig. 3.13. Systems for the control of enzyme activity found in microorganisms. (a) Controls in an unbranched pathway leading to a single product. (b) I to V. Different forms of control found in branched pathways leading to several end products. $\longrightarrow$ =reactions of the biosynthetic pathway; - - - ✕ =feedback inhibition; ⤳ =activation; A, B, etc.=intermediates; E_A=first enzyme of the common pathway. Modified from Umbarger, H. E. (1968). *Ann. Rev. Biochem.* **38**, 323.

tration of one of the products would stop growth in the absence of the other end products. Cells have developed a number of ways of controlling such pathways. In multivalent feedback inhibition, the presence of all end products is required to inhibit E_A; in some cases no effect is seen if only one product is present, in others slight inhibition is observed with each product but their concerted effect gives complete inhibition (see Fig. 3.13 (b) I). A variant of this type of control is found where each product inhibits the activity of E_A by a small amount; the inhibition is additive and synergistic when all the products of the pathway are present (see Fig. 3.13 (b) II). The inhibition by products may be partially reversed, however, by metabolites of related pathways (see Fig. 3.13 (b) III). Compounds capable of causing

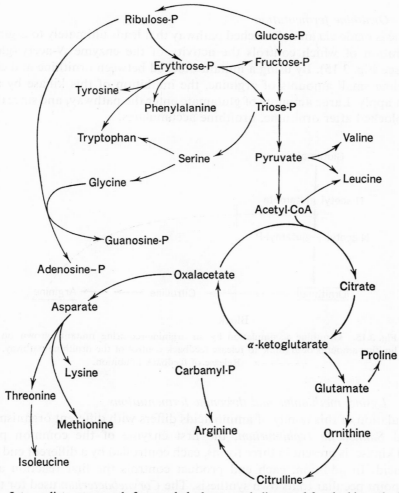

Fig. 3.14. Intermediate compounds from carbohydrate metabolism used for the biosynthesis of some amino acids and nucleotides. Modified from McFall, E., and Maas, W. K. (1967). in *Molecular Genetics* Part II, p. 262. Academic Press.

feedback inhibition are not confined to end products. Components of the pathway, and so far these have proved to be branch-point compounds, can also inhibit E_A; the concentrations of the branch-point compounds are controlled by simple feedback by the respective end products (see Fig. 3.13 (b) IV). In another control system, the first reaction of a series is catalyzed by more than one enzyme (iso- or multiple enzymes), each inhibited specifically by one of the end products (see Fig. 3.13 (b) V).

Before discussing processes for the commercial production of particular amino acids, important connections between various biosynthetic pathways leading to amino acids, purines and pyrimidines and the pathways providing carbon skeletons are summarized in Fig. 3.14.

3.3.1.1.　*Ornithine fermentation*

Ornithine is made via an unbranched pathway that leads ultimately to arginine, the concentration of which controls the activity of the enzyme *N*-acetyl-glutamate kinase (see Fig. 3.15). By using a mutant blocked between ornithine and citrulline and feeding small amounts of arginine, the inhibition of this kinase by arginine does not apply. Large amounts of glutamate enter the pathway, and since the pathway is blocked after ornithine, ornithine accumulates.

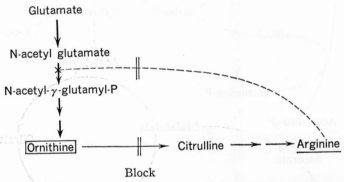

Fig. 3.15. Ornithine accumulation by an arginine-requiring mutant, grown on limiting amounts of arginine to release feedback control of the ornithine pathway.
- - -‖- - -× = Release of feedback inhibition.

3.3.1.2.　*Lysine, methionine and threonine fermentations*

The regulation of this family of amino acids differs with different organisms. In *E. coli* and *Salmonella typhimurium*, the first enzyme of the common pathway, aspartyl kinase, is present in three forms, each controlled by a different end product amino acid. In addition, each end product controls the first reaction after the branch point peculiar to its own synthesis. The *Corynebacterium* used for the commercial production of these amino acids, however, has a different control system. It has not been completely elucidated, but it is probable that here, aspartyl kinase is a single enzyme subject to multivalent inhibition by lysine and threonine. In the process for accumulating lysine, a mutant is used which requires threonine and methionine for growth because of a defective enzyme between aspartyl semialdehyde and homoserine. The mutant is then grown on low concentrations of threonine and methionine. This releases inhibition of aspartyl kinase and metabolites are diverted in large amounts to lysine. In this organism, in contrast to *E. coli*, lysine does not inhibit its own pathway after the branch point (see Fig. 3.16).

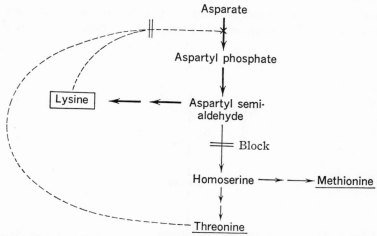

Fig. 3.16. The accumulation of lysine by a strain of *Corynebacterium* requiring threonine and methionine for growth and grown in limiting concentrations of these amino acids so that multivalent feedback cannot operate. – – –‖– – –×=Release of inhibition on aspartyl kinase. Broad arrows indicate the pathway of lysine synthesis.

Escherichia coli can be made to accumulate threonine by putting blocks in the lysine and methionine pathways and growing the cells on low concentrations of lysine and methionine. The two isoenzymes for aspartyl kinase will thus not be inhibited and metabolites will tend to flow to threonine. The lack of inhibition by threonine of its own synthesis is not fully explained since it it known that in wild type *E. coli*, threonine does inhibit the first enzyme after homoserine.

3.3.1.3. *Glutamic acid fermentation*

Monosodium glutamate is an important flavor-accentuating compound. The world production of glutamate reached 180,000 tons in 1972, 90% of which was made by fermentation. The *Corynebacterium* used for this process is biotin-requiring and lacks the enzyme α-ketoglutarate dehydrogenase which catalyzes the conversion of α-ketoglutarate to succinyl~CoA, and in the presence of ammonia, α-ketoglutarate is converted to glutamate. The efficiency of the conversion of glucose to glutamate is near 50% of theoretical when the biotin content of the medium is maintained near $2\,\mu g/l$. The pathway for this accumulation is set out in Fig. 3.17 where it can be seen that the supply of α-ketoglutarate is maintained by the operation of the glyoxylate cycle and parts of the TCA cycle.

Recent work has provided an explanation of the role of biotin in this fermentation. Biotin is not important in the biosynthesis of glutamate, but cells grown in low concentrations of biotin form a cytoplasmic membrane with high permeability to small molecules. At high biotin levels, the membrane is normal and the yields of glutamate are low because feedback mechanisms inhibit further synthesis of glutamate. Evidence for this hypothesis comes from studies of mutants of the original biotin-requiring strains which are now oleic acid-requiring. These strains accumulate glutamate at low oleic concentrations irrespective of the biotin concentration, but low glutamate yields are obtained at high concentrations of oleic acid. Biotin is

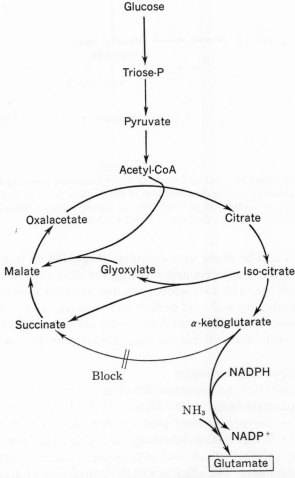

Fig. 3.17. Accumulation of glutamate. The mutant strain lacks α-ketoglutarate dehydrogenase and dicarboxylic acids to provide oxalacetate are synthesized via the glyoxylate cycle and parts of the TCA cycle.

a factor in oleic acid biosynthesis, and at low concentrations of biotin, or of oleic acid itself in the oleic-requiring strains, the cytoplasmic membrane formed is leaky, allowing glutamate to pass into the medium. Feedback mechanisms then do not operate. In addition, log phase cultures treated with penicillin or Tween have increased permeability and these also accumulate large amounts of glutamate.

For glutamate production from *n*-paraffins, a mutant of *Corynebacterium alkanolyticum* which requires glycerol is grown under conditions of glycerol limitation, to obtain high yields of glutamate. Biochemical studies of this mutant showed that it lacked glycerol-3-phosphate:NADP oxidoreductase and so the cells became deficient in phospholipid. By careful control of the glycerol in the growth medium, cells with defective membranes are produced and thus glutamate is excreted. The yields of glutamate from such cultures are not affected by the concentration of biotin or oleic acid.

3.3.2. Production of nucleotides

Certain purine nucleotides also accentuate flavor; guanosine monophosphate (GMP) is more potent than inosine monophosphate (IMP) and xanthine monophosphate (XMP), but the adenine nucleotides have no effect on flavor. Apart from the usual feedback controls that must be circumvented if fermentation processes are to be successful in accumulating nucleotides, there are further difficulties. Nucleotides are charged molecules and are, therefore, not readily excreted from the cell and tend to be hydrolyzed by intracellular phosphatases. Processes have, however, been developed where these problems have been partially solved. When mutants with metabolic blocks are grown in carefully controlled concentrations of the required end products, feedback regulation can be overcome. In addition, careful control of the components of the medium and selection of strains have enabled the permeability of the cytoplasmic membrane to be increased. In these strains, extracellular synthesis of nucleotides occurs using precursors and enzymes leaked from the cell.

Figure 3.18(a) summarizes the controls of the synthesis of the purine nucleotides. The first enzyme of the common pathway is under multivalent feedback control by AMP and GMP. When the pathway branches at IMP, the first enzyme of each pathway after IMP is controlled by AMP and GMP, respectively. The back reaction from GMP to IMP is regulated by the concentration of AMP. Figure 3.18(b) shows the accumulation of inosine and hypoxanthine using a mutant with metabolic blocks in both branches of the pathway after IMP. This mutant requires guanine and adenine for growth, and if grown in limiting concentrations of these purines, the control of the first enzyme of the common pathway is released. IMP tends to accumulate but is dephosphorylated and excreted as inosine or hypoxanthine.

Recent studies by workers at Kyowa Hakko Co. have shown that an adenine-requiring mutant grown in high concentrations of phosphate and magnesium but low concentrations of manganese and adenine, forms IMP extracellularly. It is believed that such cells become highly permeable to the intermediates inosine, hypoxanthine, ribose-5-phosphate, phosphoribosyl pyrophosphate (PRPP) and the enzymes needed to synthesize IMP extracellularly. The addition of guanine or

(a) **CONTROL OF NUCLEOTIDE BIOSYNTHESIS**

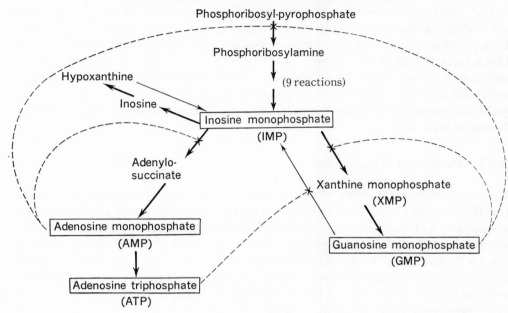

(b) **ACCUMULATION OF INOSINE AND HYPOXANTHINE**

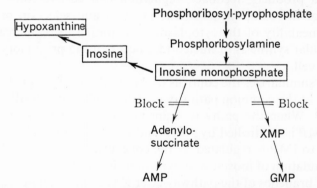

Fig. 3.18. (a) **The control of nucleotide biosynthesis.** Adenosine monophosphate (AMP) and guanosine monophosphate(GMP) exert multivalent feedback inhibition of the first enzyme of the common pathway and each nucleotide controls the first reaction from inosine monophosphate (IMP) specific to its own synthesis. In addition, adenosine triphosphate inhibits the back reaction from GMP to IMP.

(b) The accumulation of hypoxanthine and inosine by a mutant requiring adenine and guanine for growth. When grown in limiting concentrations of these two purines the feedback control of the pathway is circumvented. IMP is hydrolyzed by a phosphatase and inosine and hypoxanthine are excreted.
––––×=Feedback inhibition.

adenine to cells grown under these conditions gives increased amounts of GMP or AMP, ADP and ATP formed by extracellular, "salvage synthesis."

3.3.3. Production of antibiotics

Clearly, cells have a selective advantage if they can control the synthesis of essential metabolites such as amino acids and nucleotides. Many cells also produce secondary metabolites; these are nonessential but they too are often under cellular control. Some are commercially valuable and so it becomes important to know how the cell controls their production. Antibiotics are typical secondary metabolites and it has long been recognized that antibiotics are produced in the stationary phase when growth has almost stopped, but initiation of this secondary metabolism is not well understood, despite its acknowledged importance to fermentation management. There seem to be at least two possibilities; some regulating metabolite must accumulate to levels that stimulate the production of the enzymes for the synthesis of the antibiotic, or, alternatively, some component of the medium (or metabolite of it) acts as an inhibitor of the biosynthetic enzymes and the supply of this inhibitor must be exhausted before antibiotic production can begin. This second phenomenon is known as catabolite repression. Derepression of the synthesis of enzymes concerned in the production of antibiotics and other secondary metabolites has been shown experimentally.

In streptomycin formation, amidinotransferase is a key enzyme in the biosynthesis of the streptidine moiety of the molecule and this enzyme is found only in antibiotic-producing strains and only after active growth is over. Its production is inhibited by chloramphenicol, hence the enzyme is synthesized *de novo* and is not simply inactive early in growth. During fermentation, streptomycin and mannosido-streptomycin are produced concurrently but when glucose is exhausted, a mannosidase is formed and this hydrolyzes the mannoside to give streptomycin. In washed cells, this enzyme is produced at low, but not at high, concentrations of glucose. In addition, the concentration of phosphate in the growth medium is known to influence the yields of streptomycin. It is possible that phosphate influences the production of streptomycin by controlling the activity of the phosphatases involved at several points in the biosynthesis of the streptidine moiety.

In the penicillin fermentation, antibiotic production is inhibited by high concentrations of glucose; a less available form of carbon, or glucose fed at low concentrations, gives higher yields. The amount of acyltransferase, the enzyme which puts the side chain on 6-amino penicillanic acid, is greatly increased during penicillin production and is presumably subject to catabolite repression. There is evidence that penicillin production may be controlled in part by the concentration of lysine. In the biosynthesis of penicillin (see Fig. 3.19), a tripeptide is formed by the sequential condensation of cysteine with α-amino adipate (AAA) and with valine. This is followed by a series of reactions in which cysteine and valine form the β-lactam and 5-membered rings of the penicillin nucleus, with AAA attached as a side chain.

Penicillin itself is formed by exchanging AAA with phenyl acetate. AAA can be regarded as a branch-point compound—ultimately, it either forms lysine or penicillin. If the lysine concentration is high, lysine will inhibit its own synthesis and thus reduce the concentration of AAA; at low lysine concentrations there will be no feedback inhibition, making AAA available for penicillin synthesis.

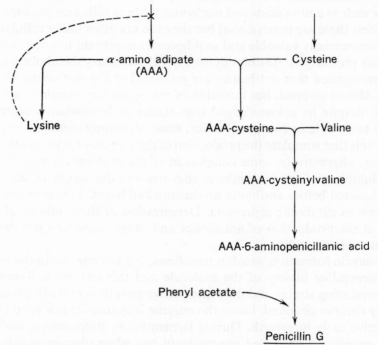

Fig. 3.19. Pathway for the synthesis of penicillin. A tripeptide of α-amino adipate (AAA), cysteine and valine is formed; the valine-cysteine portion undergoes rearrangement to form the penicillin nucleus with AAA as a side chain, then phenyl acetate exchanges with AAA to form penicillin G. Lysine controls the pathway to its own synthesis and hence controls the availability of AAA and thus the rate of synthesis of penicillin.

3.3.4. Control of amphibolic pathways

It has been emphasized in previous sections that glycolysis, the HMP, TCA and glyoxylate pathways all have the dual functions of providing energy for growth and intermediates for biosynthesis; Davis coined the adjective amphibolic for such pathways. Most purely catabolic pathways, like those of ring cleavage discussed in Section 3.2.9., are induced by the substrate, while end products commonly exert negative feedback inhibition in purely biosynthetic pathways. Amphibolic pathways have controls of both types, but the understanding of these very complex, interlocking pathways is far from complete; most work has been done with enteric bacteria and yeast but it is clear that controls in different organisms differ in detail.

The system of controls shown in Fig. 3.20 is based primarily on findings in enteric bacteria, but serves as an example of how effector metabolites within the cell may activate or inhibit key, allosteric enzymes and so control the rate of flow of metabolites through the major pathways.

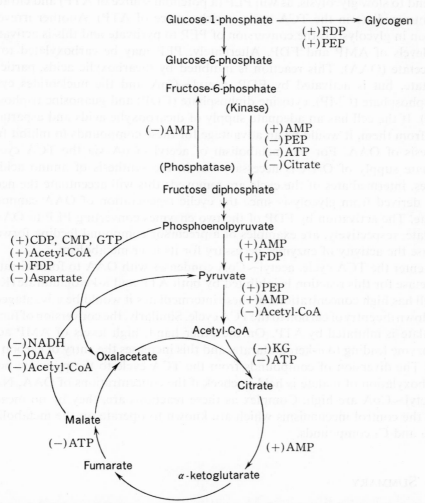

Fig. 3.20. Controls in amphibolic pathways in the enteric bacteria. Activators are shown as (+) and inhibitors as (−). AMP=adenosine monophosphate, CDP=cytosine diphosphate, FDP=fructose diphosphate, GTP=guanosine triphosphate, KG=α-ketoglutarate, OAA=oxalacetate, NADH=nicotinamide adenine dinucleotide (reduced), PEP=phosphoenolpyruvate.

In Fig. 3.20 some of the controls of glycolysis and the TCA cycle are shown. Early in glycolysis, the conversion of fructose-6-phosphate to fructose diphosphate (FDP) is virtually irreversible and is catalyzed by phosphofructokinase. This allosteric enzyme is activated by adenosine monophosphate (AMP) and inhibited by ATP, phosphoenolpyruvate (PEP) and by citrate. Notice that the rate of this reaction is increased as the level of AMP rises, which it does very rapidly if the level of ATP falls due to the demands of biosynthesis (the equilibrium constant (AMP) (ATP)/(ADP)^2=0.44). Conversely, high ATP levels inhibit the reaction and this will tend to slow glycolysis, as will PEP (a potential source of ATP) and citrate (the first intermediate in the TCA cycle, a rich source of ATP). Another irreversible reaction in glycolysis is the conversion of PEP to pyruvate and this is activated by high levels of AMP and FDP. Alternately, PEP may be carboxylated to form oxalacetate (OAA). This reaction is inhibited by dicarboxylic acids, particularly aspartate, but is activated by FDP, acetyl~CoA and the nucleotides cytosine monophosphate (CMP), cytosine diphosphate (CDP) and guanosine triphosphate (GTP). If the cell has an adequate supply of dicarboxylic acids and aspartate derived from them, it would be an advantage for these compounds to inhibit further synthesis of OAA. For the metabolism of acetyl~CoA via the TCA cycle, an adequate supply of OAA is necessary but in the synthesis of amino acids and purines, intermediates of the cycle are removed; this will accentuate the need for OAA derived from glycolysis since the cyclic regeneration of OAA cannot then operate. The activation by FDP of the two enzymes converting PEP to OAA and pyruvate, respectively, are examples of a precursor compound feeding forward to increase the activity of enzymes necessary for its later metabolism.

To enter the TCA cycle, acetyl~CoA condenses with OAA to form citrate; the synthetase for this reaction is inhibited by both ATP and α-ketoglutarate (KG). If the cell has high concentrations of these intermediates it would be advantageous to slow down the entry of carbon to the TCA cycle. Similarly, the conversion of fumarate to malate is inhibited by ATP. On the other hand, high levels of AMP activate the enzyme leading to α-ketoglutarate, and this increases the entry of carbon to the cycle. The diversion of compounds from the TCA cycle to the pyruvate pool by decarboxylation of malate is held in check if the concentrations of OAA, NADH, or acetyl~CoA are high. Complex as these reactions are, they by no means exhaust the control mechanisms which are known to operate on the metabolism of C_2, C_3 and C_4 compounds.

3.4. SUMMARY

1. Microbes of industrial importance obtain energy and small molecules for biosynthesis by oxidizing organic carbon. In general, oxidation occurs by the removal of hydrogen from the substrate rather than by insertion of oxygen, though the latter can occur in a few special reactions.

2. Energy is trapped largely as ATP either during oxidation at the substrate level, or in phosphorylations which occur as hydrogen from the substrate is transferred to oxygen. Such complete oxidations yield far more energy, and hence a greater cell mass, than anaerobic metabolism.

3. There are a number of pathways by means of which organisms metabolize carbohydrates; the most common are the glycolytic and hexose monophosphate pathways and the tricarboxylic acid cycle. The glycolytic pathway converts glucose anaerobically via fructose diphosphate and triose phosphates to pyruvate; this yields 2 ATP/glucose. The subsequent anaerobic metabolism of pyruvate functions primarily to regenerate reduced hydrogen carriers, although formation of butyric or acetic acid generates 1 mole of ATP per mole of acid produced.

4. The aerobic metabolism of pyruvate via the TCA cycle is a rich source of energy—15 ATP are formed/pyruvate oxidized by eucaryotic cells, although possibly less with bacteria. In addition, intermediates of the cycle provide the starting materials for the biosynthesis of amino acids, purines and porphyrins. There are reactions linking glycolysis with the TCA cycle, and these replenish intermediates withdrawn from the cycle for biosynthesis.

5. Growth on C_2 substrates requires the operation of the glyoxylate cycle linked with the TCA cycle.

6. Metabolism of glucose by the hexose monophosphate pathway functions largely to provides reduced pyridine nucleotides and intermediates for the biosynthesis of nucleotides and aromatic compounds.

7. Some organisms are specialists in metabolizing recalcitrant organic compounds and, in general, the special enzymes required are induced by the substrate. The products of metabolism are fed into the glycolytic pathway and the TCA cycle.

8. The cell can monitor its metabolism. Either it controls the activity of existing enzymes (inhibition or activation of allosteric enzymes), or it controls the production of enzymes (induction or repression), or both. These abilities are genetically determined and it is, therefore, possible to breed strains that lack these controls.

9. Commercial processes to accumulate intermediates essential to the cell have been developed using primarily mutants with metabolic blocks so placed that feedback inhibition and repression are bypassed and the wanted metabolite accumulates. The composition of the growth medium can influence the permeability of the cells; this may also be exploited to increase the yield of product.

10. The control of the production of secondary metabolites, such as antibiotics, is less well understood, but there is evidence that their production is also genetically controlled. Most antibiotics are produced after active growth has stopped; the presence of high concentrations of readily available substrates like glucose seems to inhibit production (catabolite repression). Further research is needed to identify controlling metabolites; but in the case of penicillin, the supply of α-aminoadipic acid and of phenyl acetic acid seems to be important, while the yield of streptomycin is influenced by phosphate.

NOMENCLATURE

AAA = α-amino adipate

ADP = Adenosine diphosphate
AMP = Adenosine monophosphate
ATP = Adenosine triphosphate
CoAS $\sim$ = Coenzyme A in the high energy form
CDP = Cytosine diphosphate
DNA = Deoxyribonucleic acid
e = Electron
E_0' = Oxidation/reduction potential
EMP = Embden-Myerhof-Parnas pathway
FDP = Fructose-1, 6-diphosphate
FAD = Flavin adenine dinucleotide
FMN = Flavin mononucleotide
GTP = Guanosine triphosphate
HMP = Hexose monophosphate pathway
IMP = Inosine monophosphate
KG = α-ketoglutarate
NAD = Nicotinamide adenine dinucleotide
NADP = Nicotinamide adenine dinucleotide phosphate
OAA = Oxalacetate
$\sim$P = High energy phosphate
i$\circledP$ = Inorganic phosphate
PEP = Phosphoenolpyruvate
P/O = moles of phosphate esterified/atom of oxygen used
RNA = Ribonucleic acid
TCA = Tricarboxylic acid cycle
XMP = Xanthine monophosphate

$$Y_{\text{ATP}} = \frac{\text{mass of cells formed (g)}}{\text{no. of moles of ATP formed/mole of substrate used}}$$

$$Y_0 = \frac{\text{mass of cells formed (g)}}{\text{atom of oxygen used for energy production}}$$

$$Y_s = \frac{\text{mass of cells formed (g)}}{\text{mole of substrate used}}$$

TEXTBOOKS AND ARTICLES FOR FURTHER READING AND REFERENCE

General Texts of Biochemistry

Conn, E. E., and Stumpf, P. K. (1966). *Outlines of Biochemistry* 2nd ed. Wiley, New York.
Lehninger, A. L. (1970). *Biochemistry* Worth Publ. Inc., New York.
Mahler, H. R., and Cordes, E. H. (1968). *Basic Biological Chemistry* Harper Int. Ed., London.

Carbohydrate Metabolism and the Production of Energy

Anderson, R. L., and Wood, W. A. (1969). "Carbohydrate metabolism in microorganisms."
Ann. Rev. Microbiol. **23**, 540.
Bauchop, T., and Elsden, S. R. (1960). "The growth of microorganisms in relation to their
energy supply." *J. Gen. Microbiol.* **23**, 457.

Gunsalus, I. C., and Stanier, R. Y. (1961). (Eds.) *The Bacteria* Vol. II: *Metabolism* Academic Press, London.

Hernandez, E., and Johnson, M. J. (1967). "Anaerobic Growth Yields of *Aerobacter cloacae* and *Escherichia coli*." *J. Bacteriol*. **94**, 991.

Mandelstam, J., and McQuillen, K. (1968). *Biochemistry of Bacterial Growth* Blackwell Sc. Pub., Oxford.

Meadow, P., and Pirt, S. J. (1969). (Eds.) *Microbial Growth (19th Symposium of the Society for General Microbiology)*.

Payne, W. J. (1970). "Energy yields and growth of heterotrophs." *Ann. Rev. Microbiol*. **24**, 17.

Pirt, S. J. (1963). "The maintenance energy of bacteria in growing cultures." *Proc. Roy. Soc. B*. **163**, 244.

Pollock, M. R., and Richmond, M. H. (1965). (Eds.) *Function and Structure in Microorganisms (15th Symposium of the Society for Microbiology)*.

Powell, E. O. *et al*. (1967). (Eds.) *Microbial Physiology and Continuous Culture* H.M.S.O., U.K.

Rainbow, C., and Rose, A. H. (1963). *Biochemistry of Industrial Microorganisms* Academic Press, London.

Rose, A. H. (1968). *Chemical Microbiology* 2nd ed. Butterworths, London.

Sokatch, J. R. (1969). *Bacterial Physiology and Metabolism* Academic Press, London.

Accumulation of Metabolites and Mechanisms of Control

Atkinson, D. E. (1966). "Regulation of enzyme activity." *Ann. Rev. Biochem*. **35**, Part I, 85.

Beckwith, J. R., and Zipser, D. (1970). (Eds.) *The Lactose Operon* Cold Spring Harbor, New York.

Datta, P. (1969). "Regulation of branched biosynthetic pathways in bacteria." *Science* **165**, 556.

Demain, A. L. (1966). "Industrial fermentations and their relation to regulatory mechanisms." *Adv. Appl. Microbiol*. **8**, 1.

Demain, A. L. (1968). "Regulatory mechanisms and the industrial production of microbial metabolites." *Lloydia* **31**, 395.

Demain, A. L. (1968). "Production of purine nucleotides by fermentation." *Prog. Appl. Microbiol*. **8**, 35.

Demain, A. L., and Innamine, E. (1970). "Biochemistry and regulation of streptomycin and mannosidostreptomycinase (α-D-mannosidase) formation." *Bact. Rev*. **34**, 1.

Gibson, F., and Pittard, A. J. (1968). "Pathways of biosynthesis of aromatic amino acids and vitamins and their control in microorganisms." *Bact. Rev*. **32**, 465.

Huang, H. T. (1964). "Microbial production of amino acids." *Prog. Ind. Microbiol*. **5**, 57.

Katz, E., and Weissbach, B. S. (1967). "Studies on the mechanism of actinomycin biosynthesis." *Prog. Ind. Microbiol*. **6**, 61.

McFall, E., and Maas, W. K. (1967). "Regulation of enzyme synthesis in microorganisms." In *Molecular Genetics*, p. 255. (Ed.) Taylor, J. H. Academic Press, London.

Martin, R. G. (1969). "Control of gene expression." *Ann. Rev. Genetics* **3**, 181.

Sanwal, B. D. (1970). "Allosteric control of amphibolic pathways in bacteria." *Bact. Rev*. **34**, 20.

Stadtman, E. R. (1963). "Enzyme multiplicity and function in the regulation of divergent metabolic pathways." *Bact. Rev*. **27**, 170.

Umbarger, H. E. (1969). "Regulation of amino acid metabolism." *Ann. Rev. Biochem*. **38**, 323.

Chapter 4

Kinetics

Kinetic studies are necessary to gain an understanding of any fermentation. As the name implies, fermentation kinetics is concerned with the rates of cell synthesis and/or fermentation product formation and the effect of environment on these rates. Kinetic studies are not necessarily limited to growing systems. They also may include dying cells.

For the most part persons involved in studying the kinetics of cellular systems have been concerned with representing or modeling the gross behavior of cells. There has been little attempt to relate this gross behavior to events that are occurring on a molecular level, i.e., induction, repression, transcription, translation, and enzyme synthesis. This chapter will deal principally with the kinetics on a gross or cellular level. However, in interpreting the gross behavior of cells, readers are suggested to refer to Chapter 2, particularly in connection with protein synthesis as viewed from the molecular level.

Two kinds of kinetic observations are made. These are the dynamic or batch culture systems and the steady state or continuous systems. Because of the ease with which shaken flask experiments can be run, most of the kinetic data are of the batch type. The difficulty with batch experiments is that the environmental conditions are continually changing; hence it is difficult to gain insight into the events on a molecular level and to evolve a real understanding of the kinetics. This is not to say, however, that kinetic models derived from batch experiments are not useful. Indeed, most commercial fermentations are run on batch. Although they are more costly and difficult to run than shaken flask experiments, continuous cultures do yield information regarding the response of cells to environmental changes (*cf.* Section 5.2., Chapter 5).

Discussions here on the gross and basic rate processes of enzymes in homogeneous systems will be followed by some remarks on oscillatory enzyme systems. Heterogeneous systems such as immobilized enzymes and their kinetics will be discussed separately in Chapter 14.

4.1. ENZYME SYSTEMS

4.1.1. Simple enzyme kinetics

A simple kinetic model for enzyme-substrate interaction is that proposed by Michaelis and Menten.[8,25] In this model the enzyme, E, combines reversibly with substrate, S, forming an enzyme-substrate complex, $E\text{-}S$, which irreversibly decomposes to form product, P, and free enzyme, E, i.e.,

$$E + S \underset{k_{-1}}{\overset{k_{+1}}{\rightleftharpoons}} E\text{-}S \xrightarrow{k_{+2}} E + P \tag{4.1}$$

where

> k_{+1} = forward reaction rate constant
> k_{-1} = reverse reaction rate constant
> k_{+2} = reaction rate constant

If the substrate concentration is much greater than the total enzyme concentration, the following equations are obtained, provided $E\text{-}S$ and S are designated as the respective concentrations:

$$\frac{dP}{dt} = k_{+2}\, E\text{-}S \tag{4.2}$$

or

$$\frac{dP}{dt} = v = \frac{k_{+2}E_0\, S}{\dfrac{k_{-1}+k_{+2}}{k_{+1}} + S}$$

$$= \frac{k_{+2}E_0\, S}{K_{\mathrm m} + S} = \frac{V_{\max}\, S}{K_{\mathrm m} + S} \tag{4.3}$$

where

$$\frac{dP}{dt} = \text{reaction rate, } v$$

$k_{+2}E_0$ = maximum rate of reaction, $V_{\max}$

E_0 = total enzyme concentration

$$K_{\mathrm m} = \frac{k_{-1} + k_{+2}}{k_{+1}} = K_{\mathrm s} + \frac{k_{+2}}{k_{+1}} \tag{4.4}$$

If the rate of product formation is controlled by the specific rate, k_{+2}, in the sequence of reactions expressed by Eq. (4.1), i.e., $k_{+2} \ll k_{+1}$, Eq. (4.4) reduces to:

$$K_m = K_s \tag{4.5}$$

Equation (4.3) is called the Michaelis-Menten equation. Note that the reaction rate is linearly related to the total enzyme concentration. The term K_m is called the Michaelis-Menten constant or simply designated as the Michaelis constant. This constant is equal to the equilibrium constant, K_s, only when the value of $k_{+2} \ll k_{+1}$. When $K_m = K_s$, the value of K_m is inversely proportional to the chemical affinity of the enzyme for the substrate. The smaller the value of K_m, the greater the affinity of the enzyme for the substrate. Equation (4.3) is shown graphically in Fig. 4.1.

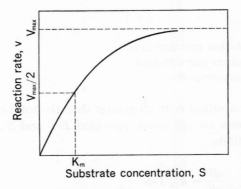

Fig. 4.1. Graphic representation of the Michaelis-Menten equation[Eq. (4.3)]. The reaction rate, v, is plotted against the substrate concentration, S. The value of V_{max} is the maximum reaction rate, and K_m in the abscissa represents the Michaelis-Menten constant.

In general, K_m values for respiratory enzymes associated with sugar metabolism are often smaller than those for hydrolytic enzymes associated with primary substrate attack. Table 4.1 gives typical K_m values in their orders of magnitude (averaged data from many publications not cited here for convenience).

TABLE 4.1

K_m VALUES FOR SOME ENZYMES.

Enzyme	Substrate	K_m (mole/l)
Maltase	Maltose	$\sim 10^{-1} \sim$
Sucrase	Sucrose	$\sim 10^{-2} \sim$
Phosphatase	Glycerophosphate	$\sim 10^{-3} \sim$
Lactic dehydrogenase	Pyruvate	$\sim 10^{-4} \sim$

4.1.2.   Lineweaver-Burk plot[8,25]

It is seen in Fig. 4.1 that the value of V_{max} cannot be determined accurately by plotting the values of v against S, because the maximum value, V_{max}, is the limit of an asymptote. An alternative to Fig. 4.1 is to take the reciprocals of both sides of Eq. (4.3).

$$\frac{1}{v} = \frac{K_m}{V_{max}} \frac{1}{S} + \frac{1}{V_{max}} \qquad (4.6)$$

If the value of $1/v$ is plotted against that of $1/S$, a straight line will be obtained as shown in Fig. 4.2. The intersection of this straight line with the ordinate represents $1/V_{max}$, while that with the abscissa is equal to $-1/K_m$. The characteristic constants (V_{max}, K_m) can then be determined from the experimental data (v, S), as plotted in Fig. 4.2, utilizing Eq. (4.6). This procedure for determining the values of V_{max} and K_m from the experimental data (v, S) is called the Lineweaver-Burk plot (1934).

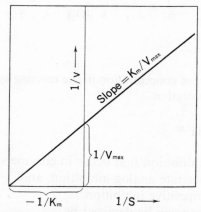

Fig. 4.2.   Graphic representation of the Lineweaver-Burk plot; the reciprocal of the reaction rate, $1/v$, is plotted against the reciprocal of the substrate concentration, $1/S$. The terms, K_m and V_{max}, in the figure are the Michaelis-Menten constant and the maximum reaction rate, respectively.

4.1.3.   Complex enzyme kinetics

Often, even an apparent single-step enzyme system does not follow Michaelis-Menten behavior. In the glucose oxidase system for enzymatic oxidation of glucose to δ-gluconolactone, an enzyme (flavoprotein) regeneration step is involved.[3] This can be represented by:

$$E_0 + S \underset{k_{-1}}{\overset{k_{+1}}{\rightleftharpoons}} E_0\text{-}S \xrightarrow{k_{+2}} E_r + \delta\text{-lactone} \qquad (4.7)$$

$$E_r + O_2 \xrightarrow{k_{+3}} E_0\text{-}P \xrightarrow{k_{+4}} E_0 + H_2O_2 \qquad (4.8)$$

where

E_0 = oxidized form of enzyme (flavoprotein)
E_r = reduced form of enzyme (flavoprotein)
S = substrate (glucose)
$k_{+1}, k_{-1}, k_{+2}, k_{+3}, k_{+4}$ = reaction rate constants

Stopped-flow experiments have shown that at pH=5.6 and 25°C, Eq. (4.7) is reduced to:

$$E_0 + S \xrightarrow{k_{+1}} E_r + \delta\text{-lactone} \qquad (4.9)$$

The observed apparent rate constant, k_a, for the appearance of lactone could then be represented from Eqs. (4.8) and (4.9) as follows:[3]

$$\frac{1}{k_a} = \frac{1}{k_{+4}} + \frac{1}{k_{+3}[O_2]} + \frac{1}{k_{+1}\,S} \qquad (4.10)$$

where

$[O_2]$ = dissolved oxygen concentration in the reacting system
S = glucose concentration

4.1.4. Enzyme inhibition[8,25]

Two common types of inhibition that occur in enzyme systems are:
   a. Competitive or substrate analog inhibition, and
   b. Reversible noncompetitive inhibition.
The competitive inhibition can be depicted by

$$
\begin{array}{c}
\pm I \nearrow \quad E\text{-}I \text{ (inactive)} \\
E \\
\pm S \searrow \quad E\text{-}S \longrightarrow E+P
\end{array}
\qquad (4.11)
$$

$$E + I \underset{k_{-i}}{\overset{k_{+i}}{\rightleftharpoons}} E\text{-}I \qquad (4.12)$$

$$K_i = \frac{k_{-i}}{k_{+i}} \qquad (4.13)$$

This model leads to a reaction rate expression of the form

$$v = \frac{V_{\max} S}{K_{\mathrm{m}} + S + \dfrac{K_{\mathrm{m}}}{K_i} I}$$

(4.14)

Rearranging Eq. (4.14),

$$\frac{1}{v} = \left(1 + \frac{I}{K_i}\right)\left(\frac{K_{\mathrm{m}}}{V_{\max}}\right)\frac{1}{S} + \frac{1}{V_{\max}}$$

(4.15)

provided:

   I = inhibitor concentration

An example of this type of inhibition is the glucose effect on the action of invertase on sucrose.

Reversible noncompetitive inhibition can be depicted by

$$
\begin{array}{c}
\pm I \nearrow \quad E\text{-}I \quad \searrow \pm S \\
E \qquad\qquad\qquad E\text{-}I\text{-}S \\
\pm S \searrow \quad E\text{-}S \quad \nearrow \pm I \\
\downarrow \\
E + P
\end{array}
$$

(4.16)

This model leads to a reaction rate expression of the form

$$v = \frac{V_{\max} S K_i}{(K_{\mathrm{m}} + S)(K_i + I)}$$

(4.17)

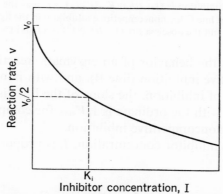

Fig. 4.3. Effect of noncompetitive inhibitor on enzyme kinetics [see Eq. (4.17)],
where v_0 = reaction rate in the absence of inhibitor.

The effect of inhibitor on the reaction rate is shown in Fig. 4.3. Noncompetitive inhibition is typical of the effect of organic acids such as acetate, propionate and lactate on hydrolytic enzymes.

Equation (4.17) can be rearranged to give:

$$\frac{1}{v} = \left(1 + \frac{I}{K_i}\right)\left(\frac{K_m}{V_{max}}\right)\frac{1}{S} + \left(1 + \frac{I}{K_i}\right)\frac{1}{V_{max}} \tag{4.18}$$

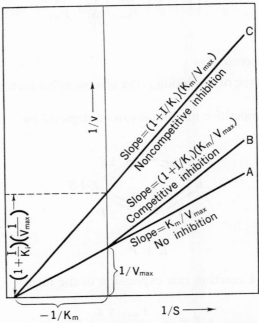

Fig. 4.4. Equation (4.15) for the competitive inhibition of enzymes is shown graphically by the Lineweaver-Burk plot; the value of K_i is assumed to be independent of the value of S; see line B. The lower straight line (line A) in the figure represents Eq. (4.3) (no inhibition). It is apparent that the slope of the straight line for competitive inhibition is $(1+I/K_i)(K_m/V_{max})$, whereas the intercept of the line with the ordinate is $1/V_{max}$. Regarding line C for noncompetitive inhibition in the light of Eq. (4.18), the intersections of the ordinate and the abscissa are $(1+I/K_i)(1/V_{max})$ and $(-1/K_m)$, respectively.

Figure 4.4 summarizes the behavior of an enzymatic reaction with no inhibition (line A), with competitive inhibition (line B), and with noncompetitive inhibition (line C). In both cases of inhibition, the slope of the lines is $(1+I/K_i)$ (K_m/V_{max}); however, the intercept with the ordinate is $1/V_{max}$ for competitive inhibition, and $(1+I/K_i)$ $(1/V_{max})$ for noncompetitive inhibition.

In some systems the inhibitor concentration, I, is proportional to the substrate concentration, S, i.e.,

$$I = \gamma S \tag{4.19}$$

where

γ = proportionality constant

In other systems the substrate will cause inhibition at high concentrations, i.e., $\gamma = 1.0$. In these cases Eq. (4.17) can be rewritten as

$$v = \frac{V_{max}}{1 + \dfrac{K_m}{S} + \dfrac{K_m}{K_i'} + \dfrac{S}{K_i'}}$$

$$= \frac{V_{max}}{1 + \dfrac{K_m}{S} + \dfrac{S}{K_i'}} \tag{4.20}$$

provided:

$$K_i' \gg K_m$$
$$K_i' = \frac{K_i}{\gamma}$$

4.1.5. Example of calculation for determining the type of enzyme reaction and the characteristic constants of the reaction

Experimental data on the hydrolysis of starch with α-amylase are given in Table 4.2.[8] The sequence of enzyme reaction is as follows:

$$\text{Starch} \longrightarrow \alpha\text{-dextrin} \longrightarrow \text{Limit dextrin} \longrightarrow \text{Maltose}$$

It is desired to determine the values of the Michaelis constant, K_m; the equilibrium constant for the dissociation of the enzyme-inhibitor complex, K_i; the maximum reaction rate, V_{max}; and the types of inhibition involved in the above sequence of reactions.

Solution
The experimental data are plotted as shown in Fig. 4.5 in terms of $1/v$ vs. $1/S$. Figure 4.5 shows that limit dextrin and maltose exhibit noncompetitive inhibition patterns, while the inhibition by α-dextrin is competitive (cf. Fig. 4.4).

However, Fig. 4.5 shows that inhibitions by limit dextrin (noncompetitive) and by α-dextrin (competitive) are clearly revealed when the substrate concentration is below 4 and 10 mg/ml, respectively.

For no inhibition in Fig. 4.5,

$-1/K_m = -0.290$ (the abscissa reading for $1/v = 0$) (see Fig. 4.4)

$K_m = 3.45$ mg/ml

$1/V_{max} = 0.0078$ (the ordinate reading for $1/S = 0$) (see Fig. 4.4)

$V_{max} = 128$ (relative hydrolysis reaction rate)

TABLE 4.2

DATA ON THE HYDROLYSIS OF STARCH.[8]

Inhibitor	Inhibitor concentration (mg/ml)	Substrate concentration (mg/ml)	Relative hydrolysis velocity
None	0.00	12.56	101
		11.24	98.2
		9.00	92.4
		8.12	90.0
		6.33	82.7
		5.61	79.1
		4.28	70.9
		3.56	65.0
		2.34	51.7
		1.00	28.8
Maltose	12.7	10.00	77.0
		7.70	71.4
		5.26	62.5
		4.55	58.9
		3.33	51.4
		2.04	38.9
		1.89	37.0
		1.67	34.2
Limit dextrin	13.35	10.00	95.3
		7.15	87.0
		6.25	83.4
		5.00	75.9
		4.17	47.6
		3.77	45.5
		3.14	41.7
		2.39	35.7
		2.10	33.0
		1.75	29.4
α-dextrin	3.34	33.30	116
		20.0	109
		12.9	102
		10.0	85.5
		6.06	71.5
		3.64	55.6
		2.82	47.6
		1.84	35.6
		1.60	32.2
		1.43	29.5

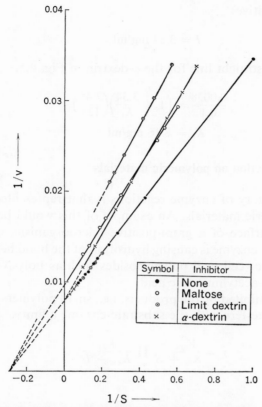

Fig. 4.5. Plot of $1/v$ vs. $1/S$ using the data given in Table 4.2 as an example of calculation.

Maltose (noncompetitive inhibition)

$$I = 12.7 \text{ mg/ml}$$

Slope of the Lineweaver-Burk plot in Fig. 4.5 for maltose,

$$\frac{0.0063}{0.190} = \left(1 + \frac{12.7}{K_i}\right)\left(\frac{3.45}{128}\right)$$

$$K_i = 56 \text{ mg/ml (see Fig. 4.4)}$$

Limit dextrin (noncompetitive inhibition)

$$I = 13.35 \text{ mg/ml}$$

Slope of the straight line for the limit dextrin in Fig. 4.5,

$$\frac{0.0053}{0.135} = \left(1 + \frac{13.35}{K_i}\right)\left(\frac{3.45}{128}\right)$$

$$K_i = 29 \text{ mg/ml}$$

α-dextrin (competitive)

$$I = 3.34 \text{ mg/ml}$$

Slope of the straight line for the *α*-dextrin in Fig. 4.5,

$$\frac{0.0038}{0.1} = \left(1 + \frac{3.34}{K_i}\right)\left(\frac{3.45}{128}\right)$$

$$K_i = 8.16 \text{ mg/ml}$$

4.1.6. Enzymatic action on polymeric materials

An interesting category of enzyme reactions is that represented by the attack of enzymes on polymeric materials. An example of this would be the attack of egg lysozyme on the surface of a gram-positive microorganism, causing the cell to lyze. In this case, the enzyme is causing hydrolysis of the bond between the 1-carbon and bridge-oxygen of cell wall polyglucosides such as poly-*N*-acetylglucosamine and oligomers of *N*-acetylmuramic acid.

The reaction is inhibited by the products, i.e., small polymers and the monomer, because they can interfere with the substrate-enzyme affinity.

$$v = \frac{V_{\max}S}{K_{\mathrm{m}} + S} \prod_i \frac{K_{Ii}}{K_{Ii} + I_i} \qquad (4.21)$$

where

S = substrate concentration (cell wall)
I_i = concentration of *i*-th inhibitor (product of low molecular weight)

Table 4.3 illustrates how the reaction rate (pH=5.5, 25°C, in phosphate buffer solution) is affected by the various molecular weight products.[12] This appears to be an interesting problem. Optimization of this reaction will become important for the recovery of protoplasmic protein and intracellular enzymes from single cells (*cf.* Section 13.2., Chapter 13).

TABLE 4.3

SPECIFIC ACTIVITY OF LYSOZYME.[12]

Substrate	Relative rate	K_{m} (mole/l)
N-acetylglucosamine (NAG)	0	$(4-6) \times 10^{-2}$
di-NAG	0.003	1.75×10^{-4}
tri-NAG	1	6.58×10^{-6}
tetra-NAG	8	9.45×10^{-6}
penta-NAG	4,000	9.35×10^{-6}
hexa-NAG	30,000	6.15×10^{-6}

4.2. ABSOLUTE REACTION RATE THEORY AS APPLIED TO KINETICS

The ideas regarding the absolute reaction rate theory, while not inconsistent with the Michaelis-Menten theory, represent a different view; namely, that the transition state described in Fig. 4.6 is not necessarily the same as the enzyme-substrate complex.

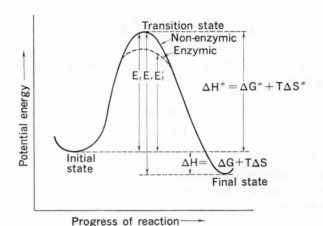

Fig. 4.6. Potential energy model in the absolute reaction rate theory;[8] enzyme reaction reduces the energy barrier from E_f to $E_{f'}$.

The schematic diagram in Fig. 4.6 shows that the enzyme reduces the potential energy barrier for forward and reverse reactions; the reduction is from E_f to E_f' for the forward reaction in the figure.

$$A + B \rightleftharpoons \text{Transition state} \longrightarrow \text{Product} \qquad (4.22)$$

The reaction is formulated in Eq. (4.22). The value of $\Delta E = E_f - E_r$ in the biological system is approximately equal to the value of ΔH (heat of reaction) (see later).

The equilibrium constant, K^*, for the formation of the transition state is obtained from Eq. (4.22) as follows:

$$K^* = \frac{C^*}{C_A C_B} \qquad (4.23)$$

where

$$C^* = \text{concentration of the transient material}$$
$$C_A, C_B = \text{concentrations of A and B in Eq. (4.22)}$$

It has been shown by the absolute reaction rate theory that

$$k_r C_A C_B = \frac{C^* kT}{h} \tag{4.24}$$

where

k_r = specific reaction rate to form the product
k = Boltzmann constant
h = Planck's constant
T = absolute temperature

From Eqs. (4.23) and (4.24),

$$k_r = \frac{kT}{h} \frac{C^*}{C_A C_B} = \frac{kT}{h} K^* \tag{4.25}$$

The equilibrium constant, $K,^*$ is related, on the other hand, to the free energy change, $\Delta G,^*$ as follows:

$$\Delta G^* = -RT \ln K^*$$
$$= \Delta H^* - T\Delta S^* \tag{4.26}$$

where

ΔH^* = heat of reaction of activation
ΔS^* = entropy change of activation
R = gas constant

From Eqs. (4.25) and (4.26),

$$\Delta G^* = -RT \ln \frac{k_r h}{kT} = \Delta H^* - T\Delta S^* \tag{4.27}$$

$$k_r = \frac{kT}{h} e^{\Delta S^*/R} e^{-\Delta H^*/RT} \tag{4.28}$$

Assuming that ΔS^* is independent of temperature,

$$\frac{d \ln k_r}{dT} = \frac{\Delta H^*}{RT^2} + \frac{1}{T} = \frac{\Delta H^* + RT}{RT^2} \tag{4.29}$$

$$\Delta H^* + RT = E \tag{4.30}$$

where

E = activation energy defined by the Arrhenius equation, which relates the change of specific rate of reaction with temperature, namely,

$$\frac{d \ln k_r}{dT} = \frac{E}{RT^2} \tag{4.31}$$

An error, due to the fact that ΔH^* in Eq. (4.28) is made approximately equal to

E, must be considered. However, this error is not necessarily grave because in most biological systems at about 300°K, the value of $RT=1.987\times300\doteqdot600$ cal/mole and, moreover, ΔH^* is usually more than 10,000 cal/mole. Then, from Eq. (4.28),

$$k_r = \frac{kT}{h}e^{\Delta S^*/R}\,e^{-E/RT} \tag{4.32}$$

The values of k_r in Eq. (4.32) in a biological system reveal an overall picture of the activities of the enzymes involved in a biological reaction. By measuring the change of k_r values with temperature, either the values of E or ΔS^* can be estimated. The values of E or ΔS^* thus determined for a specific biological reaction may give further insight into the kinetics of the whole system.

4.2.1.   Example 1:   Catalytic breakdown of hydrogen peroxide

Sizer has studied the activation and inactivation of the crystalline catalase-hydrogen peroxide system by changing the temperature of the reaction.[35] The catalase he used was extracted from beef liver and recrystallized. The activity of the enzyme was expressed by the rate of oxygen evolution (converted to standard temperature) from the catalase-hydrogen peroxide system. The rate was measured with a Warburg-Barcroft manometer at temperatures from 0° to 60°C.

The experimental data are shown in Fig. 4.7, the ordinate of which represents the log rate of oxygen evolution, mm³/min, while the abscissa is the reciprocal of

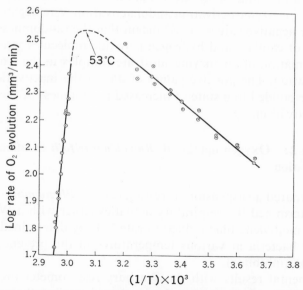

Fig. 4.7.   Catalytic breakdown rate of H_2O_2 depending on temperature.[35]

the absolute temperature. Although a definite temperature at which the rate of oxygen evolution was maximized could not be determined by the experiment, it is apparent from Fig. 4.7 that the data points fall into two groups, one of which was related to the activation of the enzyme in the temperature range up to 53°C, while the other dealt with inactivation of the enzyme at higher temperatures.

Assuming that the k_r values in Eq. (4.30) can be represented by the rate of oxygen evolution,

$$E = \frac{2.303\,R \log \frac{k_{r2}}{k_{r1}}}{\frac{1}{T_1} - \frac{1}{T_2}} \div \frac{4.6 \log \frac{k_{r2}}{k_{r1}}}{\frac{1}{T_1} - \frac{1}{T_2}} \qquad (4.33)$$

where

k_{r2} = specific activity of enzyme measured by the rate of oxygen evolution at T_2
k_{r1} = specific activity of enzyme at T_1

From the slopes of solid lines drawn through the data points in Fig. 4.7 and with Eq. (4.33),

$$E_1 = 4,200 \text{ cal/mole (below 53°C)}$$
$$E_2 = 51,000 \text{ cal/mole (above 53°C)}$$

The former value is the energy required for enzyme activation, while the sum of E_1 and E_2 (55,000 cal/mole) represents the energy for the thermal inactivation of the enzyme.

Sizer calculated the value of ΔS^* from Eq. (4.32). According to his calculations, the values of ΔS^* in enzyme activation and inactivation were negative and positive, respectively. The negative values of ΔS^* meant that the randomness decreased due to reorientation of catalase and hydrogen peroxide molecules, presumably resulting from the formation of an enzyme-substrate complex in the temperature range of enzyme activation. The positive values of ΔS^* at the inactivation temperature apparently corresponded to a state of increased randomness of the molecules during enzyme inactivation.

4.2.2. Example 2: Oxygen uptake of *Rhizobium trifolii* as affected by urethane concentration

Koffler *et al.* prepared a suspension of resting cells of *Rhizobium trifolii* in a buffer solution and determined the respiratory activities either with a Warburg respirometer or by the methylene blue reduction rate.[21] They determined the respiratory activities of the bacteria at various temperatures in the presence or absence of urethane solution (inhibitor).

Their experimental results with the Warburg respirometer are shown in Fig. 4.8; the relative rate of oxygen uptake is plotted against the reciprocal of the

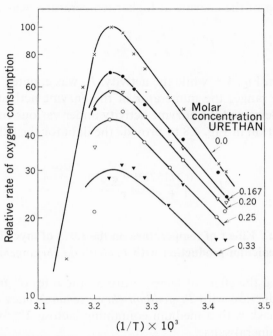

Fig. 4.8. Effects of temperature and urethane concentration on oxygen uptake rate of *Rhizobium trifolii*; the ordinate and abscissa are the relative rate of oxygen consumption measured with the Warburg respirometer and the reciprocal of the absolute temperature, respectively; the parameter is the molar concentration of urethane (inhibitor).[21]

absolute temperature. It is clear from Fig. 4.8 that the respiratory activity of the enzyme was inhibited markedly by urethane and that the enzyme, regardless of the presence or absence of the inhibitor was activated in a lower temperature range, approximately 19° to 37°C ($1/T = 3.42 \times 10^{-3}$ to 3.23×10^{-3}). The denaturation became appreciable at temperatures above 37°C ($1/T \leq 3.2 \times 10^{-3}$).

A curve based on the data obtained in the absence of urethane is shown in Fig. 4.8; the relation is represented by:

$$I' = \frac{\alpha T\, e^{-E/RT}}{1 + e^{-\Delta H/RT}\, e^{\Delta S/R}} \tag{4.34}$$

where

$$
\begin{aligned}
E &= \text{activation energy} \\
\Delta H &= \text{heat of reaction} \\
I' &= \text{enzyme activity measured by oxygen uptake rate (Warburg respirometer)} \\
R &= \text{gas constant} \\
\Delta S &= \text{entropy change} \\
T &= \text{absolute temperature} \\
\alpha &= \text{proportionality constant}
\end{aligned}
$$

The denaturation of the enzyme at higher temperatures was expressed by

$$I' = \frac{\alpha T}{1 + e^{-\Delta H/RT}\,e^{\Delta S/R}}$$

as can be seen from Fig. 4.8, while the activity, I', was equated to $\alpha Te^{-E/RT}$ in the lower temperature range; the numerator of the enzyme activation in Eq. (4.34).

Koffler et al. determined the enzyme activity under various conditions using the curve for no inhibition in Fig. 4.8 to provide the data for the calculation set out in Eq. (4.34).

$$I' = \frac{0.3775Te^{21.61}\,e^{-6,700/T}}{1 + e^{153.07}\,e^{-48,000/T}} \qquad (4.35)$$

4.2.3. Example 3: Effect of temperature on the rates of mycelial growth, respiration, and penicillin production with *Penicillium chrysogenum*

Calam et al. studied the effect of temperature on the rates of growth, respiration, and penicillin production with *Penicillium chrysogenum*.[5] They cultured the mold in a 2 l bolthead flask with a medium containing lactose 2%; C.S.L. (corn steep liquor) 2%; and mineral salts.

The mycelial growth rate, g/20 hr/l (see Fig. 4.9), the respiration rate, ml CO_2/ min/g dry mycelium (see Fig. 4.10), and the penicillin production rate, units/ 20 hr/g dry mycelium (see Fig. 4.11), were measured at various temperatures. The temperature range studied was from 13° to 35°C.

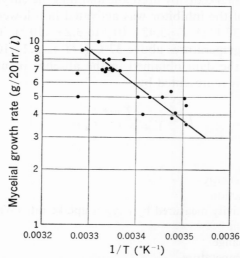

Fig. 4.9. Effect of temperature on the mycelial growth of *Penicillium chrysogenum*;[5] mycelial growth rate is plotted on a semi-logarithmic scale against the reciprocal of the absolute temperature.

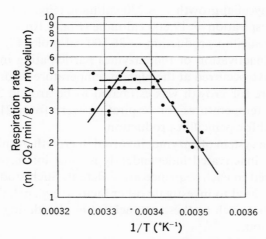

Fig. 4.10. Effect of temperature on the respiration rate of *Penicillium chrysogenum*;[5] the respiration rate, measured with the Warburg respirometer, is plotted on a semi-logarithmic scale against the reciprocal of the absolute temperature.

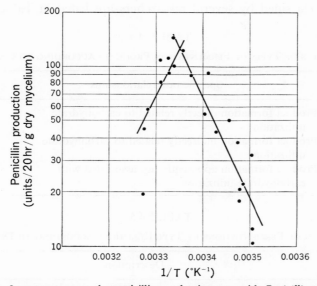

Fig. 4.11. Effect of temperature on the penicillin production rate with *Penicillium chrysogenum*;[5] the production rate is plotted on a semi-logarithmic scale against the reciprocal of the absolute temperature.

It is interesting to note from these figures that the effect of temperature on the activity of the various enzymes involved in growth, respiration, and penicillin production is appreciably different. It is also seen from Figs. 4.10 and 4.11 that the enzyme systems are inactivated in the higher temperature range (above 26°C). If Eq. (4.31) is applied for the assessment of E values,

E for mycelial growth $= 8,230$ cal/mole (Fig. 4.9)
E for respiration $= 17,800$ cal/mole (Fig. 4.10)
E for penicillin production $= 26,800$ cal/mole (Fig. 4.11)

It appears that inactivation of the enzymes participating in the production of mycelium might have occurred at the highest temperature recorded in Fig. 4.9; the optimal temperature for growth was estimated to be around 30°C. On the other hand, the optimal temperatures for respiration were from 21.7° to 28.6°C, and 24.7°C was optimal for penicillin production.

The fact that each system of enzymes involved in growth, respiration, and penicillin production was inactivated independently is most interesting, suggesting that the enzymes involved in each sequence are different. Such studies of fermentation kinetics can also be used to investigate other factors; e.g., pH, the concentration of nutrients, etc. Studies such as these are helpful in elucidating the kinetic mechanisms of fermentations.

4.3. KINETIC PATTERNS OF VARIOUS FERMENTATIONS

Fermentations classified by several researchers are listed in Tables 4.4 and 4.5.

TABLE 4.4

EXAMPLES OF TYPES OF FERMENTATION PROCESSES ACCORDING TO GADEN.[15]

Type	Specific rate relationships	Example
I	Product formation directly related to carbohydrate utilization	Ethanol
II	Product formation indirectly related to carbohydrate utilization	Citric acid
III	Product formation apparently not associated with carbohydrate utilization	Penicillin

TABLE 4.5

CLASSIFICATION OF FERMENTATIONS BY TYPE-REACTIONS ACCORDING TO DEINDOERFER.[9]

Type	Description
Simple	Nutrients converted to products in a fixed stoichiometry without accumulation of intermediates
Simultaneous	Nutrients converted to products in variable stoichiometric proportion without accumulation of intermediates
Consecutive	Nutrients converted to product with accumulation of an intermediate
Stepwise	Nutrients completely converted to intermediate before conversion to product or Nutrients selectively converted to product in preferential order

The system suggested by Gaden[15] relates the formation of product to substrate utilization. Another way of viewing this classification would be to assess the extent to which energy-producing reactions are coupled to product-forming reactions. This approach has certain advantages for studying continuous fermentation.

The classification proposed by Deindoerfer[9] is based on the course of the fermentation, i.e., consecutive, stepwise, etc. This approach is particularly useful in the study of batch fermentations.

Some examples of the latter classification follow:

Simple reactions. This type of fermentation kinetics involves two subtypes, growth and nongrowth reactions. These are shown in Figs. 4.12 and 4.13. Model

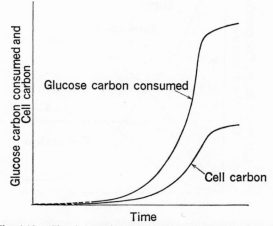

Fig. 4.12. Simple reaction: growth of *Aerobacter cloacae*; schematic representation of the data presented by Pirt.[30]

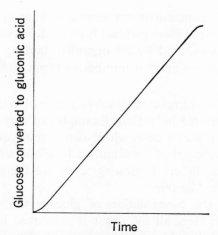

Fig. 4.13. Simple reaction: conversion of glucose to gluconic acid by resuspended *Aspergillus niger* mycelia; schematic representation of the data published by Moyer *et al.*[29]

processes for these reactions are the growth of yeast, and the production of gluconic acid using recycled mycelium.

Simultaneous reactions. Simultaneous reactions are those in which more than one product is produced and the relative rates of production of these products vary with nutrient concentration. They involve overflow or shunt metabolism. Figure 4.14 shows some results of relative synthesis of cell protein and cell fat during the growth of *Rhodotorula glutinis.*

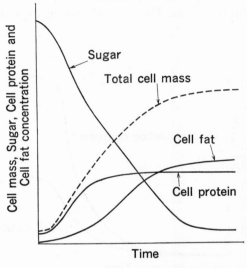

Fig. 4.14. Simultaneous reaction: conversion of sugar into cell protein and cell fat during growth of *Rhodotorula glutinis*; schematic representation of the data presented by Enebo *et al.*[11]

Consecutive reactions. Consecutive reactions are those in which an intermediate accumulates to some degree before product is formed. An example is the fermentation of glucose to gluconic acid by an organism lacking gluconolactonase (see Fig. 4.15). Antibiotic formation in a number of fermentations may fall into this category.

Stepwise reactions. The stepwise reaction is a case of two simple reactions that may be regulated by enzyme induction. Examples of this reaction are shown in Figs. 4.16 and 4.17. In the first case, where two carbohydrate substrates such as hexoses and pentoses are supplied simultaneously, *Escherichia coli* first completely utilizes the hexose sugar before beginning to utilize the pentose sugar. Monod calls this type of growth "diauxie."

Another example is the biooxidation of glucose to 5-ketogluconic acid by *Acetobacter suboxydans.* Here all the sugar is converted to gluconic acid before ketose formation begins. Similar reactions occur in the multi-point but stepwise attack on the steroid nucleus by some microorganisms.

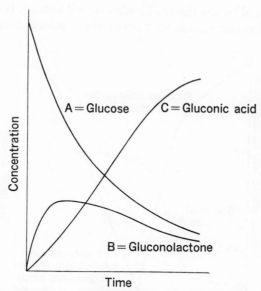

Fig. 4.15. Consecutive reaction: conversion of glucose to gluconic acid by *Pseudomonas ovalis*; schematic representation of the data presented by Humphrey *et al.*[18]

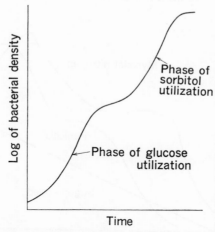

Fig. 4.16. Stepwise reaction: diauxic growth of *E. coli*; schematic representation of the data presented by Monod.[28]

Complex cases. Most fermentation processes involve a combination of reactions. Their complexity can vary tremendously. An examination of the fermentation patterns in the penicillin process suggests just such a number of reactions. The growth curve in Fig. 4.18 is typically diphasic. The penicillin production curve also exhibits a diphasic character and lags behind the growth curve. The accumulation of an intermediate product somewhere between sugar disappearance and penicillin appearance is suggested by the figure. This, as is well known, is the case in penicillin fermentation and is one reason why precursor is added during the process.

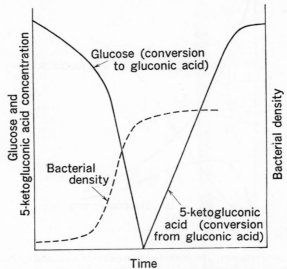

Fig. 4.17. Stepwise reaction: diphasic biooxidation of glucose to 5-ketogluconic acid by *Acetobacter suboxydans*; schematic representation of the data presented by Stubbs *et al.*[34]

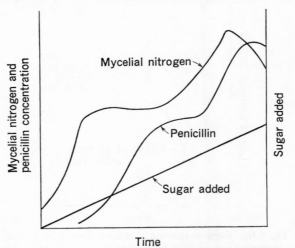

Fig. 4.18. Penicillin production in synthetic medium with continuous glucose feed (concentrations not to the same scale); schematic representation of the data presented by Hosler and Johnson.[17]

4.4. EXPRESSIONS FOR THE KINETIC PARAMETERS FOR CELLULAR ACTIVITIES

With respect to cellular activities, this section will consider only kinetics of growth and product formation. In the case of unicellular growth, the growth rate of cells can be expressed in terms of the cell concentration, X, the concentrations of a growth-limiting substrate, S, (Note: there are cases where more than a single substrate can be growth limiting) and an inhibitor, I, i.e.,

$$\frac{dX}{dt} = f(X, S, I) \tag{4.36}$$

Generally, the inhibitor, I, in Eq. (4.36) may imply a predator, X_2 (see Section 4.5 later). In many instances the variables, X, S, and I are highly coupled. Hence, expressions for cell growth rate are usually highly nonlinear.

The specific growth rate, μ, is defined as

$$\mu \equiv \frac{1}{X} \frac{dX}{dt} \tag{4.37}$$

If the value of μ is constant, Eq. (4.37) represents the so-called exponential growth, where growth is proportional to the mass of cells present. Note that the specific growth rate is related to the mass-doubling time, t_d, by

$$t_d = 0.693/\mu = \frac{\ln 2}{\mu} \tag{4.38}$$

Growth other than the exponential type has been proposed. For example, linear growth (dX/dt=constant) occurs in some hydrocarbon fermentations where limitation is due to the rate of diffusion of substrate from oil droplets, provided their surface area is constant.[13,14] In filamentous organisms where growth occurs from the tip, but nutrients diffuse throughout the filamentous cell mass, the growth rate may be proportional to the surface area of mycelia or the 2/3rds power of the cell mass.

4.4.1. Expressions for μ

The following empirical equation has been commonly used to express the specific growth rate, μ,

$$\mu = \frac{\mu_{max}S}{K_s + S} \tag{4.39}[28]$$

where

  μ_{max} = maximum specific growth rate
  K_s = saturation constant
  S = concentration of growth-limiting substrate

Equation (4.39) is analogous to Eq. (4.3) (Michaelis-Menten equation), but Eq. (4.39), the so-called Monod equation, has been derived empirically, while Eq. (4.3) is theoretical.

Obviously, Eq. (4.39) is an oversimplification of the true picture.[32,36] There are problems of inhibition which can include substrate inhibition at high concentration, product inhibition and even enzyme poisoning. Then, there are reaction lags due to the repression and induction control mechanism. Finally, there are problems of multi-substrate control, competition from other species for the substrate, commensalism, amensalism and predation.

Figure 4.19 is a typical representation of the growth-substrate relationship. For the case where $S \ll K_s$, a linear relationship exists between the specific growth rate and substrate concentration; for (b), the Monod equation applies, and (c) is a region of high substrate concentration where normally the maximum specific growth rate is achieved, but where inhibition due to either metabolic products of growth or substrate occurs.

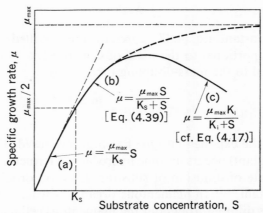

Fig. 4.19. Effect of substrate concentration on specific growth rate; though no theoretical equation has been made available for (c), the equation assumed in the figure originates from an analogy between the Michaelis-Menten and Monod equations with respect to enzymatic reaction and specific growth rate.

When no single substrate is limiting, the kinetics becomes more complicated. Often one substrate is a repressor of the enzyme system for the other substrate. When this occurs, it is possible to use a fermentation to separate or purify a system of mixed substrates, i.e., one substrate would be preferentially utilized in a continuous culture.[7]

4.4.2. Yield concepts

An overall yield, $Y_{x/s}$, is defined as

$$Y_{x/s} \equiv \frac{\Delta X}{-\Delta S} \qquad (cf.\ \text{Section 5.1., Chapter 5}) \tag{4.40}$$

It must be mentioned here that certain substrates, primarily those utilized for energy, are involved in cell maintenance. A cell can use glucose for endogenous respiration without growth. At low concentration, therefore, the yield expression must include a factor to account for utilization of the substrate for maintenance.

$$-\frac{dS}{dt} = \frac{\dfrac{dX}{dt}}{Y_G} + mX \qquad (4.41)$$

where

$$Y_G \equiv \frac{\Delta X}{(-\Delta S)_G} = \text{yield for growth}$$

m = specific rate of substrate uptake for cellular maintenance
(see Eq. (5.37), Chapter 5)

4.4.3. Structured growth models

A number of workers have suggested that growth models should be structured to include age distribution.[10,33] The concept of age is most difficult to visualize with cells undergoing binary fission, except for the idea of age in terms of time since division. On the other hand, the age of yeasts which bud and scar when the bud breaks away from the mother cell can be identified by counting the number of bud scars.

Table 4.6 compares some scar distribution data for yeast grown in continuous culture at various dilution rates (cf. Section 5.1., Chapter 5) with the ideal scar distribution.[2] The data tend to indicate that ideal distributions are nearly achieved at various growth rate conditions. Deviations from ideal distribution may be due to a physical selection mechanism that operates on the exit stream. There is very little encouragement from these results to include an age distribution as a parameter of cell culture models.

TABLE 4.6

SCAR DISTRIBUTION DATA FOR A GROWING SYSTEM.[2]

Number of scars	Scar distribution (%) D (hr^{-1})			
	Ideal*	0.11	0.2	0.3
0	50	57.94	59.08	56.45
1	25	17.33	18.18	18.84
2	12.5	10.14	9.08	11.25
3	6.25	6.05	5.53	4.98
4	3.125	3.58	3.13	3.09
> 4	3.101	4.74	5.02	4.13

* Time intervals between buddings are assumed to be all equal.

However, there are distinct compositional changes in cells depending upon the environment and the rate at which they are grown (see Fig. 2.11 in Chapter 2). This has led several workers to develop models for cellular activities that are structured.[10,33] By "structured" it is meant that various cell components can be utilized to represent different cellular capabilities. For example, RNA is indicative of the capability for enzyme synthesis. Protein composition is indicative of the cellular enzyme concentration. Such models undoubtedly will prove useful in correlating cellular activities with productivities (see Section 5.2., Chapter 5).

4.4.4. Product formation

Many models have been proposed for the production of various useful metabolites. Three of these models have been reasonably useful in expressing the behavior of simple systems. These are
    a. growth-associated model,
    b. non-growth associated model, and
    c. combined model.

When a substrate is stoichiometrically converted to a single product, P, the rate of product formation is related to the rate of growth by:

$$\frac{dP}{dt} = \alpha \frac{dX}{dt} \qquad (4.42)$$

provided:

    α = stoichiometric constant

This is the so-called growth-associated model.

For cases where the rate of product formation is only dependent on cell concentration, the cell has a constitutive enzyme system that controls the product formation rate, then,

$$\frac{dP}{dt} = \beta X \qquad (4.43)$$

provided:

    β = proportionality constant

The constant, β, is similar to enzyme activity and it can be thought of as representing so many units of product-forming activity per mass of cells.

Luedeking *et al.* proposed that Eqs. (4.42) and (4.43) be combined to express the productivity of lactic acid fermentations.[26]

$$\frac{dP}{dt} = \alpha \frac{dX}{dt} + \beta X \qquad (4.44)$$

or

$$\nu \equiv \frac{1}{X}\frac{dP}{dt} = \alpha\mu + \beta \qquad (4.45)$$

where

ν = specific rate of product formation
μ = specific growth rate

Luedeking and others have successfully applied this model to some fermentations (batch and continuous).[1,22-24]

A more complicated use of the combined model has been made by treating the constants, α and β in Eqs. (4.42) and (4.43) as variables, having discrete and particular values for each of the four phases of batch growth.[23,24] The phases of batch growth considered were lag, exponential, constant, and declining phases.

Terui has developed a series of kinetic models for enzyme production by microorganisms.[36] These models consider such factors as so-called preferential synthesis, salvage synthesis, and repression-derepression relationships.

It is considered from industrial experience with several antibiotic fermentation systems that the most useful models will be those that involve environmental factors such as temperature, pH, dissolved oxygen, concentrations of inducer and precursor, and rate of feed of energy substrate and nitrogen substrate, because these can be used in feedback loops for computer control of the process (see Section 12.7., Chapter 12).

4.5. Oscillatory System Kinetics

4.5.1. Cellular level (mixed culture)

Mixed culture systems are receiving considerable attention these days because of problems in biological waste control, where naturally occurring mixed culture system are utilized.[4,6,20,31] Also, in the behavior of estuary microflora and microfauna, prey-predation systems are at work.

The prey-predator system can be described under continuous culture conditions by the following set of equations:

$$\frac{dS}{dt} = D(S_0 - S) - \frac{\mu_1 X_1}{Y_1} \tag{4.46}$$

$$\frac{dX_1}{dt} = \mu_1 X_1 - \frac{\mu_2 X_2}{Y_2} - DX_1 \tag{4.47}$$

$$\frac{dX_2}{dt} = \mu_2 X_2 - DX_2 \tag{4.48}$$

$$\mu_1 = f(S) \tag{4.49}$$

$$\mu_2 = f(X_1) \tag{4.50}$$

where

D = dilution rate
S = concentration of growth-limiting substrate for prey (host) cells
S_0 = concentration of growth-limiting substrate in fresh medium
X_1 = dry-cell mass concentration of prey (host) population
X_2 = dry-cell mass concentration of predator population
Y_1 = overall yield factor for growth of prey $\left(= \dfrac{\varDelta X_1}{-\varDelta S} \right)$

Y_2 = overall yield factor for growth of predator $\left(= \dfrac{\varDelta X_2}{-\varDelta X_1} \right)$

μ_1 = specific growth rate of prey population
μ_2 = specific growth rate of predator population

In addition to a complete mixing assumed in Eqs. (4.46) to (4.50), it should be noted that another assumption—host organisms take soluble nutrients represented by the growth-limiting substrate, while predators take only host organisms as food with no direct metabolism of the soluble substrate—underlies the above set of equations.

If μ_2 is proportional to X_1 in Eq. (4.50), i.e.,

$$\mu_2 = \alpha' X_1 \tag{4.51}$$

provided:

α' = proportionality constant

Equations (4.47) and (4.48) are reduced to:

$$\frac{dX_1}{dt} = \mu_1 X_1 - \frac{\alpha'}{Y_2} X_1 X_2 - D X_1 \tag{4.52}$$

$$\frac{dX_2}{dt} = \alpha' X_1 X_2 - D X_2 \tag{4.53}$$

The phase-plane analysis of Eqs. (4.52) and (4.53) shows that the populations of X_1 and X_2 oscillate around a vortex point (D/α' for X_1, $Y_2(\mu_1 - D/\alpha'$ for X_2), provided α' and Y_2 are constant, respectively.[6] This is the classical Lotka-Volterra model.[27] A disadvantageous feature of this classical model is in that the oscillation depends exclusively on the initial conditions of the system.

Canale presented his phase-plane analysis by using Monod-type relations between μ_1 and S, and μ_2 and X_1, respectively, in place of the linear assumption of Eq. (4.51), i.e.,

$$\mu_1 = \mu_{1\mathrm{max}} \frac{S}{K_s + S} \tag{4.54}$$

$$\mu_2 = \mu_{2\mathrm{max}} \frac{X_1}{K_s' + X_1} \tag{4.55}$$

where

K_s' = saturation constant for predator
subscript max = maximum value[6]

An example of computations with Eqs. (4.46) to (4.48), and with Eqs. (4.52) and (4.53) (HITAC 5020, Computer Center, University of Tokyo) is shown in Fig. 4.20; numerical values used are also shown in the figure. Depending on these values, oscillations of X_1, X_2, and S are quite obvious, as illustrated in this figure. In particular, a large amplitude of oscillation in S is noted.[20]

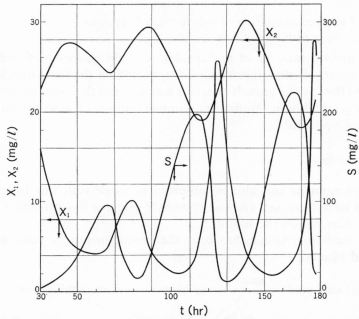

Fig. 4.20. Simulation of mixed and continuous culture (*Colpidium campylum* and *Alcaligenes faecalis*); dilution rate, D=0.04 hr^{-1}. The following values were used in the simulation: μ_{1max}=0.5 hr^{-1}, μ_{2max}=0.11 hr^{-1}, K_s=K_s'=10 mg/l, Y_1=0.14, Y_2=0.5, S_1=10 mg/l, X_{1i}=25 mg/l, X_{2i}=7 mg/l, provided: *subscript* i=initial condition.

According to Canale, the singular point on the first quadrant of the X_2-X_1 plane is claimed to be unstable (he predicted the existence of a limit cycle).[6] The marked oscillation of S as exemplified in Fig. 4.20 has hardly been recognized in mixed and continuous cultures of *Alcaligenes faecalis* (prey; X_1) and *Colpidium campylum* (predator; X_2) at 20°C, pH=6.5 to 7, using asparagine as a limiting substrate of the bacterium.[20] This implies that ample room remains for further improvement of mixed culture simulation.

In more general and rather qualitative terms, another set of equations for mixed culture growth under conditions of continuous run involve:

$$\frac{dS}{dt} = D(S_0 - S) - \frac{\mu_1 X_1}{Y_1} \tag{4.46}$$

$$\frac{dX_1}{dt} = \mu_1 X_1 - X_1 D \tag{4.56}$$

$$\vdots \qquad \vdots \qquad \vdots$$

$$\frac{dX_n}{dt} = \mu_n X_n - X_n D \tag{4.57}$$

$$\mu_n = f(S, X_1, X_3, \cdots, X_n, I) \tag{4.58}$$

provided:

X_n = cell mass concentration of n-th species of organism

μ_n = specific growth rate of n-th species

The specific growth rate, μ_n, of any species, n, could be increased, inhibited, or not affected by any of the other species. Depending on the particular model used for expressing the specific growth rate, one may observe the interesting phenomena of competition, inhibition, mutualism, commensalism, or amensalism along the lines discussed previously.

4.5.2. Enzyme level

Oscillatory phenomena can also be observed in single cells due to feedback or feed-forward control of key enzyme systems.[32] Higgins has presented a detailed theoretical discussion on the subject.[16]

One such oscillatory system has been the genetic-metabolic control suggested by Jacob and Monod.[19]

$$\tag{4.59}$$

However, this control can act directly without gene regulation by back activation or forward inhibition of an enzyme in the following reaction sequence.

$$\tag{4.60}$$

$$\tag{4.61}$$

Oscillatory phenomena on an enzymatic level in the light of Eqs. (4.59) to (4.61) will be exemplified below. By the use of a fluorescent trace of NADH₂ in a buffered yeast cell suspension, Pye[32] showed that, upon feeding 100 mM of glucose and producing an anaerobic state, slowly damped oscillations in the NADH₂ content arise. He suggested that glycolytic flux pulses probably arise from alternate activation of phosphofructokinase (PFK).

ADP (adenosine diphosphate) and F-1,6-P (fructose-1,6-diphosphte) in Fig. 4.21 activate PFK, followed by a higher flux of glucose through the yeast cell membrane, the transfer rate through which becomes the rate-limiting step. Then, concentrations of G-6-P (glucose-6-phosphate), F-6-P, and F-1,6-P decrease, and eventually the activity of PFK becomes another rate-limiting factor; an accumulation of ADP and F-1,6-P triggers the activation of PFK. This sort of cycle is reflected by the oscillation of NADH₂ content in the subsequent metabolism of F-1,6-P to ethanol and carbon dioxide through pyruvate. Due to the limited amount of glucose added, the oscillation is destined to damp out.

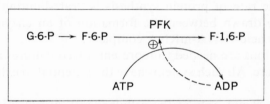

Fig. 4.21. Activation of PFK in anaerobic culture of yeast cells.[32]

Detailed studies of these oscillatory biological systems certainly present a challenge if one is to understand the control systems that operate within cellular systems (*cf.* Section 3.3.4.).

4.6. SUMMARY

1. Simple enzyme kinetics can be represented by the Michaelis-Menten equation, i.e.,

$$v = \frac{V_{max}S}{K_m + S}$$ [Eq. (4.3)]

2. The Lineweaver-Burk plot is useful for representing enzyme kinetic data.

3. Noncompetitive inhibition kinetics of the type sometimes encountered in substrate inhibition can be approximated by the following relationship:

$$v = \frac{V_{max}}{1 + \dfrac{K_m}{S} + \dfrac{S}{K_i'}}$$ [Eq. (4.20)]

4. Most cellular systems exhibit exponential growth, i.e.,

$$\mu = \frac{1}{X}\frac{dX}{dt} = \text{constant} \qquad [cf.\ \text{Eq. (4.37)}]$$

5. The doubling time, t_d, can be related to the specific growth rate, μ,

$$t_d = 0.693/\mu = \frac{\ln 2}{\mu} \qquad [\text{Eq. (4.38)}]$$

6. For growth limitation by a single substrate, the specific growth rate can be related to the concentration of substrate by the Monod equation, i.e.,

$$\mu = \frac{\mu_{max}S}{K_s + S} \qquad [\text{Eq. (4.39)}]$$

7. It is highly likely that the Monod model has some basis in molecular biology. If, as suspected, the rate of protein synthesis is "rate-limiting" in growth, then an analogy may be drawn between the formation of an enzyme-substrate complex and the amino acid-tRNA-mRNA complex.

8. A set of equations are needed to represent mixed cultures and prey-predator relationships in nature. All such systems have the potential to exhibit stable oscillations.

9. Oscillations exist not only at the cellular level, but also in enzyme systems where feedback activations and inhibitions apply.

NOMENCLATURE

A, B = materials
C_A = concentration of material A
C_B = concentration of material B
C^* = concentration of transient material
D = dilution rate, hr^{-1}
E = enzyme, enzyme concentration; activation energy defined by the Arrhenius equation
E_0 = total enzyme concentration; oxidized form of enzyme
E_f, E_r, E_f' = activation energies
E_r = activation energy; reduced form of enzyme
$E\text{-}S$ = enzyme-substrate complex or its concentration
f = function
ΔG^* = free energy change of activation
ΔG = free energy change
ΔH^* = heat of reaction of activation
ΔH = heat of reaction

h = Planck's constant, 6.62×10^{-27} erg sec

I = inhibitor or inhibitor concentration

I' = enzyme activity measured by oxygen uptake rate (Warburg respirometer)

K^* = equilibrium constant, $C_A + C_B \rightleftharpoons C^*$

$K_i' = K_i/\gamma$

K_i = equilibrium constant $(= k_{-i}/k_{+i})$ for $E\text{-}I \underset{k_{+i}}{\overset{k_{-i}}{\rightleftharpoons}} E+I$

K_m = Michaelis constant

$K_s = k_{-1}/k_{+1}$, equilibrium constant between $E\text{-}S \underset{k_{+1}}{\overset{k_{-1}}{\rightleftharpoons}} E+S$

K_s, K_s' = saturation constants

k = Boltzmann constant, 1.37×10^{-16} erg/°K

k_a = apparent reaction rate constant

k_{+i} = forward reaction rate constant, $E+I \overset{k_{+i}}{\rightleftharpoons} E\text{-}I$

k_{-i} = reverse reaction rate constant, $E+I \underset{k_{-i}}{\rightleftharpoons} E\text{-}I$

k_r = specific rate of reaction

k_{+1} = forward reaction rate constant $(E+S \overset{k_{+1}}{\rightleftharpoons} E\text{-}S)$

k_{-1} = reverse reaction rate constant $(E+S \underset{k_{-1}}{\rightleftharpoons} E\text{-}S)$

k_{+2} = reaction rate constant $(E\text{-}S \overset{k_{+2}}{\longrightarrow} E+P)$

k_{+3}, k_{+4} = reaction rate constants

m = specific rate of energy-yielding substrate uptake for cellular maintenance

P = product, product concentration; unknown and intermediate metabolite in Eq. (4.8)

R = gas constant, 1.987 cal/°K mole

S = substrate or substrate concentration, mg/ml

S_0 = initial substrate concentration, mg/ml

ΔS^* = entropy change of activation

T = absolute temperature, °K

t = time, hr

t_d = doubling time, hr

$V_{\max}$ = maximum reaction rate

v = rate of product formation

v_0 = reaction rate in the absence of inhibitor

X = cell mass concentration, mg/ml

X_1 = cell mass concentration of prey population, mg/ml

X_2 = cell mass concentration of predator population, mg/ml

Y = overall yield factor, $\Delta X/(-\Delta S)$

$Y_{x/s}$ = overall yield factor, $\Delta X/(-\Delta S)$; abbreviated as Y

Y_G = yield factor for growth, $\Delta X/(-\Delta S)_G$

Y_1 = overall yield factor for growth of prey $[= \Delta X_1/(-\Delta S)]$

Y_2 = overall yield factor for growth of predator $[= \Delta X_2/(-\Delta X_1)]$

subscripts

1, 2, ..., *i*, ... = species for substrate or inhibitor; 1 = prey (host), 2 = predator, unless otherwise noted

G = growth without taking maintenance into account

i = *i*-th product or inhibitor

Greek letters

α = stoichiometric constant; proportionality constant

α' = proportionality constant

β = proportionality constant

γ = proportionality constant

μ = specific growth rate, hr^{-1}

μ_{max} = maximum value of specific growth rate, hr^{-1}

ν = specific rate of product formation, hr^{-1}

REFERENCES

1. Aiyar, A. S., and Luedeking, R. (1966). "A kinetic study of the alcoholic fermentation of glucose by *Saccharomyces cerevisiae.*" *C.E.P. Symposium Series* **62**, 55.
2. Beran, K., Málek, I., Streiblová, E., and Lieblová, J. (1967). "The distribution of the relative age of cells in yeast populations." *Microbial physiology and continuous culture* (Eds.) Powell, E. O., Evans, C. G. T., Strange, R. E., and Tempest, D. W. p. 57. Her Majesty's Stationary Office, London.
3. Bright, H. J., and Appleby, M. (1969). "The pH dependence of the individual steps in the glucose oxidase reaction." *J. Biol. Chem.* **244**, 3625.
4. Bungay, H. R., and Krieg, N. R. (1966). "Growth in mixed culture processes." *C.E.P. Symposium Series* **62**, 68.
5. Calam, C. T., Driver, N., and Bowers, R. H. (1951). "Studies in the production of penicillin, respiration and growth of *Penicillium Chrysogenum* in submerged culture, in relation to agitation and oxygen transfer." *J. Appl. Chem.* **1**, 209.
6. Canale, R. P. (1970). "An analysis of models describing predator-prey interaction." *Biotech. & Bioeng.* **12**, 353.
7. Chian, S. K., and Mateles, R. I. (1968). "Growth of mixed cultures on mixed substrates. I. Continuous culture." *Appl. Microbiol.* **16**, 1337.
8. Dawes, E. A. (1967). *Quantitative problems in biochemistry* 4th Ed. pp. 106–174. E. & S. Livingstone, Edinburgh and London.
9. Deindoerfer, F. H. (1960). "Fermentation kinetics and model process." *Adv. Appl. Microbiol.* (Ed.) Umbreit, W. W. **2**, 321. Academic Press, New York.
10. Eakman, J. M., Fredrickson, A. G., and Tsuchiya, H. M. (1966). "Statistics and dynamics of microbial cell populations." *C.E.P. Symposium Series* **62**, 37.
11. Enebo, L., Anderson, L. G., and Lundin, H. (1946). "Microbiological fat synthesis by means of *Rhodotorula* yeast." *Arch. Biochem.* **11**, 383.
12. *Enzyme Handbook* (1969). (Ed.) Barman, T. E. **2**, 572. Springer-Verlag, New York, Berlin, Heidelberg.
13. Erickson, L. E., Humphrey, A. E., and Prokop, A. (1969). "Growth models of cultures with two liquid phases. I. Substrate dissolved in dispersed phase." *Biotech. & Bioeng.* **11**, 449.

14. Erickson, L. E., and Humphrey. A. E. (1969). "Growth models of cultures with two liquid phases. II. Pure substrate in dispersed phase." *Biotech. & Bioeng.* **11**, 467.
15. Gaden, E. L., Jr. (1955). "Fermentation kinetics and productivity." *Chem. & Ind.* p. 154. London.
16. Higgins, J. (1967). "The theory of oscillating reactions." *Ind. Eng. Chem.* **59**, 18.
17. Hosler, P., and Johnson, M. J. (1953). "Penicillin from chemically defined media." *Ind. Eng. Chem.* **45**, 871.
18. Humphrey, A. E., and Reilly, P. J. (1965). "Kinetic studies of gluconic acid fermentations." *Biotech. & Bioeng.* **7**, 229.
19. Jacob, F., and Monod, J. (1961). "Genetic regulatory mechanisms in the synthesis of proteins." *J. Mol. Biol.* **3**, 318.
20. Kobayashi, K. (1972). Master's thesis. Dept. Chem. Eng., Univ. of Tokyo. "Some experiments and analysis on predator-prey models; interaction between *Colpidium campylum* and *Alcaligenes faecalis* in a continuous and mixed culture." *Progress Rept.* No. 77. (1972). Biochem. Eng. Lab., Inst. Appl. Microbiol., Univ. of Tokyo.
21. Koffler, H., Johnson, F. H., and Wilson, P. W. (1947). "Combined influence of temperature and urethane on the respiration of *Rhizobium.*" *J. Am. Chem. Soc.* **69**, 1113.
22. Koga, S., Burg, C. R., and Humphrey, A. E. (1967). "Computer simulation of fermentation systems." *Appl. Microbiol.* **15**, 683.
23. Kono, T., and Asai, T. (1969). "Kinetics of continuous cultivation." *Biotech. & Bioeng.* **11**, 19.
24. Kono, T., and Asai, T. (1969). "Kinetics of fermentation processes." *Biotech. & Bioeng.* **11**, 293.
25. Laidler, K. J. (1958). *The chemical kinetics of enzyme action* pp. 30–93. Oxford Univ. Press.
26. Luedeking, R., and Piret, E. L. (1959). "A kinetic study of the lactic acid fermentation. Batch process at controlled pH." *J. Biochem. Microbiol. Tech. & Eng.* **1**, 393.
27. Lotka, A. J. (1925). *Elements of physical biology* Williams and Wilkins, Baltimore, Md.
28. Monod, J. (1949). "The growth of bacterial cultures." *Ann. Review of Microbiol.* **3**, 371.
29. Moyer, A. J., Umberger, E. J., and Stubbs, J. J. (1940). "Fermentation of concentrated solutions of glucose to gluconic acid—Improved process—." *Ind. Eng. Chem.* **32**, 1379.
30. Pirt, S. J. (1957). "The oxygen requirement of growing cultures of an *Aerobacter* species determined by means of the continuous culture technique." *J. gen. Microbiol.* **16**, 59.
31. Proper, G., and Garver, J. C. (1966). "Mass culture of the protozoa *Colpoda steinii.*" *Biotech. & Bioeng.* **8**, 287.
32. Pye, E. K. (1969). "Biochemical mechanisms underlying the metabolic oscillations in yeast." *Can. J. Botany* **47**, 271.
33. Ramkrishna, D., Fredrickson, A. G., and Tsuchiya, H. M. (1967). "Dynamics of microbial population: Models considering inhibitors and variable cell composition." *Biotech. & Bioeng.* **9**, 129.
34. Stubbs, J. J., Lockwood, L. B., Roe, E. T., Tabenkin, B., and Ward, G. E. (1940). "Ketogluconic acids from glucose—Bacterial production—." *Ind. Eng. Chem.* **32**, 1626.
35. Sizer, I. W. (1944). "Temperature activation and inactivation of the crystalline catalase-hydrogen peroxide system." *J. Biol. Chem.* **154**, 461.
36. Terui, G., Okazaki, M., and Kinoshita, S. (1967). "Kinetic studies on enzyme production by microbes. (I) On kinetic models." *J. Ferm. Tech.* (*Japan*) **45**, 497.

Chapter 5

Continuous Cultivation

Continuous culture was initiated in the early 20's of this century, when fodder yeast was produced by a continuous process. Since then, continuous culture of microorganisms has been carried out by a large number of workers. Especially during the past decades, considerable attention has been focused on the continuous cultivation of microbes. Various international symposia have been devoted specifically to the topic of continuous cultivation.[11,19,34]

So far, most publications on continuous culture have been related to the physiological behavior of microorganisms. Specific devices—the "chemostat," "turbidostat," etc., have been utilized for continuous culture.[9,10,15,37] Continuous culture has been favored for physiological studies, primarily because environmental conditions can be kept constant, in sharp contrast with batch cultivation, where microorganisms are subject to continual change.

TABLE 5.1

REPRESENTATIVE GENERA OF ORGANISMS GROWN IN CONTINUOUS CULTURE.[24]

Organisms	Genera
Actinomycetes	*Streptomyces*
Algae	*Chlorella*
	Euglena
	Scenedesmus
Bacteria	*Aerobacter*
	Azotobacter
	Bacillus
	Brucella
	Clostridium
	Salmonella
Fungi	*Ophiostoma*
	Penicillium
Protozoa	*Tetrahymena*
Yeast	*Saccharomyces*
	Torula
Mammalian cells	Embryo rabbit kidney

TABLE 5.2

REPRESENTATIVE CHEMICALS PRODUCED BY CONTINUOUS FERMENTATION.[24]

Growth-associated	Non-growth associated
Acetic acid	Acetone
Butanediol	Butanol
Ethanol	Glycogen
Gluconic acid	Subtilin
Hydrogen sulfide	Chloramphenicol
Lactic acid	Penicillin
	Streptomycin
	Vitamin B_{12}

Tables 5.1 and 5.2 list some of the organisms and products which have been studied with continuous systems. Most of these systems have been studied in laboratories rather than on a commercial scale. Exceptions are *Chlorella*, yeast, and activated sludge; the last contains a variety of microorganisms extending from protozoa to bacteria.

Although many studies of yeast, bacteria, and fungi in continuous cultivation in laboratories have been published, full-scale production in continuous cultivation systems has been surprisingly scarce.[8,39] Exceptions are as follows: continuous fodder yeast production from sulfite paper-mill waste; continuous baker's yeast culture from molasses in England, and the activated sludge process to purify sewage and industrial wastes.

This lack of large-scale continuous processes may be ascribed to the following difficulties: *a*. the possibility that the microbial species used will undergo deleterious mutations, *b*. technical difficulties of running aseptically for a long period of time, and *c*. lack of knowledge on the dynamic aspects of microbial behavior.

As will be shown later, the productivity of continuous cultivation is usually higher than that for batch culture, so even a semi-continuous cultivation, which avoids a high incidence of microbial mutation, may be advantageous. Technical difficulties with aseptic control may also be avoided to some extent in a semi-continuous operation.

Figure 5.1 exemplifies the dynamic response of bacteria in continuous culture. By changing the value of pH sinusoidally, the changes in the concentrations of bacterial cells, X, and of glucose, Z, were studied by Fuld *et al.*[14] using *Lactobacillus delbrueckii*. The figure also shows that the bacterial concentration changes in direct response to the sinusoidal change of the pH values of the incoming broth, but that the glucose concentration of the medium was unchanged. This sort of study is important to the future development of practical continuous cultivations.

This chapter will be devoted primarily to a consideration of continuous cultivations from an engineering viewpoint. The discussions, moreover, will be restricted to the basic principles underlying the theory of the continuous stirred reactor, with-

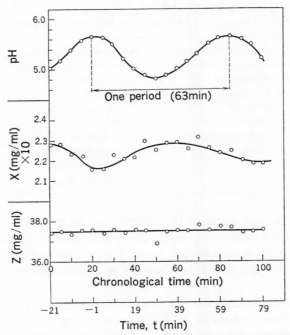

Fig. 5.1. Response of different variables with time (*Lactobacillus delbrueckii*).[14] The ordinate represents pH, cell concentration, *X* mg/ml, and glucose concentration, *Z* mg/ml, in the broth, while the abscissa is time, *t* min.

out considering particularly the theory of variations such as tubular or cylindrical types of reactors. However, the well-defined character of continuous cultivation is of considerable help in gaining an understanding of the physiological and kinetic activities of microbes.

The nomenclature used in this chapter will follow as closely as possible the recommendations made by a committee organized at the Symposium on Continuous Cultivation held in Czechoslovakia in 1962.[13]

5.1.   STEADY-STATE CONTINUOUS CULTIVATION THEORY

5.1.1.   Mass balance in a series of vessels

To provide a general model, a series of equivolume vessels, *n* in total number, will be considered. Denoting (Fig. 5.2) the rate of medium flow through the series of vessels as *F* l/hr, and the volume of each vessel as *V* l, the mass balance equations regarding concentrations of cell mass, *X*, product, *P*, and limiting substrate, *S* can be expressed respectively as follows:

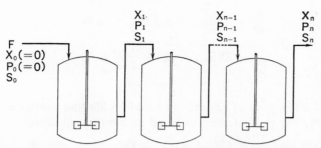

Fig. 5.2. *n* vessels in series; $F =$ flow rate of medium; $X =$ cell mass concentration; $P =$ product concentration; $S =$ limiting substrate concentration; subscript, 1,2,..., *n* are for the 1st, 2nd, ..., *n*-th vessels, whereas the subscript 0 is for the incoming medium. If a fresh medium is charged into the 1st vessel, $X_0 = 0$ and $P_0 = 0$.

$X:$

$$V\frac{dX_n}{dt} = FX_{n-1} - FX_n + V\left(\frac{dX_n}{dt}\right)_{\text{Growth in } n\text{-th vessel}} \tag{5.1}$$

$$= FX_{n-1} - FX_n + V\mu_n X_n$$

$$\frac{dX_n}{dt} = D(X_{n-1} - X_n) + \mu_n X_n \tag{5.2}$$

where

$D = F/V =$ dilution rate, hr^{-1}

$= 1/(V/F) = 1/\bar{t}$

$=$ reciprocal of mean holding time (or retention time) of flowing medium in each vessel

$\mu_n = \dfrac{1}{X_n}\dfrac{dX_n}{dt} =$ specific growth rate of cells at *n*-th vessel

subscripts $n, (n-1) = n$-th, $(n-1)$-th vessels, respectively

Similarly,

$P:$

$$\frac{dP_n}{dt} = \frac{F}{V}(P_{n-1} - P_n) + \left(\frac{dP_n}{dt}\right)_{\text{Production in } n\text{-th vessel}} \tag{5.3}$$

$$\frac{dP_n}{dt} = D(P_{n-1} - P_n) + Y_{P/X}\mu_n X_n \tag{5.4}$$

where

$Y_{P/X} = \Delta P/\Delta X =$ yield of product based on cell mass

For $S:$

$$\frac{dS_n}{dt} = \frac{F}{V}(S_{n-1} - S_n) + \left(\frac{dS_n}{dt}\right)_{\text{Consumption in } n\text{-th vessel}} \tag{5.5}$$

$$\frac{dS_n}{dt} = D(S_{n-1} - S_n) - \frac{1}{Y_{X/S}} \mu_n X_n$$

$$= D(S_{n-1} - S_n) - \frac{Y_{P/X}}{Y_{P/S}} \mu_n X_n \tag{5.6}$$

where

$Y_{X/S} = -\Delta X/\Delta S$ = yield of cell growth based on limiting substrate
$Y_{P/S} = -\Delta P/\Delta S$ = yield of product based on limiting substrate

$$= \left(\frac{\Delta P}{\Delta X}\right)\left(\frac{\Delta X}{-\Delta S}\right) = Y_{P/X} \cdot Y_{X/S}$$

It is assumed that the medium flowing from the $(n-1)$-th vessel into the n-th vessel is mixed instantaneously and completely with the contents of the n-th vessel; this is a necessary assumption for the derivation of Eqs. (5.1) to (5.6). Values of $Y_{X/S}$, $Y_{P/X}$, and $Y_{P/S}$ in these equations are all assumed to be constant, regardless of the number of vessels under consideration. Discussions of these assumptions will appear later.

At steady-state conditions, the left-hand side of all equations from Eqs. (5.1) to (5.6) is zero. Then, from Eq. (5.2),

$$X_n = \frac{DX_{n-1}}{D - \mu_n} \quad (n \neq 1) \tag{5.7}$$

Accordingly,

$$X_{n-1} = \frac{DX_{n-2}}{D - \mu_{n-1}}$$

$$\vdots \qquad \vdots$$

$$X_n = \frac{DX_{n-1}}{D - \mu_n} = \frac{D^2 X_{n-2}}{(D - \mu_n)(D - \mu_{n-1})} = \cdots$$

$$= \frac{D^{n-1} X_1}{\prod_{i'=2}^{n} (D - \mu_{i'})} \tag{5.8}$$

Equation (5.8) is useful for estimating the value of X_n in the n-th vessel or, conversely, the value of μ_n in the steady state can be estimated from the cell concentrations, X_n, X_{n-1}, etc. For single vessels $(n=1)$, the value of X_0 in the right-hand side of Eq. (5.2) or (5.1) is zero. Then,

$$\mu_1 = D = \mu_{\max} \frac{S_1}{K_s + S_1} \quad \text{[see Eq. (4.39)]} \tag{5.9}$$

Note that since $S_1/(K_s+S_1) < 1$, the maximum growth rate attainable in the first reactor is always less than the maximum growth rate theoretically possible, i.e.,

$D < \mu_{max}$. This is not true, however, when the reactor is not perfectly mixed, when there are concentration effects in the exit line, or when there is recycling.

The steady-state equation for P is given by:

$$P_n = \frac{DP_{n-1} + Y_{P/X}\mu_n X_n}{D} \tag{5.10}$$

Since the value of P_0 is zero in a single reactor or in the first of a series of continuous reactors, then

$$P_1 = \frac{Y_{P/X}\mu_1 X_1}{D} = Y_{P/X}X_1 \tag{5.11}$$

Lastly, the steady state for S is given by:

$$S_n = S_{n-1} - \frac{1}{D} \cdot \frac{Y_{P/X}}{Y_{P/S}}\mu_n X_n$$

$$= S_{n-1} - \frac{1}{D} \cdot \frac{1}{Y_{X/S}}\mu_n X_n \tag{5.12}$$

For $n = 1$,

$$S_1 = S_0 - \frac{\frac{Y_{P/X}}{Y_{P/S}}}{D}\mu_1 X_1$$

$$= S_0 - \frac{Y_{P/X}}{Y_{P/S}}X_1$$

$$= S_0 - \frac{1}{Y_{X/S}}X_1 \tag{5.13}$$

5.1.2. Mass balance in single vessels with recycling[28]

An interesting consideration is the case of single vessels with reuse or recycle of some of the microbial mass (see Fig. 5.3). It is evident from the figure that

$$F_a = (1 + \omega)F \tag{5.14}$$

$$F = F_e + F_{ex} \tag{5.15}$$

Mass balance equation for X in the reactor of complete mixing in Fig. 5.3 is:

$$V\frac{dX}{dt} = X_X\omega F - F_a X + V\left(\frac{dX}{dt}\right)_{Growth}$$

Unless otherwise noted, the subscript 1 for single vessels is omitted hereafter in this chapter for simplicity.

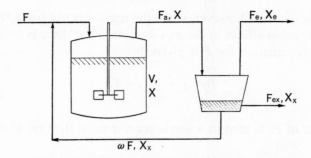

Fig. 5.3. Single vessels with recycling; F=flow rate of fresh medium; F_a=flow rate of cell suspension from reactor; F_e=flow rate of effluent from separator; F_{ex}=flow rate of concentrate cell suspension from separator; ω=recycle ratio; ωF=flow rate of cell suspension recycled; X=cell mass concentration in reactor; X_e=cell mass concentration in effluent from separator; X_x=cell mass concentration in recycled suspension; V=reactor (working) volume.

At steady state,

$$\frac{1}{X}\left(\frac{dX}{dt}\right)_{\text{Growth}} = \frac{F_a}{V} - \frac{\omega F}{V}\cdot\frac{X_x}{X}$$ (5.16)

From Eqs. (5.16) and (5.14),

$$\mu = (1 + \omega)D - \omega D\frac{X_x}{X}$$

$$= D\left\{1 + \omega\left(1 - \frac{X_x}{X}\right)\right\}$$ (5.17)

provided:

$$D = \frac{F}{V}$$

Another mass balance of cells regarding the separator (Fig. 5.3) in steady state gives:

$$F_a = F_e\frac{X_e}{X} + \frac{X_x}{X}(F_{ex} + \omega F)$$

$$\frac{X_x}{X} = \frac{F_a - F_e\left(\dfrac{X_e}{X}\right)}{F_{ex} + \omega F}$$ (5.18)

From Eqs. (5.14) and (5.15), Eq. (5.18) is rearranged as follows:

$$\frac{X_x}{X} = \frac{1 + \omega - \dfrac{F_e}{F}\cdot\dfrac{X_e}{X}}{1 + \omega - \dfrac{F_e}{F}}$$ (5.19)

From Eqs. (5.17) and (5.19),

$$\mu = D\left\{1 + \omega\left(1 - \frac{1 + \omega - \dfrac{F_e}{F}\cdot\dfrac{X_e}{X}}{1 + \omega - \dfrac{F_e}{F}}\right)\right\}$$

(5.20)

If $X_e/X \doteq 0$, Eq. (5.20) is reduced to:

$$\mu = D\left\{1 + \omega\left(1 - \frac{1 + \omega}{1 + \omega - \dfrac{F_e}{F}}\right)\right\}$$

(5.21)

The relationship between μ and D in recycling in single vessels can be obtained from Eq. (5.21), as shown in Fig. 5.4; in the figure, μ/D is plotted against ω, parameters being F_e/F. It is noted from this diagram that steady-state operation can be realized, even if the value of dilution rate, D, in single vessels is rather larger than the specific growth rate, μ, of the cells, provided the recycle ratio, ω, is selected ap-

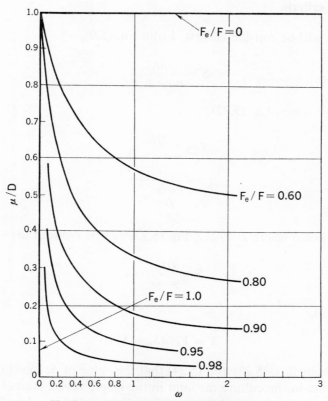

Fig. 5.4. μ/D vs. ω for recycling in steady state.

propriately, depending on the value of F_e/F. An aeration basin of the activated sludge process for sewage and industrial waste treatment is a typical example, because the dilution rate, D, in this example must necessarily be large, as compared with the specific growth rate, μ, of the sludge, to enhance the capacity of the treatment.

For the case of $F_e/F=0$, Eq. (5.21) reduces to nonrecycling, where the value of μ must always be equal to D in single vessels, if steady-state operation is considered. For $F_e/F=1.0$, as another extreme in Fig. 5.4 [or Eq. (5.21)], the value of μ/D becomes zero; unless a nongrowth and blank system is concerned, this extreme case is unlikely.

Suppose that F_e/F in Fig. 5.4 be taken as 0.98, for instance, in the activated sludge process for municipal sewage treatment. Then, the figure suggests, if the assumptions underlying Eq. (5.21) are acceptable in this example, that even sludge whose specific growth rate, μ, is around 1/10 the value of D (reciprocal of the mean retention time of flowing sewage in the aeration basin), warrants steady-state cultivation, provided a recycle ratio, $\omega \doteqdot 0.20-0.25$ is secured in this continuous treatment.

5.1.3. Design criteria

A single vessel will be considered first. From Eq. (5.9),

$$S = \frac{\mu K_s}{\mu_{max} - \mu} \tag{5.22}$$

Substituting $D=\mu$ into Eq. (5.22),

$$S = \frac{DK_s}{\mu_{max} - D}$$
$$= \frac{K_s}{\dfrac{\mu_{max}}{D} - 1} \tag{5.23}$$

For the special case where $D \ll \mu_{max}$, Eq. (5.23) can be reduced to:

$$S = \frac{DK_s}{\mu_{max}} \tag{5.24}$$

From Eq. (5.13),

$$X = Y_{x/s}(S_0 - S) \tag{5.25}$$

It is noted from Eqs. (5.23) and (5.24) that the value of S, the concentration of limiting substrate in the culture medium in the single vessel, starts from zero and increases in proportion to the increase of D as long as $D \ll \mu_{max}$ [see Eq. (5.24)]. As the value of D approaches that of μ_{max}, S increases sharply [see Eq. (5.23)]. Although Eq. (5.23) indicates that the value of S becomes infinite when $D=\mu_{max}$, such a situation cannot be realized in practice since $S<S_0$.

The cell concentration, X, from the vessel is approximately $Y_{X/S}S_0$ at low values of D. However, as D approaches μ_{max} where $S \to S_0$, the value of X approaches zero. This situation is referred to as "wash out" and occurs because the growth rate can no longer keep up with the dilution rate. These various conditions are depicted in Fig. 5.5.

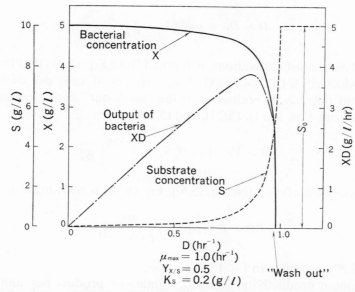

Fig. 5.5. S vs. D and X vs. D (calculated).[18,19] The abscissa represents dilution rate, D, while the ordinates of the figure are limiting substrate concentration, S, cell concentration, X, and the product of XD. These curves are calculated using the values of $\mu_{max}=1.0$ hr^{-1}, $Y_{X/S}=0.5$, $S_0=10$ g/l, and $K_s=0.2$ g/l.

Curves in Fig. 5.5 were prepared from Eqs. (5.23), (5.24), and (5.25), into which the following values were substituted:

$$\mu_{max} = 1.0 \text{ hr}^{-1}$$
$$Y_{X/S} = 0.5$$
$$S_0 = 10 \text{ g/}l$$
$$K_s = 0.2 \text{ g/}l$$

By and large, it is difficult to maintain a steady state near conditions of "wash out," since a slight change of D will result in a large variation of either S or X. The curve of bacterial output in the figure was obtained from a curve for $XF/V=XD$, i.e., from Eqs. (5.25) and (5.23),

$$XD = DY_{X/S}\left(S_0 - \frac{DK_s}{\mu_{max} - D}\right) \qquad (5.26)$$

To find the particular value of D at which the value of XD becomes maximum,

$$\frac{d(XD)}{dD} = 0 \qquad (5.27)$$

Solving Eq. (5.27) for D with Eq. (5.26), if $Y_{X/S}$ remains unaffected by D [cf. Eq. (5.37)],

$$D = D_m = \mu_{max}\left(1 - \sqrt{\frac{K_s}{K_s + S_0}}\right) \qquad (5.28)$$

Since $K_s \ll S_0$ in most cultivations, it is noted from Eq. (5.28) that the maximum output (productivity), $(XD)_{max} = (XF/V)_{max} = $ (mass of cells per unit volume of broth per unit time)$_{max}$, is realized near the "wash out" condition (see Fig. 5.5).

Similarly, from Eqs. (5.11), (5.23), and (5.25),

$$PD = Y_{P/X}Y_{X/S}D\left(S_0 - \frac{DK_s}{\mu_{max} - D}\right) \qquad (5.29)$$

From the definition of $Y_{P/S}$ [cf. Eq. (5.6)], Eq. (5.29) is rearranged to:

$$PD = DY_{P/S}\left(S_0 - \frac{DK_s}{\mu_{max} - D}\right) \qquad (5.30)$$

Equation (5.30) is similar to Eq. (5.26).

The maximum productivity, $(PD)_{max} = $ (mass of product per unit volume of broth per unit time)$_{max}$, is then expected to be at the value of $D = D_m$ expressed by Eq. (5.28). Substituting Eq. (5.28) into Eq. (5.30), the concentration of product, P_m, at the maximum productivity is given by:

$$P_m = Y_{P/S}\{S_0 + K_s - \sqrt{K_s(S_0 + K_s)}\} \qquad (5.31)$$

Similarly, the cell concentration, X_m, at the maximum output will be:

$$X_m = Y_{X/S}\{S_0 + K_s - \sqrt{K_s(S_0 + K_s)}\} \qquad (5.32)$$

All of the above equations from (5.1) to (5.32) are useful in the design of continuous cultivations. Figure 5.5 is useful in determining the values of D which are most economical for a specific continuous cultivation.

It should be remembered that the above equations present design criteria for continuous cultivations which are based on constant values of $Y_{X/S}$, $Y_{P/X}$, and $Y_{P/S}$ irrespective of the values of D or μ.

The material balance [Eq. (5.1)] and the experimental data (dX/dt vs. X) obtainable from batch cultivation provide another approach to the design of continuous culture. Readers should be cognizant that, in this case, batch data may not be reliable, since they are obtained under continuously varying conditions. From Eq. (5.1) for steady state, considering that $X_0 = 0$,

$$\frac{dX_1}{dt} = 0 = \left(\frac{dX_1}{dt}\right)_{\text{Growth}} - \frac{F}{V} X_1$$

from which

$$X_1 = \frac{\left(\dfrac{dX_1}{dt}\right)_{\text{Growth}}}{F/V} \tag{5.33}$$

Likewise,

$$\frac{dX_2}{dt} = 0 = \left(\frac{dX_2}{dt}\right)_{\text{Growth}} - \frac{F}{V}(X_2 - X_1)$$

$$X_2 - X_1 = \frac{\left(\dfrac{dX_2}{dt}\right)_{\text{Growth}}}{F/V} \tag{5.34}$$

$$X_n - X_{n-1} = \frac{\left(\dfrac{dX_n}{dt}\right)_{\text{Growth}}}{F/V} \tag{5.35}$$

Typical batch data in terms of dX/dt vs. X are represented by curve abc in Fig. 5.6. The value of X_1 for the first reactor can be graphically determined by the intersection at A of curve abc with a straight line whose slope is F/V, as shown by Eq. (5.33). By repeating such procedures, the values of X_2, X_3,$\cdots$ will be determined successively, as indicated by Eqs. (5.34) and (5.35) and as shown in the figure.[12,23,38] The intersection labeled A′ in the figure is not the steady-state value of (dX/dt); it is unstable since it occurs in the lag phase of growth before cell division begins.

If a fermentation is considered in which a product other than cell mass is produced, the values of P_1, P_2,$\cdots$ in each vessel in the series can be obtained graphically following the same procedure as that in Fig. 5.6; i.e., (dP/dt) is plotted against P using data from batch experiments [refer to Eq. (5.3) and Eqs. (5.33) to (5.35)].

It must be emphasized that the design procedure mentioned above is acceptable only if the microbial characteristics expressed by (dX/dt vs. X) or (dP/dt vs. P) in batch runs represent those in continuous operation. Many questions relating to this controversial procedure are left open for discussion and further experimentation. However, graphical estimations of X_1, X_2,$\cdots$ or P_1, P_2,$\cdots$ from batch fermentation data are not totally worthless; they can be a valuable guide for the design and operation of a continuous culture plant. For the rational design of a large-scale

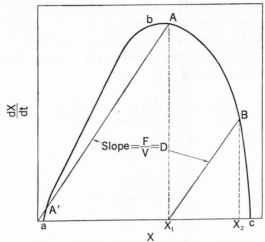

Fig. 5.6. Graphical solution of design equation in continuous fermentation;[23] cell growth rate, dX/dt, is plotted against cell concentration, X, both of which are obtained from batch experiment.

plant, however, there is no substitute for data obtained from continuous cultivation experiments in pilot-plant equipment.

If energy source is a limiting substrate in the first reactor vessel, no growth is expected in the subsequent vessels, unless additional substrate is provided to these vessels. It must also be emphasized that the graphical estimations of X_1, X_2, ··· take no account of the substrate limiting growth. This point must be remembered throughout this section and Section 5.2.

5.2.   MICROBIAL DYNAMICS IN CHEMOSTAT CULTURE (SINGLE VESSELS); SOME EXAMPLES

Here, microbial dynamics deals with the way in which, and the extent to which, the mass balance eqations for X, S, and P [cf. Eqs. (5.1) to (5.6)] can describe microbial behavior in an unsteady state.

The unsteady state applies in the following cases: the transient period at the start of continuous culture on initiating the flow of fresh medium into the batch culture, the period after a change in one of the operating variables such as dilution rate, rotation speed of impeller, etc. which disrupt the steady state, the change from chemostat to batch, and finally, batch culture itself.

The microbial dynamics will be discussed for each case, followed by a short general treatment of the well-known theory and practice of automatic control (cf. Section 4.4.3., Chapter 4).[21,29,30] In connection with microbial dynamics, "balanced" and "unbalanced" cell growth will be discussed from the mass balance concept, which is useful from an engineering viewpoint.

From Eqs. (5.2) and (5.9), taking $n=1$ in Eq. (5.2), and remembering the omission for convenience of the subscript $n=1$, the mass balance is then,

$$\frac{dX}{dt} = \left(\frac{\mu_{max}S}{K_s + S} - D\right)X \tag{5.36}$$

The following equation has been derived from another mass balance with limiting substrate,[1,27,31,35] and confirmed experimentally by many workers with the carbon source as the limiting substrate (cf. Eq. (4.41), Chapter 4)[2,17,18]

$$\frac{1}{Y_{X/S}} = \frac{m}{\mu} + \frac{1}{Y_G} \tag{5.37}$$

where

m = specific rate of consumption of carbon for cellular maintenance
Y_G = yield factor for cell growth

From Eqs. (5.37) and (5.6), taking $n=1$,

$$\frac{dS}{dt} = D(S_0 - S) - \left(\frac{1}{Y_G} \cdot \frac{\mu_{max}S}{K_s + S} + m\right)X \tag{5.38}$$

Equations (5.36) and (5.38) are the basic equations to describe one aspect of the microbial dynamics. Koga *et al.* propose that X and S can show one of six patterns of transient response, depending on the initial conditions. The transient response may show a single maximum and/or a single minimum in either X or S before they become stabilized. So far as the values of Y_G, m, K_s, and μ_{max} can be regarded as constant, they pointed out from phase-plane analysis that the solutions for X and S converge ultimately at one point on the phase plane; no oscillation is expected.[21] In other words, the Monod-type equation [Eq. (5.9)] which is involved in the right-hand sides of Eqs. (5.36) and (5.38) is fairly stable. These arguments of Koga *et al.* apply to the transient period associated with transfer from batch to continuous cultivation.[2]

Values of μ_{max}, K_s, Y_G, and m, in Eqs. (5.36) and (5.38) can be determined with the chemostat culture (steady state) by observing the values of $Y_{X/S}$ and S at each dilution rate, $D (=\mu)$.[2]

Emphasis in this discussion, however, will be placed on the extent to which the solutions for X and S derived from Eqs. (5.36) and (5.38), using the values of μ_{max}, K_s, Y_G, and m assessed from the chemostat culture, do indeed represent the dynamic behavior. Before proceeding to discuss some examples relevant to another aspect of microbial dynamics, the points worthy of note are as follows:

1. The dynamics of intracellular components will provide the key to elaborating the overall behavior of microbes.
2. It is difficult to define the dynamic pattern of intracellular components

rigidly; the pattern may vary from case to case; for instance, it may be first-
or second-order (*cf.* Eqs. (5.39) and (5.47) later).
3. Steady-state kinetic patterns are generally representative of the data on microbial dynamics in both continuous and batch cultivation.
4. Conceptually, the examples below originate partly from the pioneering work
   on microbial dynamics such as that shown in Fig. 5.1.

5.2.1. Response to change in environment of chemostat culture (*Azotobacter vinelandii*; unsteady state)[2]

Figure 5.7 shows the response of *Azotobacter vinelandii* to a new environment, in
which the dilution rate was changed stepwise from $D=0.180$ to $D=0.266$ hr^{-1}. The
nitrogen-fixing bacterium was cultivated in a chemostat (limiting substrate=
glucose) (working volume=13 *l*) at 30°C, pH=7.0 before the dilution rate was
elevated. In the lower half of this diagram, the responses of S and X and the change
in the RNA content, N_R, of the cells are plotted against time, t hr. In the upper
diagram, each specific rate which was calculated (see later) from the solid curves in
the lower diagram for S, X, and N_R is also plotted against time.

Solid curves for X and S were calculated by solving Eqs. (5.36) and (5.38) with an
analog computer. Another solid curve for N_R was calculated from the following
equation.

$$\frac{dN_R}{dt} = \beta \, (N_{Rm} - N_R) \tag{5.39}$$

where

    β = empirical constant
    N_{Rm} = final value of RNA content (fraction) of cells

It is observed from Fig. 5.7 that Eqs. (5.36) and (5.38) do represent, in general,
the responses of X and S, even though the values of various constants required for
the calculation were determined from the steady-state condition (chemostat culture). It is interesting to note from the figure that the RNA fraction shows apparently a first-order response, different from those of X and S. It is necessary here to
demonstrate the mutual relationship between the responses of X and S and that of
N_R, before discussing the upper diagram in Fig. 5.7.

The value of cell mass concentration, X, is shown generally by the following
equation.

$$X = \sum_{i=1}^{n} C_i \tag{5.40}$$

Dividing both sides of Eq. (5.40) by X,

$$1 = \sum_{i=1}^{n} (C_i)_f \tag{5.41}$$

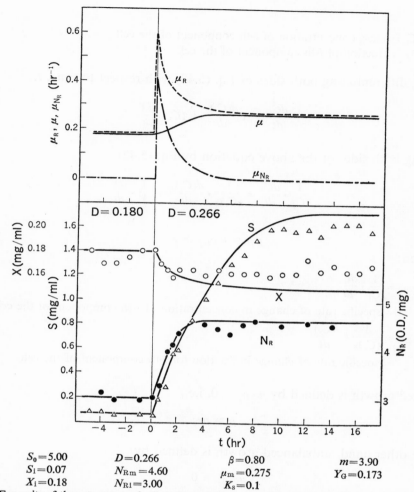

$S_0=5.00$	$D=0.266$	$\beta=0.80$	$m=3.90$
$S_i=0.07$	$N_{Rm}=4.60$	$\mu_m=0.275$	$Y_G=0.173$
$X_i=0.18$	$N_{Ri}=3.00$	$K_s=0.1$	

For units of these terms see the list of nomenclature section at the end of this chapter.

Fig. 5.7. Patterns of substrate, S, cell concentration, X, RNA fraction in cells, N_R, and the values of specific rates vs. time in unsteady state when the value of dilution rate, D, was changed stepwise from $D=$ .0180 to $D=0.266$ hr^{-1}.[2] Solid curves in the lower half of the figure were calculated by using an analog computer from Eqs. (5.36) and (5.38) (for X and S), and from Eq. (5.39) (for N_R); values of respective terms required for the computation are summarized (see the table above).

Solid, dotted and chain-like curves in the upper half of the figure are calculated from solid curves in the lower half of the figure. Provided: μ_R=specific rate of change in RNA concentration, XN_R, in medium; μ_{NR}=specific rate of change in RNA content (fraction) of cell.

$$C_t = X(C_i)_f \tag{5.42}$$

Differentiating Eq. (5.41) with respect to time, t,

$$\sum_{i=1}^{n} \frac{d(C_i)_f}{dt} = 0 \tag{5.43}$$

where

C_i = mass concentration of i-th conponent of the cell

$(C_i)_f$ = fraction of i-th component of the cell

Now, differentiating both sides of Eq. (5.42) with respect to time, t,

$$\frac{dC_i}{dt} = X\frac{d(C_i)_f}{dt} + (C_i)_f\frac{dX}{dt}$$

Dividing both sides of the above equation by Eq. (5.42),

$$\frac{1}{C_i}\cdot\frac{dC_i}{dt} = \frac{1}{(C_i)_f}\cdot\frac{d(C_i)_f}{dt} + \frac{1}{X}\cdot\frac{dX}{dt}$$

Then,

$$\mu_{c_i} = \mu_{(c_i)_f} + \mu \qquad\qquad (5.44)$$

provided:

$$\mu_{c_i} = \frac{1}{C_i}\cdot\frac{dC_i}{dt}$$

= specific rate of change in concentration of i-th component of the cell

$$\mu_{(c_i)_f} = \frac{1}{(C_i)_f}\cdot\frac{d(C_i)_f}{dt}$$

= specific rate of change in fraction for i-th component of the cell

Balanced growth is defined by $\mu_{(c_i)_f} = 0$, i.e.,

$$\mu = \mu_{c_i} \qquad\qquad (5.45)$$

On the other hand, unbalanced growth is defined by

$$\mu_{(c_i)_f} \neq 0 \qquad\qquad (5.46)$$

In other words, the value of specific growth rate, μ, in balanced growth is equal to that of specific rate of change, μ_{c_i}, in concentration of a particular cellular component, C_i. On the other hand, in unbalanced growth one cannot equate the value of μ to μ_{c_i} since, by definition, $\mu_{(c_i)_f} \neq 0$, but even in unbalanced growth Eq. (5.43) must be satisfied; namely, if the content of any one cellular component is increasing with time, there will be other components decreasing such that the mass balance equation [cf. Eq. (5.43)] is satisfied at all times. For a practical example, consult ref. 2.

In Fig. 5.7, μ_R is the specific rate of change in RNA concentration, $X N_R$, and it should be noted that balanced growth is realized far ahead of the establishment of the new steady state at $D = 0.266$ hr^{-1}. It is apparent during the short period of unbalanced growth after D was raised from 0.180 to 0.266 hr^{-1}, that μ_R is not equal to μ, but it approaches the value of μ as balanced growth is attained.

Logarithmic growth of cells in batch cultivation approximates balanced growth defined by Eq. (5.45); for bacteria and yeasts[3] this has been confirmed experimentally, and while the relatively sparse data for fungi are also in agreement,[16] additional experimental evidence is needed.

Another study of microbial dynamics is shown in Fig. 5.8.[25] This unsteady-state behavior was observed when the steady state of *Azotobacter vinelandii* (limiting substrate=glucose, steady-state concentration=2 to 3μg/ml) was disrupted by adding a pulse of glucose to give a final concentration of glucose of 5 to 7 mg/ml, and transferring the chemostat to batch operation. During the unsteady state which continued till complete exhaustion of the glucose, cell concentration, X, residual

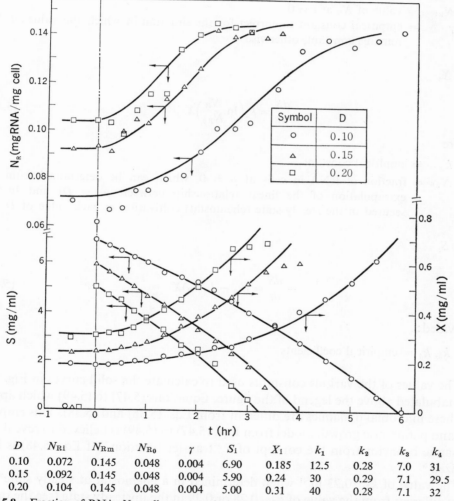

D	N_{Ri}	N_{Rm}	N_{R0}	γ	S_i	X_i	k_1	k_2	k_3	k_4
0.10	0.072	0.145	0.048	0.004	6.90	0.185	12.5	0.28	7.0	31
0.15	0.092	0.145	0.048	0.004	5.90	0.24	30	0.29	7.1	29.5
0.20	0.104	0.145	0.048	0.004	5.00	0.31	40	0.28	7.1	32

Fig. 5.8. Fraction of RNA, N_R, cell concentration, X, and glucose concentration, S vs. t in unsteady growth of *Azotobacter vinelandii*. Parameters are dilution rates, D, in the preceding steady-state growth.[25]

glucose concentration, S, and RNA content, N_R were plotted against time, t hr. The parameters are the dilution rates, D, applying at the former steady states.

Solid curves drawn through the data points in Fig. 5.8 were calculated from the following equations, using an analog computer for the curve-fitting procedure. For N_R, autocatalytic-type reaction kinetics were applied.

$$\frac{dN_R}{dt} = k_1(N_{Rm} - N_R)(N_R - N_{Ri} + \gamma) \qquad (5.47)$$

where

k_1 = coefficient
N_{Rm} = maximum value of N_R
N_{Ri} = value of N_R at $t = 0$
γ = empirical constant to correct for the situation in which the value of N_R remains constant, otherwise

For X,

$$\frac{dX}{dt} = k_2\left(\ln \frac{N_R}{N_{R0}}\right)X \qquad (5.48)$$

where

k_2 = empirical constant
N_{R0} = fraction of RNA in cells at $\mu = 0$, which can be determined from an extrapolation of the linear relationship between $\mu (= D)$ and $\ln N_R$, secured in the steady-state (chemostat) cultivation at each value of D [25]

For S,

$$-\frac{dS}{dt} = (k_3 - k_4 N_R)X \qquad (5.49)$$

provided:

k_3, k_4 = empirical coefficients

The values of the various constants used to calculate the solid curves in Fig. 5.8 are tabulated above the legend of the figure. Equations (5.47) to (5.49) which apply to these microbial dynamics are different from Eqs. (5.36) and (5.38). This empirical and parametric growth model from Eqs. (5.47) to (5.49) is believed to reveal the dynamic behavior from the concept of a "transfer function," if Eq. (5.48) is linearized.[5]

The value of $k_2 = 0.23$ hr^{-1} when determined in a chemostat at steady state[25] is very different from the value of $k_2 = 0.28$ or 0.29 in the unsteady state (see the table accompanying Fig. 5.8). This is worthy of further discussion (see later).

5.2.2. Calculation vs. observation of anaerobic fermentation of baker's yeast (unsteady state)

The following kinetic equations for an anaerobic fermentation by baker's yeast were derived from a chemostat culture[4,6] (limiting substrate=glucose).

$$\frac{dX}{dt} = \frac{\mu_0}{1 + \dfrac{P}{K_p}} \cdot \frac{S}{K_s + S} X \qquad (5.50)$$

$$\frac{dP}{dt} = \frac{\nu_0}{1 + \dfrac{P}{K_p'}} \cdot \frac{S}{K_s' + S} X \qquad (5.51)$$

$$-\frac{dS}{dt} = \frac{1}{Y_{X/S}} \cdot \frac{dX}{dt} = \frac{1}{Y_{P/S}} \cdot \frac{dP}{dt} \qquad (5.52)$$

where

$$P = \text{ethanol concentration}$$
$$S = \text{glucose concentration}$$
$$K_s, K_s' = \text{saturation constants}$$
$$K_p, K_p' = \text{empirical constants}$$
$$\mu_0 = \text{specific growth rate at } P = 0$$
$$\nu_0 = \text{specific rate of ethanol production at } P = 0$$

Figure 5.9 shows anaerobic fermentation by the yeast cells in a batch reactor. Solid curves in the figure were calculated from Eqs. (5.50) to (5.52) by a digital computer. The values of K_s, K_s', $Y_{P/S}$, and $Y_{X/S}$ required for the computation were determined from the chemostat at steady state, whereas the values of K_p, and K_p' were assessed either in shaken flasks or in a Warburg respirometer.[6]

Although the arbitrary selection of these values might be questioned, the point to be emphasized is that Eqs. (5.50) to (5.52) represent, in general terms, the dynamic behavior in batch, though there are no accepted ways of arriving at the values of the various constants involved in these equations (see Fig. 5.9)

Assuming that the kinetic equations remain unchanged in both batch and continuous, the difference between unsteady- and steady-state cultivations may be established via the numerical values for some of the constants.[7,26]

The previous examples (*cf.* Figs. 5.7 and 5.8) show indeed that the various constants determined from the chemostat culture could be used to depict approximately the dynamic behavior. A better agreement in Fig. 5.7 between calculation and observation would require values other than those adopted.

Though the examples discussed here suggest that the Monod-type equation (μ vs. S) is applicable to the unsteady state, it is still early to draw a general conclu-

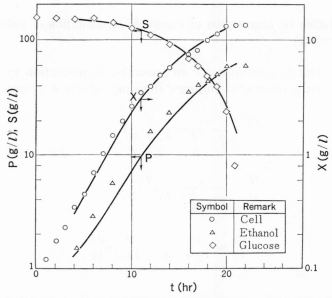

Fig. 5.9. Batch alcohol fermentation observed, and calculated. Numerical values substituted into respective coefficients of the kinetic equations are: $K_p=16.0,$[6] $K_p'=71.5,$[7] $\mu_0=0.408,$ $\nu_0=1.0,$[4] $K_s=0.22,$[4] $K_s'=0.44,$[4] $Y_{P/S}=0.35$ $(Y_{X/S}=0.10).$[4] For units of these constants, see the nomenclature section at the end of this chapter.

sion as to the way to formulate the dynamic behavior of microorganisms. Further studies of the dynamic behavior of chemostat cultures are required.

5.3.   COMPARISON BETWEEN BATCH AND CONTINUOUS CULTIVATIONS

The upper graph of Fig. 5.10 illustrates schematically an operating cycle for batch cultivation. For simplicity, the curve is divided into four periods—logarithmic growth, harvest of cells, preparation for another batch run, and the lag following inoculation. These periods are designated as t_m, t_0, t_1, and t_2, respectively (see Fig. 5.10). The total period of time, t_c, required for one cycle of the batch cultivation is:

$$\begin{aligned} t_c &= t_m + t_0 + t_1 + t_2 \\ &= t_m + t_l \\ &= \frac{1}{\mu_{max}} \ln \frac{X_m}{X_i} + t_l \end{aligned}$$
(5.53)

where

$$t_l = t_0 + t_1 + t_2$$
X_m = maximum cell concentration
X_i = initial cell concentration
μ_{max} = maximum value of specific growth rate

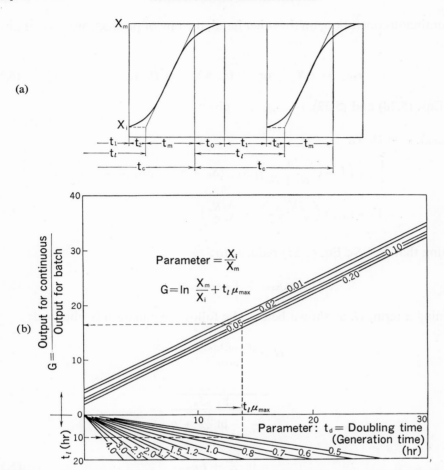

Fig. 5.10. Comparison between batch and continuous cultivations; the upper graph of the figure shows schematically the cycle of batch cultivation expressed by the cell concentration, X, and the time, t, of cultivation. The lower graph represents how much gain, G, is obtained by the continuous cultivation compared with the batch. For symbols and use of the figure, see the text.

Designating the substrate concentration of fresh medium as S_0 and introducing the yield, $Y_{X/S}$ from Eq. (5.6),

$$X_m - X_i = Y_{X/S}S_0 \tag{5.54}$$

Consequently, rate of production of cells in batch, r_{batch}, will be:

$$r_{batch} = \frac{Y_{X/S}S_0}{\frac{1}{\mu_{max}}\ln\frac{X_m}{X_i} + t_l} \tag{5.55}$$

In continuous operation, on the other hand, the rate of production, $r_{cont.}$, is given by:

$$r_{cont.} = DX \quad \text{or} \quad (r_{cont.})_{max} = D_m X_m \tag{5.56}$$

From Eqs. (5.28) and (5.32), $(r_{cont.})_{max}$ is given by:

$$(r_{cont.})_{max} = D_m X_m$$

$$= \left\{ \mu_{max}\left(1 - \sqrt{\frac{K_s}{K_s + S_0}}\right) \right\} Y_{X/S}\left\{ S_0 + K_s - \sqrt{K_s(S_0 + K_s)} \right\}$$

$$= Y_{X/S}\mu_{max}S_0\left(\sqrt{\frac{K_s + S_0}{S_0}} - \sqrt{\frac{K_s}{S_0}}\right)^2 \tag{5.57}$$

Assuming that $K_s \ll S_0$, Eq. (5.57) reduces to:

$$(r_{cont.})_{max} \doteq Y_{X/S}\mu_{max}S_0 \tag{5.58}$$

Defining a term, G, as shown below, the following equation is obtained:

$$G = \frac{(r_{cont.})_{max}}{r_{batch}}$$

$$= \frac{Y_{X/S}\mu_{max}S_0}{\dfrac{Y_{X/S}S_0}{\dfrac{1}{\mu_{max}}\ln\dfrac{X_m}{X_i} + t_i}}$$

$$= \ln\frac{X_m}{X_i} + t_i\mu_{max} \tag{5.59}$$

Equation (5.59) is shown graphically in Fig. 5.10, parameters being inoculum size fraction, X_i/X_m, and doubling time, t_d. The use of this figure will be exemplified; suppose that a specific microbe, whose doubling time $t_d = 0.5$ hr, is to be cultivated batchwise with the value of $t_i = 10$ hr and inoculum size fraction, $X_i/X_m = 0.05$. If this batch cultivation is carried out continuously, what is the value of the gain, G, obtained?

Draw a line parallel to the abscissa in the lower part of Fig. 5.10 from $t_i = 10$ and from the intersection between the line and the parameter for $t_d = 0.5$, draw another line parallel to the ordinate in the lower half of Fig. 5.10. The intersection between the vertical line and another parameter for $X_i/X_m = 0.05$ gives the value of $G \doteq 17$ in this example. For another example of $X_i/X_m = 0.01$, the values of G expected for various doubling times are shown in Table 5.3.

It is clear that the value of G is affected markedly by the value of t_d and t_i, but is less sensitive to the value of X_i/X_m. It is also evident that continuous cultivation is

TABLE 5.3

COMPARISON BETWEEN BATCH AND CONTINUOUS CULTIVATIONS
(SEE FIG. 5.10).

t_d (hr)	G	
	$t_l = 10$ hr	$t_l = 20$ hr
3.0	7	9
1.5	9	14
1.0	12	18

Note: Values of gain, G, the productivity ratio of continuous to batch cultivations are affected by the doubling time, t_d, of microorganisms and the time, t_l required between successive batch runs; provided the ratio of inoculum size fraction, X_i/X_m is taken as 0.01 (*cf.* Fig. 5.10).

more advantageous with fast-growing cells; although Fig. 5.10 illustrates this fact for the production of cells, the advantage also applies to the formation of product, P, in a growth-associated fermentation.

It must be remembered that Eq. (5.59) or Fig. 5.10 is derived assuming that the values of yield, $Y_{x/s}$, remain unchanged for a specific organism, irrespective of whether it is in batch or continuous cultivation. This assumption is not necessarily true. However, this fact does not invalidate the use of Fig. 5.10 as a basis of comparison between batch and continuous cultivations.

5.4. EXAMPLES OF CONTINUOUS CULTIVATION

5.4.1. Yeast

Figure 5.11 is an example of baker's yeast production in continuous culture, as conducted by Humphrey with a 5-l jar-type reactor.[20] The ordinate of Fig. 5.11 is the value of yield, $Y_{x/s}$, while the abscissa represents the dilution rate, D (or specific growth rate, μ). The various data points on which the curve is based were obtained by maintaining the continuous cultivation in a steady state for a couple of days.

It is seen from Fig. 5.11 that the value of $Y_{x/s}$ is not constant even over a narrow range of D. The variation in $Y_{x/s}$ was considered by Humphrey to be due to a specific physiological state at the low dilution rates where substrate concentration is low, and to the difficulties in maintaining a true steady state near "wash out" at high dilution rates.

An interesting observation was that after several days of continuous operation at relatively high dilution rates, mycelial forms of the yeast begin to appear. These were not contaminants because when transferred to shaken flasks with poorer aeration, the elipsoidal or normal forms reappeared. This mycelial yeast form, al-

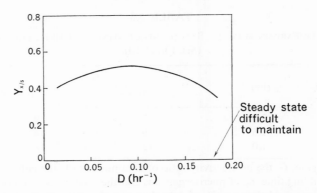

Fig. 5.11. Continuous yeast cultivation;[20] yield, $Y_{x/s}$, is plotted against dilution rate, D.

though having a very high respiratory activity, would not make good baker's yeast because of its grainy character.

5.4.2. Bacteria

Figure 5.12 is an example of continuous steady-state culture of *Aerobacter aerogenes*.[19] This example was subject to conditions where NH_3 was the growth-limiting substrate, S (μg nitrogen/ml). Both values of cell mass, X, and substrate, S, were measured only after at least 2 to 3 days at steady state at each value of D. The experimental curves in Fig. 5.12 are in accord with the shapes predicted from Eqs. (5.23) and (5.25). The yield, $Y_{x/s}$ (not shown in the figure) was constant except in the particular range of D near "wash out."

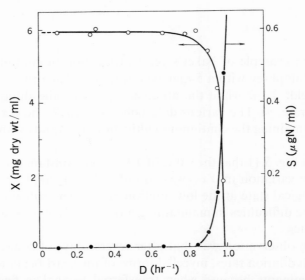

Fig. 5.12. Continuous culture of *Aerobacter aerogenes*;[19] cell concentration, X, and limiting substrate concentration, S, in single vessels are plotted against dilution rate, D.

5.4.3. Fungi

The continuous growth of *Actinomyces* and fungi in small-scale reactors in series has been studied by several workers.[8,32,36] In these cases, the transportation of mycelia from vessel to vessel is an additional problem not encountered when other microbes such as yeast and bacteria are continuously cultivated. Because of this problem, there are not as many extensive studies on the continuous culture of fungi as there are on yeast and bacteria.

Sikyta *et al.*[33] studied the continuous culture of *Streptomyces aureofaciens* with a 10-*l* reactor using synthetic medium. Sucrose was selected as the limiting substrate. Two levels of sucrose concentration were studied. Steady states at each dilution rate were maintained for 1 to 3 days before the values of S and X were measured.

Figure 5.13 is an example of the results obtained with *Streptomyces aureofaciens*. Again, these results are similar to those predicted by Eqs. (5.23) and (5.25). In these

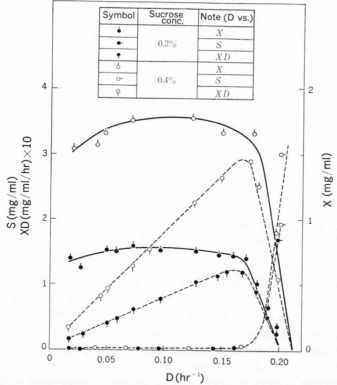

Fig. 5.13. Continuous culture in a single vessel of *Streptomyces aureofaciens*;[33] the ordinate represents limiting substrate concentration (sucrose), S, cell concentration, X, and productivity of cells, XD, while the abscissa is dilution rate, D.

experiments the value of $Y_{x/s}$ was found not to be constant (see Fig. 5.14), as was the case with the continuous culture of yeast in Fig. 5.11. The maximum outputs of mycelia for both levels of sucrose concentration in the feed were observed to occur near "wash out." This experimental fact is in agreement with Eq. (5.28).

5.5. PRACTICAL PROBLEMS WITH CONTINUOUS OPERATION

There are certain practical problems inherently associated with continuous operation. These often detract from the utility of continuous cultivations.

5.5.1. Lack of homogeneity in the continuous-reactor vessel

The cause of poorer yields at low dilution rates, as shown in Figs. 5.11 and 5.14 (also see Eq. (5.37)) may be accentuated by a lack of homogeneity in the reactor.

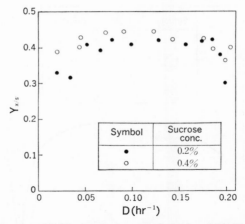

Fig. 5.14. $Y_{x/s}$ vs. D (*Streptomyces aureofaciens*); yield, $Y_{x/s}$ is plotted against dilution rate, D; data points were calculated from those in Fig. 5.13.[33]

The problem of ensuring homogeneity of nutrients throughout a reactor becomes more serious at low dilution rates where the limiting substrate concentration is low, especially in the case of thick and viscous media and for large-scale equipment. In order to minimize the problem, good mixing of the broth, either with mechanical agitation or with high aeration, becomes significant. Research projects on the mixing of highly viscous or non-Newtonian fluids are lacking, but are necessary if this problem is to be understood and solved in large-scale continuous cultivations.

Convenient and reliable techniques are needed for measuring the dissolved oxygen concentration in fermentation broth (*cf.* Section 12.3.3., Chapter 12) and for following changes in viscosity with time. These would allow the aeration to be varied to control the dissolved oxygen at a selected concentration, and in addition, the broth could be diluted to provide a constant viscosity. These measures would assist in maintaining homogeneous conditions during continuous operation.

5.5.2. Maintenance of sterility

A second problem is that of maintaining a continuous cultivation under aseptic conditions for long periods of time, though not required for the activated sludge process. The maintenance of such conditions places a serious burden on the operation of sterilizing media and air. Hence, in continuous cultivation, these systems must be designed and operated with extreme care.

However, even with the most appropriately designed fibrous air filter there exists a finite chance of contaminating microbes eventually passing through the filter into the medium (*cf.* Chapter 10). Also in the medium sterilization, even if the operation is performed with the utmost care, the probability of undesirable microbes being charged into the broth cannot be reduced absolutely to zero (*cf.* Chapter 9).

To cope with such eventualities, in a continuous glutamic acid fermentation Ueda advocated the use of a mixture of drugs which suppressed the growth of undesirable microorganisms without hampering the activities of the glutamic acid-producing strain.[39] The strain he used was acclimatized to the drug mixture. This sort of procedure is an excellent example of an alternate, and perhaps more practical, solution to the problem of reducing contamination in continuous cultivations.

5.5.3. Stability

Stability is a third problem. There are two kinds of stability involved; one concerned with the microbial strain, the other with mechanical operations. As was apparent from Figs. 5.11, 5.12, and 5.13, operation at dilution rates near "wash out" is not recommended due to the instability experienced in practical operation. The inherent instability of feed devices and the sensitivity of the cell mass, X, and of the substrate concentration, S, near "wash out" to slight changes in dilution rate, D should be remembered. Although the constancy of the feeding device is of prime importance in the case of small-scale laboratory experiments, the problem in large reactors will not be so serious.

In long continuous cultivations the microbes are inevitably subject to mutations. When the mutant form has a selective advantage over the desired strain, the mutant will dominate the cultivation. Knowledge of ways to suppress undesirable mutations or maintain the desired strain in continuous cultivations is needed. One technique is that of semi-continuous cultivation in which the first vessel of the series is periodically reseeded. In passing it should be noted that continuous culture is an excellent tool for studying mutations of microbial populations.

5.6. EXAMPLES OF DESIGN CALCULATION

5.6.1. Recycling in single vessels

The activated sludge process accomplishes essentially two operations. It supplies oxygen to the microbial population and separates the activated sludge floc from the mixed liquor. Usually, part of the sludge is fed back to the aeration unit.

Data recorded (at 27°C) in steady-state operation at a sewage treatment plant are as follows:

Flow rate of raw sewage, $F = 310,000$ m³/day
Flow rate of excess sludge, $F_{ex} = 6,700$ m³/day
Working volume of aeration unit, $V = 68,000$ m³
Recycle ratio, $\omega = 0.197$
Concentration of a specific species of microbe ("marker") in aeration basin, $X = 21,000$/ml
Concentration of the "marker" microbe in effluent from separator, $X_e = 52$/ml

Assuming complete mixing in the aeration unit and that no growth of the sludge occurs in the separator, estimate the value of specific growth rate, μ, of the sludge.

Solution
From Eq. (5.15), flow rate of effluent from separator,

$$F_e = F - F_{ex}$$
$$= 310,000 - 6,700$$
$$= 303,300 \, \text{m}^3/\text{day}$$

In addition,

$$\frac{X_e}{X} = \frac{52}{21,000} = 2.37 \times 10^{-3}$$

Then, from Eq. (5.20),

$$\mu = \frac{F}{V}\left\{1 + \omega\left(1 - \frac{1 + \omega - \frac{F_e}{F} \cdot \frac{X_e}{X}}{1 + \omega - \frac{F_e}{F}}\right)\right\}$$

$$= \frac{310,000}{68,000}\left\{1 + 0.197\left(1 - \frac{1 + 0.197 - \frac{303,300}{310,000} \cdot \frac{52}{21,000}}{1 + 0.197 - \frac{303,300}{310,000}}\right)\right\}$$

$$= 0.59 \, \text{day}^{-1}$$

5.6.2. Two vessels in series

Batch data on growth of *Lactobacillus delbrueckii* with a medium containing glucose 5%, yeast extract 3%, and mineral salts at 45°C and pH = 6.0 are given in Table 5.4 (two columns from the left in the table).[22] For this example, the organism is to be transferred to grow continuously in two vessels in series, each with a work-

TABLE 5.4

BATCH DATA ON GROWTH OF *Lactobacillus delbrueckii*.[22]

t (hr)	$N\left(\dfrac{\text{U.O.D.}}{\text{ml}}\right)$	$\dfrac{dN}{dt}\left(\dfrac{\text{U.O.D.}}{\text{ml hr}}\right)$	t (hr)	$N\left(\dfrac{\text{U.O.D.}}{\text{ml}}\right)$	$\dfrac{dN}{dt}\left(\dfrac{\text{U.O.D.}}{\text{ml hr}}\right)$
1.00	0.12	0.04	7.50	2.21	1.06
1.50	0.14	0.05	8.00	2.79	1.33
2.00	0.17	0.06	8.50	3.46	1.63
2.50	0.22	0.09	9.00	4.31	1.99
3.00	0.28	0.13	9.50	5.35	2.09
3.50	0.35	0.16	10.00	6.30	1.70
4.00	0.45	0.21	10.50	7.00	1.30
4.50	0.57	0.27	11.00	7.60	1.06
5.00	0.72	0.35	11.50	8.14	0.89
5.50	0.91	0.44	12.00	8.51	0.74
6.00	1.15	0.55	12.50	8.84	0.62
6.50	1.39	0.67	13.00	9.13	0.54
7.00	1.76	0.85	13.50	9.40	0.44

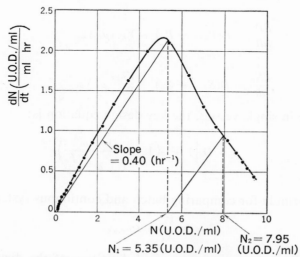

Fig. 5.15. Plot of dN/dt against N from the data given in Table 5.4.

ing volume of 100 l. It is desired to adjust the cell concentration in terms of optical density so that the effluent from the second vessel will have a concentration, $N_2 = 8$ U.O.D./ml. The necessary flow rate, F, and the values of N_1 and N_2 in each vessel are to be determined.

Solution
From the batch data (t vs. N in Table 5.4), each value of dN/dt is calculated and plotted against N as shown in Fig. 5.15. Utilizing the graphical method to determine the cell concentration in each vessel, as shown in Fig. 5.6, the value of F required for the desired condition of $N_2 = 8$ U.O.D./ml can be determined by a trial-and-error method (see Fig. 5.15). The answers to this problem are:

$$F/V = 0.40 \text{ hr}^{-1}$$
$$F = 0.40 \times 100 = 40 \text{ } l/\text{hr}$$
$$N_1 = 5.35 \text{ U.O.D./ml}$$
$$N_2 = 7.95 \text{ U.O.D./ml}$$

5.7. SUMMARY

1. Continuous cultivation of microorganisms is a powerful tool for revealing microbial metabolic responses. Continuous culture systems can be used both for studying steady-state and unsteady-state behavior.

For steady-state operation, key equations are:

For X:
$$V\frac{dX_n}{dt} = FX_{n-1} - FX_n + V\mu_n X_n \qquad \text{[Eq. (5.1)]}$$

For P:
$$\frac{dP_n}{dt} = D(P_{n-1} - P_n) + Y_{P/X}\mu_n X_n \qquad \text{(Eq. (5.4)]}$$

For S:
$$\frac{dS_n}{dt} = D(S_{n-1} - S_n) - \frac{Y_{P/X}}{Y_{P/S}}\mu_n X_n \qquad \text{[Eq. (5.6)]}$$

2. For recycle in single vessels, the key design equation is:

$$\mu = D\left\{1 + \omega\left(1 - \frac{1+\omega}{1+\omega - \frac{F_e}{F}}\right)\right\} \qquad \text{[Eq. (5.21)]}$$

3. A useful formula for comparing batch and continuous system is:

$$G = \ln\frac{X_m}{X_i} + t_l\mu_{max} \qquad \text{[Eq. (5.59)]}$$

4. For unsteady-state operation, a rigid definition of the dynamic pattern of

intracellular components is difficult; the RNA fraction shows an apparent first- or second-order response with time.

5. The various constants in the kinetic equation (derived from steady-state chemostat culture) can be used to describe unsteady-state performance.

6. Dynamic equations have been developed to deal with the following cases: a. transient periods associated with initiation of continuous cultivation, b. transfer from continuous to batch, c. disruption of steady state caused by changing the concentration of limiting substrate, either in a step- or delta-function, d. batch cultivation.

7. In balanced growth, the specific rate of growth, μ, is equal to the specific rate of change in any cellular component concentration, μ_{c_i} in the medium, where-as in unbalanced growth the value of μ is not equal to μ_{c_i}, i.e., the value of specific rate of change in the fraction, $\mu_{(c_i)_f}$ of cellular component is not equal to zero.

8. Some practical problems associated with chemostat culture are: a. lack of homogeneity in the reactor, b. maintenance of sterility, c. cellular stability.

NOMENCLATURE

$C_i = (C_i)_f X$ = mass concentration of i-th component of the cell, mg/ml

$(C_i)_f$ = fraction of i-th component of the cell

D = dilution rate, hr^{-1}

D_m = dilution rate at which either XD or PD becomes maximum

F = flow rate of fresh medium, l/hr

F_a = flow rate of cell suspension from reactor, l/hr

F_e = flow rate of effluent from separator, l/hr

F_{ex} = flow rate of concentrate cell suspension from separator, l/hr

G = gain, productivity ratio of continuous to batch cultivation

k_1, k_2, k_3, k_4 = empirical coefficients

K_p, K_p' = empirical constants, g/l

K_s, K_s' = saturation constants, g/l

m = specific consumption rate of carbonaceous material for cellular maintenance, hr^{-1}

N = concentration of microorganisms, 1/ml or U.O.D./ml

N_R = RNA content (fraction) of cells

N_{Ri} = value of N_R at $t = 0$

N_{Rm} = final value of RNA content (fraction) of cells

N_{R0} = fraction of RNA in cells at $\mu = 0$

P = product concentration, ethanol concentration, mg/ml

r_{batch} = cell productivity in batch, mg/ml hr

$r_{cont.}$ = cell productivity in continuous run, mg/ml hr

S = limiting substrate concentration in single vessels unless otherwise noted, mg/ml

S_0 = limiting substrate concentration in fresh medium, mg/ml

t = time, hr

t_d = doubling time (generation time) = $0.693/\mu$, hr

t_m = time for exponential growth in batch, hr

t_1 = time of preparation for another batch run, hr

t_2 = lag time, hr

t_0 = time required for harvest in batch, hr

$t_l = t_0 + t_1 + t_2$, hr

$t_c = t_0 + t_1 + t_2 + t_m$, hr

$\bar{t}$ = holding time of flowing medium (= V/F), hr

V = volume of broth, l

X = cell mass concentration in single vessels unless otherwise noted, mg/ml

X_m = maximum cell mass concentration, mg/ml

X_e = cell mass concentration in effluent from separator, mg/ml

X_0 = cell mass concentration in fresh medium, mg/ml

X_x = cell mass concentration in pipe for recycling, mg/ml

$Y_G = \dfrac{X}{(-\varDelta S)_G}$ — yield factor for cell growth [$(-\varDelta S)_G$ = carbon source, per unit volume of liquid, devoted to cell growth]

$Y_{X/S} = -\varDelta X/\varDelta S$, overall yield for cell growth

$Y_{P/X} = \varDelta P/\varDelta X$, yield

$Y_{P/S} = -\varDelta P/\varDelta S$, yield

Z = substrate concentration in Fig. 5.1, mg/ml

subscripts

1, 2, ..., $(n-1)$, n = 1st, 2nd, ..., $(n-1)$-th, and n-th vessels

0 = value regarding fresh medium

m; max = maximum value

Greek letters

β = empirical constant

γ = empirical constant, Eq. (5.47)

μ = specific growth rate, hr^{-1}

μ_0 = value of μ at $P = 0$

$\mu_{c_i} = \dfrac{1}{C_i} \cdot \dfrac{dC_i}{dt}$ = specific rate of change in concentration of i-th component of cell, hr^{-1}

μ_R = specific rate of change in RNA concentration, XN_R, in medium, hr^{-1}

$\mu_{(c_i)_f} = \dfrac{1}{(C_i)_f} \cdot \dfrac{d(C_i)_f}{dt}$ = specific rate of change in fraction for i-th component of cell, hr^{-1}

μ_{N_R} = specific rate of change in RNA content (fraction) in cell, hr^{-1}

μ_{max} = maximum value of specific growth rate, hr^{-1}

ν_0 = specific rate of ethanol production at $P = 0$

ω = recycle ratio, Fig. 5.3

REFERENCES

1. Aiba, S., Nagai, S., Nishizawa, Y., and Onodera, M. (1967). "Energetic and nucleic analyses of a chemostatic culture of *Azotobacter vinelandii*." *J. Gen. Appl. Microbiol.* **13**, 73.

2. Aiba, S., Nagai, S., Nishizawa, Y., and Onodera, M. (1967). "Nucleic approach to some response of chemostatic culture of *Azotobacter vinelandii." J. Gen. Appl. Microbiol.* **13**, 85.

3. Aiba, S., Nagatani, M., and Furuse, H. (1967). "Some analysis of lag phase in the growth of microbial cells." *J. Ferm. Tech. (Japan)* **45**, 475.

4. Aiba, S., Shoda, M., and Nagatani, M. (1968). "Kinetics of product inhibition in alcohol fermentation." *Biotech. & Bioeng.* **10**, 845.

5. Aiba, S., Nagai, S., Endo, I., and Nishizawa, Y. (1969). "A dynamical analysis of microbial growth." *A I Ch E Journal* **15**, 624.

6. Aiba, S., and Shoda, M. (1969). "Reassessment of the product inhibition in alcohol fermentation." *J. Ferm. Tech. (Japan)* **47**, 790.

7. Aiba, S., and Shoda, M. (1972). "Bioenergetic analysis of yeast cell growth." *Proc. 1st International Symposium of Chemical Reaction Engineering. Series 109*, Am. Chem. Soc., Washington, D.C.

8. Bartlett, M. C., and Gerhardt, P. (1959). "Continuous antibiotic fermentation—Design of a 20 liter, single-stage pilot plant and trials with two contrasting processes." *J. Biochem. Microbiol. Tech. & Eng.* **1**, 359.

9. Borzani, W., Falcome, M., and Vario, M. L. R. (1960). "Kinetics of the continuous alcoholic fermentation of blackstrap molasses." *Appl. Microbiol.* **8**, 136.

10. Contois, D. E. (1959). "Kinetics of bacterial growth: Relationship between population density and specific growth rate of continuous cultures." *J. gen. Microbiol.* **21**, 40.

11. *Continuous cultivation of microorganisms.* (1969). (Eds.) Málek, I., Beran, K., Fencl, Z., Munk, V., Řicica, J., and Smrcková, H. Academia, Prague and Academic Press, New York and London.

12. Deindoerfer, F. H., and Humphrey, A. E. (1959). "A logical approach to—Design of multistage systems for simple fermentation processes." *Ind. Eng. Chem.* **51**, 809.

13. Fencl, Z. (1963). "A uniform system of basic symbols for continuous cultivation of microorganisms." *Folia Microbiol.* **8**, 192.

14. Fuld, G. J., Mateles, R. I., and Kusmierek, B. W. (1961). "A method for the study of the dynamics of continuous fermentation." p. 54. *S.C.I. Monograph No. 12* (Soc. of Chem. Industry), London.

15. Golle, H. A. (1953). "Theoretical considerations of a continuous culture system." *Agr. & Food Chem.* **1**, 789.

16. Gottlieb, D., and van Etten, J. L. (1964). "Biochemical changes during the growth of fungi. I. Nitrogen compounds and carbohydrate changes in *Penicillium atrovenetum." J. Bacteriol.* **88**, 114.

17. Harte, M. J., and Webb, F. C. (1967). "Utilization of mixed sugars in continuous fermentation. II." *Biotech. & Bioeng.* **9**, 205.

18. Herbert, D., Elsworth, R., and Telling, R. C. (1956). "The continuous culture of bacteria; a theoretical and experimental study." *J. gen. Microbiol.* **14**, 601.

19. Herbert, D. (1961). "A theoretical analysis of continuous culture systems." p. 21. *S.C.I. Monograph No. 12* (Soc. of Chem. Industry), London.

20. Humphrey, A. E. (1963). "Some observations of continuous fermentation." p. 215. *Proc. 5th Symposium, Inst. Appl. Microbiol. Univ. of Tokyo.*

21. Koga, S., and Humphrey, A. E. (1967). "Study of the dynamic behavior of the chemostat system." *Biotech. & Bioeng.* **9**, 375.

22. Luedeking, R., and Piret, E. L. (1959). "A kinetic study of the lactic acid fermentation. Batch process at controlled pH." *J. Biochem. Microbiol. Tech. & Eng.* **1**, 393.

23. Luedeking, R., and Piret, E. L. (1959). "Transient and steady states in continuous fermentation. Theory and experiment." *J. Biochem. Microbiol. Tech. & Eng.* **1**, 431.

24. Maxon, W. D. (1960). "Continuous fermentation." *Adv. Appl. Microbiol.* (Ed.) Umbreit, W. W. **2**, 335.
25. Nagai, S., Nishizawa, Y., Endo, I., and Aiba, S. (1968). "Response of a chemostatic culture of *Azotobacter vinelandii* to a delta type of pulse in glucose." *J. Gen. Appl. Microbiol.* **14**, 121.
26. Nagatani, M., Shoda, M., and Aiba, S. (1968). "Kinetics of product inhibition in alcohol fermentation. Part 1. Batch experiments." *J. Ferm. Tech.* (*Japan*) **46**, 241.
27. Pirt, S. J. (1965). "The maintenance energy of bacteria in growing culture." *Proc. Roy. Soc.* **B163**, 224.
28. Pirt, S. J. (1970). Private communication.
29. Ramkrishna, D., Fredrickson, A. G., and Tsuchiya, H. M. (1966). "The dynamics of microbial growth. (I) A distributed, unstructured model." *J. Ferm. Tech.* (*Japan*) **44**, 203.
30. Ramkrishna, D., Fredrickson, A. G., and Tsuchiya, H. M. (1966). "The dynamics of microbial growth. (II) A distributed, structured model." *J. Ferm. Tech.* (*Japan*) **44**, 210.
31. Schulze, K. L., and Lipe, R. S. (1964). "Relationship between substrate concentration, growth rate, and respiration rate of *Escherichia coli* in continuous culture." *Archiv. für Mikrobiol.* **48**, 1.
32. Sikyta, B., Doskocil, J., and Kaspavora, J. (1959). "Continuous streptomycin fermentation." *J. Biochem. Microbiol. Tech. & Eng.* **1**, 379.
33. Sikyta, B., Slezak, J., and Herold, M. (1961). "Growth of *Streptomyces aureofaciens* in continuous culture." *Appl. Microbiol.* **9**, 233.
34. Spicer, C. C. (1955). "The theory of bacterial constant growth apparatus." *Biometrics* **11**, 225.
35. Tempest, D. W., Hunter, J. R., and Sykes, J. (1965). "Magnesium-limited growth of *Aerobacter aerogenes* in a chemostat." *J. gen. Microbiol.* **39**, 355.
36. Terui, G., Konno, N., and Okazaki, M. (1963). "Studies on continuous culture for enzyme production through preferential synthesis." p. 259. *Proc. 5th Symposium, Inst. Appl. Microbiol. Univ. of Tokyo.*
37. *Theoretical and methodological basis of continuous culture of microorganisms* (1966). (Eds.) Màlek, I., and Fencl, Z. Academia, Prague and Academic Press, New York and London.
38. Ueda, K. (1956). "Studies on continuous fermentation. Part 6. Fermentation cycle in multi-stage system." *J. Agr. Chem. Soc.* **30**, 335.
39. Ueda, K., Takahashi, H., and Oguma, T. (1963). "Production of glutamic acid by continuous multi-stage fermentation." p. 232. *Proc. 5th Symposium, Inst. Appl. Microbiol. Univ. of Tokyo.*

Chapter 6

Aeration and Agitation

The purposes of aeration and agitation in fermentors are, firstly to supply micro-organisms with oxygen, and secondly, to mix fermentation broths in such a way that a uniform suspension of microbes is achieved and the mass-transfer rate of the metabolic product accelerated. Many fermentors are equipped with impellers for the mechanical agitation of broths to disintegrate air bubbles and to intensify the turbulence of the liquid, but fermentors without mechanical agitators are also used, as in some aerated, activated sludge processes.

In principle, aeration and agitation can be viewed from two aspects; the demand for oxygen by the microbes, and the supply of oxygen from air bubbles to the liquid. A sequence of enzymatic reactions underlies the oxygen requirements of microbes, while the oxygen supply from air bubbles is a purely physical operation.

Except for viscous fungal fermentations, aeration and agitation can provide oxygen for most fermentations without difficulty. However, knowledge of microbial oxygen demand, which is liable to change with change in the dissolved oxygen concentration in the broth, will be useful for a better understanding of the process.

The discussions which follow in this chapter are concerned both with bubble aeration and with mechanical agitation. In 1944, Cooper et al.[13] published an experimental procedure for measuring the coefficient of oxygen absorption in a bubble aerator using the sulfite-oxidation method. Many studies on the subject of oxygen transfer have appeared since then. The rational design of aerators for Newtonian liquids is now possible, but oxygen transfer to non-Newtonian liquids such as concentrated mycelial broths still involves many problems.

After briefly reviewing mass-transfer theories and microbial respiration, some significant data regarding bubble motion in liquids and several concepts associated with liquid agitation by impellers, will be discussed. The correlation between the oxygen-transfer coefficient and operating variables will then be elaborated.

6.1. Mass Transfer and Microbial Respiration

6.1.1. Mass-transfer resistance

The rate of mass transfer is expressed by:

163

$$\text{mass-transfer rate} = \text{driving force/resistance} \tag{6.1}$$

The unidirectional rate of mass (oxygen) transfer, n_{O_2}, for the gas component (oxygen) across the unit interfacial area between air bubbles and liquid in a steady state is obtained from Eq. (6.1) by the following equation, if the liquid-film resistance is controlling oxygen transfer in bubble aeration.[2]

$$n_{O_2} = Hk_L(\bar{p} - p^*) \tag{6.2}$$

$$= k_L(C^* - \bar{C}) \tag{6.3}$$

where

H = Henry's constant
k_L = mass-transfer coefficient in liquid film
$\bar{C}$ = dissolved oxygen concentration in bulk liquid
$\bar{p}$ = partial pressure of oxygen in bulk gas phase
C^*, p^* = hypothetical values which are in equilibrium with $\bar{p}$ in bulk gas phase and $\bar{C}$ in bulk liquid phase, respectively

The volumetric oxygen-transfer rate, n_{O_2}', is expressed by:

$$n_{O_2}' = Hk_La(\bar{p} - p^*) \tag{6.4}$$

$$= k_La(C^* - \bar{C}) \tag{6.5}$$

where

a = interfacial area between liquid and gas (air) per unit volume of liquid

It is evident from Eqs. (6.1), (6.2), and (6.3) that the reciprocal of the value of k_L (or Hk_L) is the resistance to oxygen transfer in bubble aeration (two-film theory). Further, k_L is proportional to the square root of the reciprocal either of the contact time between air bubbles and liquid (penetration theory) or of the replacement period of the liquid film surrounding the air bubbles (surface-renewal theory).

The rationale for the management of aeration and agitation is to achieve a value of k_La such that the oxygen supply can meet the oxygen demand of the microbes. The value of k_La is a function of aeration-agitation intensity and of the physical and chemical properties of broth.

So far as the physical transfer of oxygen is concerned, the value of the volumetric coefficient, k_La, around air bubbles is usually limiting.[1,4]

6.1.2. Physical vs. enzymatic considerations

Suppose that oxygen is supplied from air bubbles into a well-dispersed suspension of either bacteria or yeast in an agitated reactor. If only the physical nature of oxygen transfer from the liquid to the microbes is considered, the transfer rate of oxygen can be calculated from the following equation.[24,34]

$$N_{\text{Sh}} = 2 + \alpha''(N_{\text{Re}})^{\beta'}(N_{\text{Sc}})^{\delta'} \tag{6.6}$$

where

N_{Sh} = Sherwood number ($=k_L d/D$)

d = representative diameter of microbe

D = molecular diffusivity of oxygen in liquid

N_{Re} = Reynolds number based on relative velocity between liquid and microbe ($=dv\rho/\mu$)

N_{Sc} = Schmidt number ($=\mu/\rho D$)

Since the relative velocity between microbe and liquid is negligible,[35] the second term on the right-hand side of Eq. (6.6) vanishes.

The maximum rate, n_{O_2}', of oxygen transfer from the liquid to the microbial surface is:

$$n_{O_2}' = k_L \bar{a} n \bar{C} \tag{6.7}$$

provided:

dissolved oxygen concentration at the microbial surface is assumed zero

$\bar{a}$ = surface area of individual cell (assumed spherical in shape)

n = number concentration of microbial cells

Then, using Eq. (6.6), the above equation can be rearranged as follows:

$$n_{O_2}' = (2D/d)\bar{a} n \bar{C} \tag{6.8}$$

As representative of each term on the right-hand side of Eq. (6.8), the following values may be taken, i.e.,

$$D = 10^{-5} \text{ cm}^2/\text{sec}$$
$$d = 2 \times 10^{-4} \text{ cm}$$
$$n = 10^8 \text{ cells/ml}$$
$$\bar{C} = 6 \text{ ppm}$$

The value of n_{O_2}' becomes of the order of 10^{-2} mole O_2/l/min.

Under the above conditions, the maximum oxygen uptake rate, measured either polarographically or with an oxygen sensor, usually ranges from 10^{-4} to 10^{-6} mole O_2/l/min.

The wide discrepancy between the two figures emphasizes that the purely physical interpretation of oxygen uptake by microbes in a medium is not acceptable. In other words, aeration and agitation have a dual purpose; the physical transfer of oxygen from air bubbles into the liquid, and the provision of oxygen for enzymatic uptake by microbes.

6.1.3. Critical value of dissolved oxygen concentration and Q_{O_2}

In designing an aeration system, it is necessary to consider the microbial demand for oxygen due to respiration. The enzymatic analysis of specific respiration (oxygen uptake) rate by microbes, Q_{O_2}, yields a Michaelis-Menten type of equation, as shown in Fig. 6.1 (a).[27] The specific rate at which cells respire increases with increase in the dissolved oxygen concentration, $\bar{C}$ up to a point which is often referred to as $C_{crit.}$. Average figures for $C_{crit.}$ range from 0.1 to 1.0 ppm[18] for bacteria, yeast and fungi growing at 20° to 30°C, provided they are well dispersed in the medium. As is also shown in the figure, by a broken curve, further increase in the dissolved oxygen concentration beyond $C_{crit.}$ may entail a deterioration of the value of Q_{O_2} due to an inhibitory effect of C on the enzyme sequence of reactions.[7]

If the value of Q_{O_2} is plotted against the specific growth rate, $\mu\ (=D)$ of the microbe, provided: $D=$dilution rate, a linear correlation is obtained [see Fig. 6.1 (b)]; an intersection of the straight line with the ordinate, if extrapolated, is designated as $(Q_{O_2})_m$, required for cellular maintenance. This type of correlation is

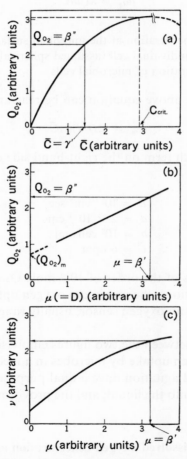

Fig. 6.1. Schematic diagrams showing relationship between specific rates of growth, respiration, product formation and dissolved oxygen concentration in liquid.

justified from an energetic viewpoint and supported experimentally with respect to various microorganisms.[26]

Further, the specific rate of product formation, ν, may be correlated with μ by a specific function,

$$\nu = f(\mu) \tag{6.9}$$

The functional form is either linear, hyperbolic or quadratic, depending on the microbial species and the fermentation conditions. A hyperbolic function is assumed in Fig. 6.1 (c).

The background of each diagram of Fig. 6.1 is different, even with respect to the same microorganism; Fig. 6.1 (a) is the course usually taken by an exhaustion experiment (batch run), and Fig. 6.1 (c) is also taken empirically from batch. On the other hand, the bioenergetic aspects of the microbe [Fig. 6.1 (b)] are generally observed with chemostat culture (continuous). Accordingly, discussion of these three diagrams may yield an insight into the microbial demand for oxygen if and only if the batch performance [Fig. 6.1 (a) and (c)] can also reflect, without grave error, that for continuous runs. The following discussion stems from this premise.

Suppose that a value of ν corresponding to a specific value of $\mu(=\beta')$ is achieved. This presumption requires through Fig. 6.1 (b) and (a) that the value of $\bar{C}$ be maintained at $\bar{C} = \gamma'$, as is apparent from these diagrams. In practice, the value of $\bar{C}$ is monitored continuously (cf. Section 12.3.3., Chapter 12) and as the demand for oxygen by the culture approaches the maximum oxygen-transfer capability of a fermentor, the value of $\bar{C}$ tends to decrease, eventually to zero; this is clear from the following material balance.

$$\frac{d\bar{C}}{dt} = k_L a(C^* - \bar{C}) - Q_{O_2}X \tag{6.10}$$

where

t = time
X = cell mass concentration

In a steady state,
$$\bar{C} = C^* - \frac{Q_{O_2}X}{k_L a} \tag{6.11}$$

To cope with this situation, a portion of the broth can be withdrawn and replaced by fresh medium, reducing effectively the value of $Q_{O_2}X$, and permitting the value of Q_{O_2} to be maintained at the specific value of $Q_{O_2}(=\beta'')$ required [see Fig. 6.1 (b)]. The rate at which regrowth occurs can then be controlled by manipulating a nutrient (usually sugar) in the replacement medium. With repeated withdrawals, it is possible to achieve a greatly prolonged period at the desired level of product formation.

6.1.4. Respiration of mycelial pellet[6,40]

Mycelial pellets are observed frequently in fungal fermentations and it is considered that dissolved oxygen is most likely to become limiting in such cases. To make the limiting situation clear, the distribution of dissolved oxygen within the pellets will be considered. In other words, the physical interpretation of oxygen transfer will now be discussed for cells not well dispersed.

Suppose that a mycelial pellet is spherical (radius$=R$), mycelial density, ρ_m g/cm^3, uniform throughout, as shown in Fig. 6.2. Suppose also that the reference value of the specific respiration rate, Q_{O_2} is determined, if the pellet is disassembled into a pulpy suspension.

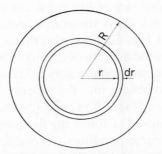

Fig. 6.2. Schematic diagram of a mold pellet.

Flow rate of oxygen with respect to a shell of radii r and $r + dr$ in Fig. 6.2 will be:

$$\text{In} = -(4\pi r^2)D\frac{\partial C}{\partial r}$$

$$\text{Out} = -\{4\pi(r + dr)^2\}D\left(\frac{\partial C}{\partial r} + \frac{\partial^2 C}{\partial r^2}dr\right)$$

$$\text{Respiration} = (4\pi r^2 dr)\rho_m Q_{O_2}{}'$$

$$\text{Accumulation} = (4\pi r^2 dr)\frac{\partial C}{\partial t}$$

Oxygen balance shows:

$$\frac{\partial C}{\partial t} = D\left(\frac{\partial^2 C}{\partial r^2} + \frac{2}{r}\cdot\frac{\partial C}{\partial r}\right) - \rho_m Q_{O_2}{}' \tag{6.12}$$

where

$$D = \text{molecular diffusivity of oxygen}$$
$$Q_{O_2}{}' = \text{specific rate of respiration at } C = C$$

Referring to Fig. 6.1 (a),

$$Q_{O_2}{}' = Q_{O_2}{}''\frac{C}{K_m + C}$$

$$= Q_{O_2}\left(\frac{K_m + \bar{C}}{\bar{C}}\right)\left(\frac{C}{K_m + C}\right) \tag{6.13}$$

provided:

$$Q_{O_2}'' = Q_{O_2}\frac{K_m + \bar{C}}{\bar{C}}$$

$= $ maximum value of specific respiration rate
$Q_{O_2} = $ specific respiration rate at $C = \bar{C}$
$\bar{C} = $ dissolved oxygen concentration in bulk medium
    ($=$ dissolved oxygen concentration at pellet surface, $r = R$) (assumed)
$K_m = $ Michaelis constant

Assuming also that the values of K_m and $\bar{C}$ are of the same order of magnitude ($\sim 10^{-4}$ mole O_2/l),[37] Eq. (6.13) can be rearranged to:

$$Q_{O_2}' = 2Q_{O_2}\frac{C}{K_m + C} \tag{6.14}$$

In the steady state, where $\partial C/\partial t = 0$, oxygen balance within the pellet can be shown by Eqs. (6.12) and (6.14) as:

$$D\left(\frac{d^2C}{dr^2} + \frac{2}{r}\cdot\frac{dC}{dr}\right) = 2\rho_m Q_{O_2}\frac{C}{K_m + C} \tag{6.15}$$

Introducing the dimensionless terms, $y = C/\bar{C}$ and $x = r/R$, Eq. (6.15) is normalized i.e.,

$$\frac{d^2y}{dx^2} + \frac{2}{x}\cdot\frac{dy}{dx} = 12\left(\frac{R}{\sqrt{\frac{6\bar{C}D}{\rho_m Q_{O_2}}}}\right)^2\frac{y}{1+y} \tag{6.16}$$

In deriving Eq. (6.16) the previous assumption that $K_m \approx \bar{C}$ is used. Defining again that

$$12\left(\frac{R}{\sqrt{\frac{6\bar{C}D}{\rho_m Q_{O_2}}}}\right)^2 = a$$

Eq. (6.16) reduces to:

$$\frac{d^2y}{dx^2} + \frac{2}{x}\cdot\frac{dy}{dx} = \frac{ay}{1+y} \tag{6.17}$$

Boundary conditions are:

$$x = 0, \quad dy/dx = 0$$
$$x = 1, \quad y = 1$$

The apparent value of specific respiration rate, Q_{O_2} for the pellet is defined by the following equation.

$$\frac{4}{3}\pi R^3 \rho_m \bar{Q}_{o_2} = \int_0^R \frac{4}{3}\pi \{(r + dr)^3 - r^3\} \rho_m Q_{o_2}{}'$$

$$= \frac{8}{3}\pi \rho_m Q_{o_2} \int_0^R \{(r + dr)^3 - r^3\} \frac{C}{K_m + C} \qquad [cf.\ Eq.\ (6.14)]$$

$$\bar{Q}_{o_2}/Q_{o_2} = 6 \int_0^1 \frac{x^2 y}{1 + y}\, dx \qquad (6.18)$$

Since

$$\frac{d}{dx}\left(x^2 \frac{dy}{dx}\right) = \frac{ax^2 y}{1 + y} \qquad [cf.\ Eq.\ (6.17)]$$

$$\bar{Q}_{o_2}/Q_{o_2} = \frac{6}{a}\left[x^2 \frac{dy}{dx}\right]_0^1$$

$$= \frac{6}{a} \cdot \frac{dy}{dx}\bigg|_{x=1} \qquad (6.19)$$

The solid curve shown in Fig. 6.3 ($\bar{Q}_{o_2}/Q_{o_2}$ vs. $R/\sqrt{6\bar{C}D/\rho_m Q_{o_2}}$) represents the result of calculation [Eqs. (6.17) and (6.19)] by a trial-and-error method using the numerical procedure of Runge-Kutta-Gill,[6] whereas the broken curve in the figure is reproduced from the work of Yano et al.,[40] who calculated $\bar{Q}_{o_2}/Q_{o_2}$ by assuming that the right-hand side of Eq. (6.15) is independent of C (zero-order reaction), and by assuming the critical radius, R_c, provided $C = 0$ for $0 \le r \le R_c$ and $Q_{o_2}{}' = Q_{o_2}$ for $R_c \le r \le R$.

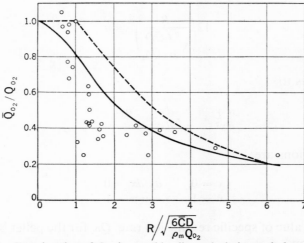

Fig. 6.3.   Deterioration of Q_{o_2} in a mold pellet; calculation and observation.[6,40]

Open circles in Fig. 6.3 are also reproduced from the work of Yano *et al.* who determined the values of $\bar{Q}_{O_2}/Q_{O_2}$ for *Asp. niger* pellets.[40] It is evident from the comparison between calculation and observation that Eq. (6.15) represents more realistically the distribution of dissolved oxygen within the pellet.

6.2. BUBBLE AERATION AND MECHANICAL AGITATION

6.2.1. Bubble aeration

6.2.1.1. *Single bubbles*

The size, d_B, of single bubbles emerging from an orifice into water was confirmed by van Krevelen *et al.* to be proportional to the cube root of the orifice diameter, $d^{1/3}$, and independent of the gas-flow rate, F, over a range from 0.02 to 0.5 cm^3/sec.[39] This relationship between bubble size and orifice diameter does not hold true when the value of F exceeds the above range.

In aeration practice in the fermentation industry, the following, purely empirical equation is applied in estimating the value of d_B:

$$d_B \propto F^\alpha, \qquad \alpha = 0.2 \sim 1.0 \quad [16] \tag{6.20}$$

All the experimental data reported by the many workers who have studied the relation between bubble diameter, d_B, and terminal ascending velocity, v_B, in water are summarized in Fig. 6.4.[39] Various symbols in the figure correspond to data obtained by many workers, whose names and original references are omitted. A characteristic type of curve is drawn through the data points in Fig. 6.4. It was urged by van Krevelen *et al.*[39] that the value of v_B was approximately proportional to d_B^{1-2} when $d_B \leq 1.5$ mm. The bubbles were nearly spherical in shape in this region. As the value of d_B increased from 1.5 to 6 mm, the shape of the bubbles began to deform appreciably, eventually mushrooming when the value of d_B exceeded 6 mm.

Peebles *et al.*[30] conducted an extensive experiment in which the relation between d_B and v_B was studied with air and 22 different kinds of liquid. The viscosities and surface tensions of the liquids used were from 0.233 to 59 cp and from 71.2 to 15.9 dyne/cm, respectively. Their experimental data were correlated in terms of the drag coefficient vs. the modified Reynolds number of bubbles. From the correlation (not shown) they found some differences in the motion of air bubbles depending upon the liquid used.

The motion of air bubbles in liquids other than in water, especially in non-Newtonian liquids, may deviate appreciably from the pattern shown by the solid curves in Fig. 6.4. To demonstrate the deviation, the experimental data reported by Calderbank *et al.*[11] on the value of v_B for carbon dioxide in an aqueous solution of polyox (pseudoplastic) are also cited in the figure (see the broken curve; data points are omitted).

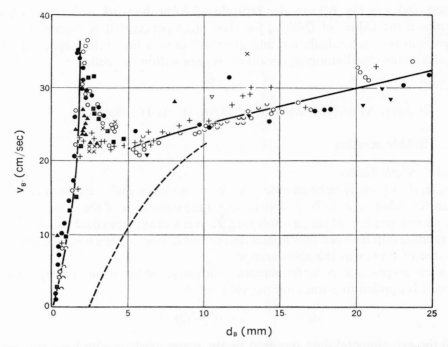

Fig. 6.4. Ascending terminal velocity of single bubbles in water; the ordinate is the ascending terminal velocity, v_B, of single bubbles, while the abscissa represents bubble diameter, d_B. Various symbols used in the figure indicate the sources of original data.[39] The broken curve in the figure represents the ascending velocity of carbon dioxide bubbles in 1.0% aqueous polyox solution (pseudoplastic; non-Newtonian liquid, consistency index, $K=0.877$ g cm^{-1} sec$^{n'-2}$, flow behavior index, $n'=0.540$ at 25°C).[11]

It is interesting to note from the broken curve that the value of v_B was virtually zero unless d_B exceeded nearly 2.5 mm, the peculiar discontinuous region for the air-water system vanished, and in addition, v_B approached the value for the air-water system when d_B was larger than 10 mm.

Air bubbles used in the fermentation industry range from about 1.5 to 10 mm in diameter. The values of v_B may then range from 20 to 30 cm/sec in the case of single bubbles in dilute fermentation media whose characteristics are nearly similar to those of water. However, the values of v_B may be retarded appreciably depending on the characteristic values (consistency index, and flow behavior index, *cf.* Section 7.3.1., Chapter 7) of non-Newtonian liquids, when mycelial fermentation broths exhibiting the non-Newtonian behavior are involved (see Fig. 6.4).

6.2.1.2. *Swarms of bubbles*
Single bubbles are rarely encountered in aeration practice in the fermentation industry; hence it is necessary to deal with swarms of bubbles. If there is no coalescence of bubbles in aeration and if no appreciable distribution is seen in bubble diameter, the values of v_B for single bubbles approximate those for a swarm of

bubbles. Aeration in the fermentation industry is sometimes accompanied by mechanical agitation, though air-lift or tower-type fermentors without mechanical agitation are becoming popular. (*cf.* Fig. 11.2, Chapter 11). The physical properties of fermentation broths change markedly with the progress of a fermentation, especially for mycelial cultivation. The motion of bubbles is affected by the turbulence caused by impellers: bubbles deform and change in diamter considerably during the course of fermentation. The estimation of v_B from Fig. 6.4 for a swarm of bubbles seems to become unacceptable under the above circumstances.

Hence it is proposed to approximate the values of v_B with the following equation for a swarm of bubbles in a fermentation process:

$$v_B = \frac{FH_L}{H_0 V} \tag{6.21}$$

where

$\quad v_B$ = ascending velocity of bubbles in a swarm
$\quad H_0$ = hold-up of bubbles
$\quad F$ = aeration rate
$\quad H_L$ = liquid depth
$\quad V$ = liquid volume

6.2.2. Mechanical agitation

6.2.2.1. *Power number vs. Reynolds number*

Rushton *et al.* developed the concept of the power number. They measured power requirements of liquid agitation with various types of impellers. The power characteristics of impellers with ungassed and Newtonian liquids will be discussed in this section.

A representative velocity, v, of liquid in an agitated vessel will be proportional to the tip velocity of the impeller:

$$v \propto nD_i \tag{6.22}$$

where

$\quad n$ = rotation speed of impeller
$\quad D_i$ = impeller diameter (see Fig. 6.5)

The ratio of external to inertial forces per unit volume of liquid is defined as the power number.

$$\frac{\text{External force}}{\text{Inertial force}} = \frac{\dfrac{Pg_c}{nD_i} \cdot \dfrac{1}{D_i{}^3}}{\rho n^2 D_i}$$

$$= \frac{Pg_c}{n^3 D_i{}^5 \rho} = N_P$$

$$= \text{power number (dimensionless)} \tag{6.23}$$

where

P = power requirements of agitation

g_c = conversion factor

Liquid motions in agitated vessels are depicted by the ratio of inertial to viscous forces per unit volume of liquid; namely, by the modified Reynolds number of the impellers.

$$\frac{\text{Inertial force}}{\text{Viscous force}} = \frac{\rho n^2 D_i}{\mu \dfrac{n}{D_i}} = \frac{n D_i^2 \rho}{\mu} = N_{Re}$$

= modified Reynolds number (dimensionless)   (6.24)

provided:

μ = liquid viscosity

The power characteristics of liquid agitation with various types of impellers have been established by Rushton *et al.*[32] as shown in Fig. 6.5. In the figure, N_P is plotted against N_{Re}. The experimental determination of the values of P by varying the values of n, D_i, ρ, and μ with each type of impeller in a series of geometrically

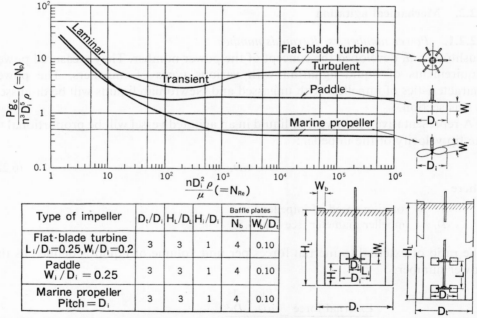

Fig. 6.5. N_P vs. N_{Re}; the power characteristics are shown by the power number, N_P, and the modified Reynolds number, N_{Re}, of single impellers on a shaft.[32] For more detailed information on each dimensionless term, see the text. The lower part of the figure shows the geometrical ratios of representative types of impellers: turbine, paddle, and marine propeller. The right-hand side of the lower part indicates another impeller configuration on the same shaft.

similar systems will yield the plot in Fig. 6.5, in which the data points are omitted. The types of impellers shown in Fig. 6.5 are most commonly used in industry. If the geometrical ratios of each system deviate from those which are indicated in the figure, each curve in Fig. 6.5 is shifted from case to case, though the general shape of each curve remains unchanged. If more detailed information is needed, the original study[32] may be consulted.

It is significant that the geometrical ratios described in Fig. 6.5 are regarded as standard, especially in the case of a flat-blade turbine. It is also important that the values of N_P at higher values of N_{Re} (i.e., in the turbulent region) become constant for each type of impeller.

It is not unusual to install multiple impellers (2 to 3) on the same shaft [see the right-hand diagram in the lower part of Fig. 6.5; cf. Fig. 11.1(b), Chapter 11]. Although power requirements with a multiple impeller are strongly affected by the ratio of the distance between impellers to the diameter, L/D_i, a consistent analysis to formulate power requirements is still needed.

However, the values of the power number for 2 to 3 impellers, N_{P2} and N_{P3} may be approximated as $2N_{P1}$ and $3N_{P1}$ respectively, provided $L/D_i \approx 1.0$ and $N_{P1}=$ power number for single impellers.[19,36]

6.2.2.2. Decrease of power requirements in aeration

Power requirements of impellers for liquid agitation in a gassed system decrease considerably compared with those of an ungassed system. The gassed system is usual in the fermentation industry. The power decrease is principally due to the fact that values of liquid density, around the impeller in particular, will apparently decrease because of the existence of bubbles. Therefore, even in a gassed system, power requirements for agitation will be affected only to a slight extent if the bubble ascending through the liquid in an agitated vessel do not effectively meet the impeller.

The degree of power decrease in a gassed as compared with an ungassed system, P_g/P, extends from 0.3 to 1.0 (unaffected), depending on the type of impeller and the rate of aeration. As is apparent from the above, the term P_g/P will be correlated with a criterion which represents the degree of dispersion of bubbles around the impeller and in the vessel. Ohyama et al. presented the following formula:[29]

$$P_g/P = f(N_a) \tag{6.25}$$

where

$$N_a = \frac{\dfrac{F}{D_i{}^2}}{nD_i}$$

$$= \frac{\text{Apparent velocity of gas (air) through sectional area of vessel}}{\text{Tip velocity of impeller}}$$

$$= \frac{F}{nD_i{}^3} \quad \text{(dimensionless)}$$

They applied a term N_a to the aeration number, which, they claimed, would provide a clue for judging the degree of dispersion of bubbles throughout the system. They determined experimentally with an air-water system the form of the function f in Eq. (6.25). This function was also dependent on the type of impeller used (see Fig. 6.6). The values of N_a in their experiments ranged from 0 to 12×10^{-3}, while those of P_g/P were in a range from 1.0 to 0.3.

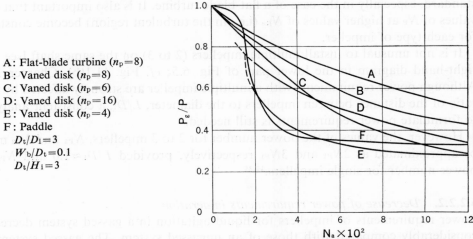

A: Flat-blade turbine ($n_p=8$)
B: Vaned disk ($n_p=8$)
C: Vaned disk ($n_p=6$)
D: Vaned disk ($n_p=16$)
E: Vaned disk ($n_p=4$)
F: Paddle
$D_t/D_i=3$
$W_b/D_t=0.1$
$D_t/H_i=3$

Fig. 6.6. Power requirements for agitation in a gassed system. The ordinate and abscissa are the degree of power decrease, P_g/P, and the aeration number, N_a. Parameters are the types of impellers, whose representative geometrical ratios in agitated vessels are also shown in the figure.[29]

The experimental correlation between P_g and operating variables was presented by Michel et al. as follows:[23]

$$P_g \propto \left(\frac{P^2 n D_i^3}{F^{0.56}}\right)^{0.45} \tag{6.26}$$

They studied the values of P_g with a standard flat-blade turbine in an air-water (liquid) system. The physical properties of the liquids used were as follows:

liquid density, $\rho = 0.8 \sim 1.65$ g/cm³
liquid viscosity, $\mu = 0.9 \sim 100$ cp
surface tension, $\sigma = 27 \sim 72$ dyne/cm

It is interesting to note from the correlation (not shown) that geometrically dissimilar systems were also expressed by Eq. (6.26). The physical properties of the liquids used did not explicitly affect the correlation with Eq. (6.26), the effects of the aeration devices (ring sparger and open nozzle) being also implicit. According to Michel et al.,[23] the data of Ohyama et al.[29] agreed well with the curves extrapolated from Eq. (6.26).

6.2.2.3. *Power requirements in non-Newtonian liquids*[10,22]
According to Metzner *et al.*,[21] the average value of shear rate, $(dv/dr)_{\mathrm{av.}}$, in an agitated vessel of non-Newtonian liquid (carboxymethyl cellulose) was simply expressed as follows:

$$(dv/dr)_{\mathrm{av.}} = \gamma n \qquad (6.27)$$

where

    γ = proportionality constant which depends on the geometrical conditions of the agitated system and the type of non-Newtonian liquid
    n = rotation speed of impeller

For non-Newtonian liquids

$$\left. \begin{aligned} \mu_{\mathrm{a}} &= \frac{\tau g_{\mathrm{c}}}{dv/dr} \\ \tau g_{\mathrm{c}} &= f(dv/dr) \end{aligned} \right\} \qquad (6.28)$$

provided:

    μ_{a} = apparent viscosity of the non-Newtonian liquid

Since it is difficult to define the value of μ_{a} for a non-Newtonian liquid in an agitated vessel, the procedure proposed by Metzner *et al.*[22] for evaluating the value of μ_{a} is:

a. First, by varying the rotation speed, n, of the impeller, measure the power requirement, P, in an agitated vessel of laboratory scale filled with a non-Newtonian liquid. Values of the power number, N_{P}, are then calculated.
b. Replace the non-Newtonian liquid with a highly viscous Newtonian liquid whose values (ρ, μ) are known; then, using the same impeller and vessel, experimentally measure the power consumed in agitating the liquid. For the Newtonian liquid, determine the relation between the power number, N_{P}, and the modified Reynolds number, N_{Re} (see Fig. 6.5).
c. The modified Reynolds number, $nD_{\mathrm{i}}^2\rho/\mu_{\mathrm{a}}$, incorporates the apparent viscosity, μ_{a}, for each value of rotation speed, n, of the impeller. It can be determined by equating both values of N_{P} — the one calculated in accordance with Step *a* above and the other in accordance with Step *b*, using the relation between N_{P} and N_{Re} — provided the latter values of $N_{\mathrm{Re}} = nD_{\mathrm{i}}^2\rho/\mu$ are those for the Newtonian liquid examined as a reference.

It is now necessary to determine the value of γ in Eq. (6.27):

a'. Values of μ_{a} determined in Step *c* are substituted into Eq. (6.28) to assess the values of dv/dr for each value of n, provided that the function $f(dv/dr)$ is given.
b'. By equating the values of dv/dr to $(dv/dr)_{\mathrm{av.}}$ in Eq. (6.27), the value of γ can be determined.

If the value of γ is determined, estimation of P can be made as follows:

a''. At a given rotation speed, n, of the impeller, the value of $(dv/dr)_{av}$. is determined with Eq. (6.27).

b''. The value of μ_a is determined with Eq. (6.28), followed by calculation of the Reynolds number, $nD_1{}^2\rho/\mu_a$.

c''. With the correlation secured earlier between the power number, N_P, and the modified Reynolds number, N_{Re}, for the Newtonian liquid, the value of N_P for the non-Newtonian liquid is determined for each value of the modified Reynolds number, $nD_1{}^2\rho/\mu_a$.

In the above approximate procedure for calculating the power requirements for agitating a non-Newtonian liquid, the step of equating the value of $(dv/dr)_{av}$. in Eq. (6.27) to that of dv/dr in Eq. (6.28) is for convenience only. The former value is related to the flow patterns of the non-Newtonian liquid in an agitated vessel, while the latter is related to that assessed with a viscometer. However, if the coupling of Eq. (6.28) with Eq. (6.27) to determine the value of γ is done only to estimate the power in this particular problem, such a procedure may be acceptable.

6.2.2.4.   *Hold-up of bubbles in an aeration vessel*
Richards *et al.*[31] studied the hold-up of air bubbles in water-filled agitated vessels. The abscissa of Fig. 6.7 represents hold-up, $H_0(\%)$, while the ordinate represents

$$(P/V)^{0.4} \times v_s{}^{0.5}$$

provided:

P/V = power input per unit volume of ungassed liquid, HP/m^3

v_s = linear velocity of air based on the empty cross-sectional area of the vessel, m/hr

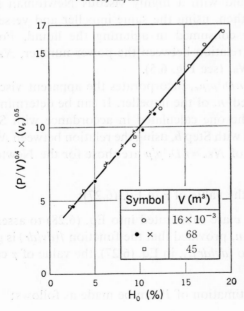

Symbol	V (m³)
o	16×10^{-3}
• ×	68
□	45

Fig. 6.7. Estimation of hold-up, H_0, with power input per unit volume of ungassed liquid, P/V, and nominal linear velocity, v_s, of air.[31]

The empirical relationship shown in Fig. 6.7 is a simplification from a publication of Calderbank, who correlated the hold-up in an aerated vessel with operating variables including the physical properties of liquids.[9] According to Richards et al.,[31] a straight line drawn through the data points in Fig. 6.7 approximates Calderbank's correlation. They also claimed that the values of H_0 were affected slightly by the geometrical dissimilarities of agitating vessels, because the data points in Fig. 6.7 were taken using widely different vessels.

6.3. CORRELATION BETWEEN OXYGEN-TRANSFER COEFFICIENTS AND OPERATING VARIABLES

6.3.1. Bubble aeration

Using the data on the values of k_L published by the many workers who have dealt with bubbles rising through water, Eckenfelder recalculated and replotted the values of both the Sherwood and the Reynolds numbers of the bubbles. These data on k_L in the bubble aeration of water were determined with the sulfite oxidation method by varying the water depth. H_L.

The sizes of bubbles in these experiments ranged from $d_B = 0.5$ to 2.0 mm.[16] An interesting fact, that the values of N_{Sh} were also dependent on the values of H_L (not shown here), is primarily due to the experimental procedure for determining the values of k_L, which involves oxygen transfer both from the surface of bubbling water and in the transient state associated with the generation of bubbles. These ambiguous factors apparently contributed to the larger values of k_L (and N_{Sh}) as the liquid depth, H_L, was lowered.[16]

Excluding the effects of the transient state and the bubbling water surface, the effect of bubble size extending from $d_B = 0.4$ to 8.8 mm on the value of k_L is shown in Fig. 6.8.[5]

The Boussinesq equation seems to represent the data points as seen in the figure. The large disagreement between calculation and observation, especially for smaller bubbles, might have resulted from difficulties in experimentation.

According to Calderbank et al.,[11] the Boussinesq equation was found applicable to the prediction of k_L values for CO_2—90.6% aqueous glycerol solution, $d_B \doteq 3$ to 55 mm; for CO_2—99.0% aqueous glycerol solution, $d_B \doteq 13$ to 55 mm; for CO_2—1.0% aqueous polyox solution, pseudoplastic, $d_B \doteq 3$ to 5 mm, provided the coefficient 1.13 in the Boussinesq equation was replaced by 0.65.

The value of a, interfacial area between the bubbles and liquid per unit volume of liquid, is shown by the following relation.

$$a \propto \frac{FH_L}{d_B v_B V} \tag{6.29}$$

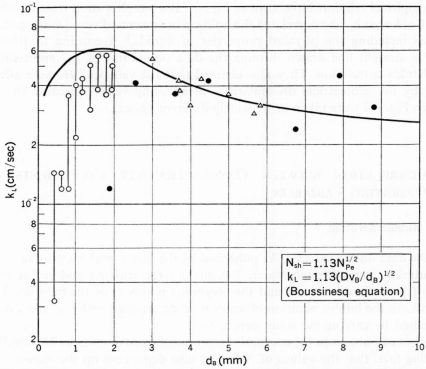

Fig. 6.8. Liquid (water)-film mass (oxygen)-transfer coefficient, k_L cm/sec, as affected by air-bubble diameter, d_B mm. Various symbols indicate the authors. For details, see reference[5]; short bars with open circles demonstrate the fluctuation of data. Solid curve represents Boussinesq equation by taking $D = 2.3 \times 10^{-5}$ cm²/sec for oxygen in water at 20°C, i.e., $k_L = 5.4 \times 10^{-3} (v_B/d_B)^{1/2}$; provided: v_B as affected by d_B was taken, referring to Fig. 6.4 for water (solid curve).

where

F = air flow rate
H_L = liquid depth
d_B = bubble diameter
v_B = ascending terminal velocity of bubbles
V = liquid volume

Since $k_L \propto (v_B/d_B)^{1/2}$ (*cf.* Fig. 6.8), disregarding the effect of the Schmidt number,

$$k_L a \propto \frac{F H_L}{d_B^{3/2} v_B^{1/2} V} \qquad (6.30)$$

the following equation is derived from Eqs. (6.30) and (6.20).

$$k_L a V = \beta F^\alpha \qquad (6.31)$$

provided:

β = empirical constant
α' = empirical exponent

By and large, the experimental work with the sulfite-oxidation method published by Eckenfelder supports the general form of Eq. (6.31) (see Fig. 6.9).[15]

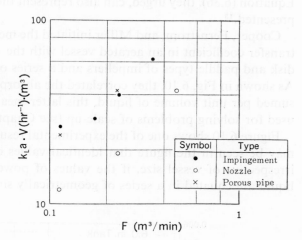

Fig. 6.9. Effect of aeration rate, F, on the value of the volumetric oxygen-transfer coefficient, k_La, times the volume of the aeration vessel, V. The symbols indicate types of aeration devices. Since the value of V in each aeration device was constant ($V = 8.5$ m³), this figure shows the effect of F on k_La. This sort of representation is only for ease of design.[15]

Symbol	Type
•	Impingement
o	Nozzle
×	Porous pipe

F (m³/min)

6.3.2. Bubble aeration with mechanical agitation

Dimensional analysis of this problem shows:[33]

$$k_L D_i / D \propto (n D_i^2 \rho / \mu)^\alpha (\mu / \rho D)^\gamma \tag{6.32}$$

Equation (6.32) can be simplified by keeping constant the values of liquid viscosity, μ, liquid density, ρ, and molecular diffusivity of gas (oxygen), D:

$$k_L D_i \propto (n D_i^2)^\alpha$$

The literature shows:

$$\alpha = 0.5 \quad [31]$$

Then,

$$k_L \propto n^{0.5} \tag{6.33}$$

According to Calderbank,[9] the value of term a in a stirred vessel is:

$$a \propto \left\{ \frac{(P_g/V)^{0.4} \rho^{0.2}}{\sigma^{0.6}} \right\} (v_s/v_B)^{0.5} \tag{6.34}$$

If the values of liquid density, ρ, surface tension of the liquid, σ, and ascending bubble velocity, v_B, are regarded as constant, Eq. (6.34) is reduced as follows:

$$a \propto (P_g/V)^{0.4} v_s{}^{0.5} \qquad (6.35)$$

Richards et al.[31] derived the following equation from Eqs. (6.33) and (6.35):

$$k_L a \propto (P_g/V)^{0.4} v_s{}^{0.5} n^{0.5} \qquad (6.36)$$

They correlated their experimental data on $k_L a$ in agitated vessels using Eq. (6.36). Equation (6.36), they urged, can also represent the data on $k_L a$ which Cooper et al. presented.[13]

Cooper, Fernstrom, and Miller initiated the measurement of volumetric oxygen-transfer coefficient in an aerated vessel with the sulfite-oxidation method. Vaned-disk and paddle types of impellers and a series of different size vessels were used. As shown in Fig. 6.10, they correlated the absorption number with the power consumed per unit volume of liquid, this latter measurement having frequently been used for solving problems of scale-up (see Chapter 7).

Figure 6.10 shows one of the experimental results published by Cooper et al.[13] It may be seen in the figure that identical values of the absorption number result, irrespective of vessel size, if the values of power consumed per unit volume of liquid are equated in a series of geometrically similar systems.

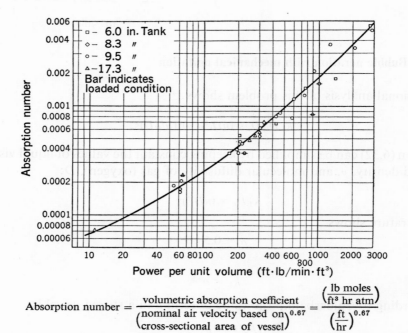

$$\text{Absorption number} = \frac{\text{volumetric absorption coefficient}}{\left(\begin{array}{c}\text{nominal air velocity based on}\\ \text{cross-sectional area of vessel}\end{array}\right)^{0.67}} = \frac{\left(\dfrac{\text{lb moles}}{\text{ft}^3 \text{ hr atm}}\right)}{\left(\dfrac{\text{ft}}{\text{hr}}\right)^{0.67}}$$

Fig. 6.10. Experimental data of Cooper et al.[13] (vaned disks). The ordinate is absorption number, defined as described in the figure, while the abscissa is the value of power requirement for agitation per unit volume of aerated liquid (water). Various symbols in the figure indicate different sizes of agitated vessels. The aeration device used was a single orifice located at the bottom of each vessel.

To avoid the inconsistency of units in Fig. 6.10, the values of K_v and other factors were converted to metric units to obtain the following equations.

For a vaned-disk impeller:

$$K_v = 0.0635(P_g/V)^{0.95}v_s^{0.67} \tag{6.37}$$

where

K_v = volumetric oxygen-transfer coefficient, kg mole/hr m³ atm
P_g/V = power input per unit volume of liquid in gassed system, HP/m³
v_s = nominal (superficial) air velocity based on empty cross-sectional area of vessel, m/hr

It is recommended that Eq. (6.37) be applied under the following conditions:[14]

For 1 set of impellers, $v_s < 90$ m/hr
For 2 sets of impellers, $v_s < 150$ m/hr

provided:

$P_g/V > 0.1$ HP/m³
$H_L/D_t = 1.0$

For a paddle impeller:

$$K_v = 0.038(P_g/V)^{0.53}v_s^{0.67} \tag{6.38}$$

provided:[14]

v_s < 21 m/hr
$P_g/V > 0.06$ HP/m³
$H_L/D_t = 1.0$

Values of K_v for $H_L/D_t = 2$ to 4 are about 50% larger than those for $H_L/D_t = 1.0$, provided the value of P_g/V is kept unchanged.[13]

In designing fermentors when no appropriate data on the volumetric coefficient of oxygen transfer, k_La or K_v, are available, Eqs. (6.37) and (6.38), or Fig. 6.10 (similar data presented by Yoshida et al.[41]) may be consulted.

6.4.   OTHER FACTORS AFFECTING THE VALUES OF OXYGEN-TRANSFER COEFFICIENTS

6.4.1.   Temperature

According to a semi-theoretical analysis by O'Conner, the values of k_La at 30° and 10°C are about 15% larger and smaller than those at 20°C, respectively. His analysis originates from the study of the transfer of oxygen during bubble aeration in the activated sludge process for the biological treatment of waste.[28] The analysis requires further study before it can be applied under actual working conditions in the fermentation industry.

6.4.2. Organic substance

The value of k_L in bubble aeration of water decreased to about one-third the initial value when 1 % peptone was added.[2,17] The bubble diameter also decreased in this instance by about 15 %. The value of $k_L a$ then became about 40 % of that without the organic substance.[17]

Zieminski et al.[42] studied the effect of various kinds of alcohol, ketone, and ester on the value of $k_L a$ in bubble aeration of distilled water. They found that the value of $k_L a$ in distilled water increased conversely by 50 to 100 % with the addition of these substances in water to the extent of about 20 ppm in each case. This trend was most clear for n-butyl and isoamylacetate. A marked increase of the interfacial area between bubbles and liquid per unit volume of liquid, a, predominating over the decrease of k_L values, accounts for the experimental results here.

6.4.3. Surface active agents

Many workers measured the effect of surface active agents on the value of $k_L a$. Figure 6.11 is an example demonstrating how a surface active agent adversely affects the values of k_L and $k_L a$ in bubble aeration.[17] Addition to water of a small amount of sodium lauryl sulfate resulted in a sharp decrease of the value of k_L in this example. The value of $k_L a$, on the other hand, first decreased sharply and then gradually recovered with increased addition of the agent as shown in Fig. 6.11. This might have been caused by the fact that air bubbles decreased in size with the addition of $NaLSO_4$ (see lower section of Fig. 6.11).

Table 6.1 summarizes the results of experiments by several workers who have studied the effect of surfactants. As seen in the table, in the case of each surfactant the value of k_L drops sharply when a small amount is added to the water, the degree of addition being far below the critical micelle concentration, C_s (see the seventh column of the table).

The critical micelle concentration, C_s, implies that the surface tension of the liquid (water) decreases with increased addition of surfactant into liquid, leveling off, however, beyond a critical value of C_s (critical micelle concentration).

Several different explanations have been put forward to account for this adverse effect of surfactant on the value of k_L. For example, in the surface-drag theory proposed by Timson et al.,[38] it is inferred that surfactant adsorbed onto air-liquid interphase does not necessarily cause resistance to oxygen diffusion from gas into the liquid phase, but that the principal controlling factor may be a calming effect of surfactant in dampening complicated movement of the interphase.

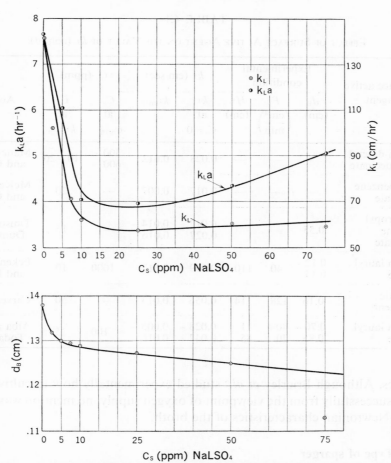

Fig. 6.11. Effect of surface active agent on $k_L a$ and k_L; the upper section of the figure shows that the value of $k_L a$ is decreased by adding sodium lauryl sulfate (a surfactant); the lower section indicates the effect of surfactant on bubble size, d_B. The abscissae of both sections are the concentrations, C_s, of sodium lauryl sulfate in water.[17]

6.4.4. Mycelium

It can be inferred that physical properties, especially viscosity, μ, or apparent viscosity, μ_a, of a fluid affect the value of $k_L a$. This inference holds true for mycelial fermentation broth. Figure 6.12, presented by Brierley *et al.*,[8] shows how a mold concentration of *Aspergillus niger* markedly affected the value of $k_L a$. They suspended newly prepared mycelia of *Aspergillus niger* in a medium of sucrose, salts, and corn steep liquor using a 8-*l* Waldhof-type fermentor. They measured the value of $k_L a$ polarographically. The abscissa of Fig. 6.12 represents the mycelium concentration in dry weight. It is surmised that the mold suspension, especially for more than 1 % concentration, might have exhibited strongly non-Newtonian char-

TABLE 6.1

Effect of Surface Active Agent on the Value of k_L (cont'd).

Surface active agent	Experimental condition			k_L (cm/sec)		C_s (ppm)		Author
	d_B (cm)	F $\left(\dfrac{cm^3}{min}\right)$	H_L (cm)	k_L at $C_s=0$	k_{Lmin}	C_s at σ_{min}	C_s at k_{Lmin}	
Sodium dioctyl sulfosuccinate	—	—	—	0.028	0.012	300 – 600	10 – 20	Mancy and Okun[20]
Alkyl benzene sulfonate	—	—	—	0.017	0.007	—	20	McKeown and Okun[25]
Pentapropyl benzene sulfonate	0.55	—	—	0.017 – 0.027	0.014 – 0.016	—	15	Timson and Dunn[38]
Sodium lauryl sulfate	0.12 – 0.14	40	110	0.039	0.017	1000	10	Eckenfelder and Barnhart[17]
Synthetic detergent	0.16	128	114	0.056	0.013	—	50	Carver[12]
Sodium lauryl sulfate	0.70 – 0.85	40 – 70	11 – 85	0.028 – 0.042	0.005 – 0.016	100	25	Aiba and Toda[3]

acteristics. Although Brierley et al.[8] studied experimentally how to cultivate such mycelia successfully from the viewpoint of oxygen supply, no mention was made of the non-Newtonian characteristics of the broth.

6.4.5. Type of sparger

Many workers have experimentally studied the oxygen-transfer capabilities of various types of aerators in their search for one which would produce a rapid rate of oxygen transfer from air bubbles to liquid. However, it is still difficult to designate the most effective device for a particular fermentor or aeration basin because of the necessity of taking into consideration such aspects of the different devices as their durability or their susceptibility to damage (principally clogging) caused by microbial cells. The basis of assessing such aspects differs from case to case.

According to Eckenfelder, the values of k_La measured with various types of aerators by the sulfite-oxidation method did not differ appreciably when compared on the basis of power consumption per unit volume of liquid (see Fig. 6.13).[15] His purpose was to determine the most effective type of aeration basin in the activated sludge process. For details of designations of sparger type in the figure, see his original paper.[15] No mechanical agitators were used in the categories from "impingement" to "porous disk" as shown in the figure. The power requirements for all of the spargers except the turbine are primarily the power consumed in

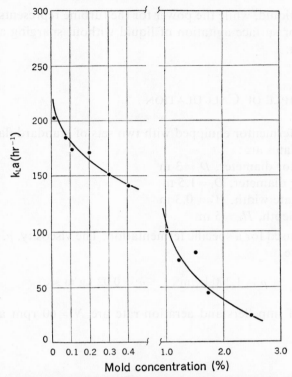

Fig. 6.12. Decrease of $k_L a$ values depending on the mycelial concentration in broth; the value determined polarographically is plotted against mold concentration % (*Aspergillus niger*).[8]

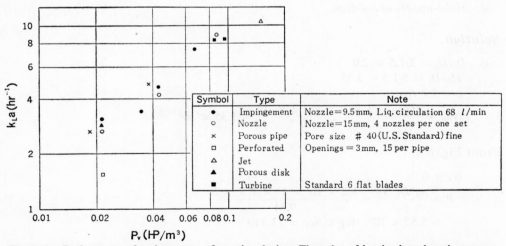

Fig. 6.13. Performance of various types of aeration devices. The value of $k_L a$ is plotted against power, P_v, consumed per unit volume of liquid (water). Symbols indicate various types of aeration devices. The value of P_v was principally supplied by air compressors with the exception of the inclusion of turbine power.[15]

sparging air into liquid, while the power for the turbine represents presumably the figure required for surface agitation of liquid without sparging air bubbles from the vessel bottom.

6.5.   AN EXAMPLE OF CALCULATION

Dimensions of a fermentor equipped with two sets of standard flat-blade turbines and four baffle plates are:

Fermentor diameter, $D_t = 3$ m

Impeller diameter, $D_i = 1.5$ m

Baffle plate width, $W_b = 0.3$ m

Liquid depth, $H_L = 5$ m

The fermentor is used for a specific fermentation. The viscosity, μ, and the density, ρ, of the broth are:

$$\rho = 1{,}200 \text{ kg/m}^3, \qquad \mu = 0.02 \text{ kg/m sec}$$

Rotation speed of impellers and aeration rate are $N = 60$ rpm and 0.4 vvm, respectively.

Calculate

a. Power requirements, P, for ungassed system,

b. Power requirements, P_g, when aerated,

c. Volumetric coefficient, K_v, of oxygen transfer, and

d. Hold-up, H_0 of bubbles.

Solution

a. $D_t/D_i = 3/1.5 = 2.0$

$H_L/D_i = 5/1.5 = 3.33$

$n = 1.0$ rps

$$N_{Re} = \frac{nD_i^2\rho}{\mu} = \frac{1 \times 1.5^2 \times 1.2 \times 10^3}{2 \times 10^{-2}} = 1.35 \times 10^5$$

From Fig. 6.5,

$N_P = 6$

$$P = \frac{\rho n^3 D_i^5 N_P}{g_c} = \frac{1.2 \times 10^3 \times 1^3 \times 1.5^5 \times 6}{9.81}$$

$$= 5.57 \times 10^3 \quad \text{Kg m/sec} = 73.3 \text{ HP}$$

Since the geometrical ratios, $(D_t/D_i)^*$ and $(H_L/D_i)^*$, of this problem deviate from D_t/D_i and H_L/D_i in Fig. 6.5, a correction factor, f_c, which is approximately expressed as shown below will be calculated.

$$f_c = \sqrt{\frac{(D_t/D_i)^*(H_L/D_i)^*}{(D_t/D_i)(H_L/D_i)}} = \sqrt{\frac{2.0 \times 3.33}{3.0 \times 3.0}} = 0.86$$

Then,

$$P^* = Pf_c = 73.3 \times 0.86 = 63 \text{ HP}$$

If power requirements, P, with two sets of impellers can be estimated by multiplying the value of P for one set of impellers by 2 (cf. 6.2.2.1.), the total power requirement, P^{**} will be:

$$P^{**} = P^* \times 2 = 63 \times 2 = 126 \text{ HP}$$

b. The aeration number, N_a, is calculated as follows:

$$N_a = \frac{F}{nD_i{}^3} = \frac{0.4 \times (\pi/4) \times 3^2 \times 5 \times (1/60)}{1 \times 1.5^3}$$

$$= 6.95 \times 10^{-2}$$

Assuming that curve A in Fig. 6.6 can be used,

$$P_g/P^{**} = 0.65, \qquad P_g = P^{**} \times 0.65 = 126 \times 0.65 = 82 \text{ HP}$$

c. $$F = 0.4 \times (\pi/4) \times 3^2 \times 5 = 14.1 \text{ m}^3/\text{min}$$

$$v_s = \frac{14.1 \times 60}{(\pi/4) \times 3^2} = 119.7 \text{ m/hr}$$

From Eq. (6.37).

$$K_v = 0.0635 \times \left\{\frac{82}{(\pi/4) \times 3^2 \times 5}\right\}^{0.95} \times 119.7^{0.67}$$

$$= 3.45 \text{ kg mole/m}^3 \text{ hr atm}$$

Supposing that the coefficient, 0.0635 be halved in the case of a flat-blade turbine (cf. Section 7.4., Chapter 7),

$$K_v = 1.72 \text{ kg mole/m}^3 \text{ hr atm}$$

$$H_L/D_t = 5/3 = 1.67$$

The correction factor, f_c, is assumed as follows: (cf. Section 6.3.2.)

$$f_c = 1.3$$

Then, the volumetric coefficient, $K_v{}^*$, of oxygen transfer is:

$$K_v{}^* = K_v f_c = 1.72 \times 1.3 = 2.24 \text{ kg mole/m}^3 \text{ hr atm}$$

It must be remembered that the above value of volumetric coefficient is maximum in terms of oxygen transfer, because Eq. (6.37) based on the sulfite-oxidation experiment is applied in the calculation.

d. Fig. 6.7 is used to calculate the hold-up, H_0.

From Fig. 6.7,

$$\left(\frac{P}{V}\right)^{0.4} v_s^{0.5} = \left\{\frac{126}{(\pi/4) \times 3^2 \times 5}\right\}^{0.4} \times 119.7^{0.5}$$

$$= 18.1$$

From an extrapolation of the solid line in Fig. 6.7,

$$H_0 = 21\%$$

6.6. SUMMARY

1. A purely physical interpretation of oxygen uptake rate by discrete microbial cells well dispersed in liquid is an incomplete description of the process.

2. For mycelia one must consider the distribution of oxygen uptake rates within clumps.

3. The effect of aeration and agitation on oxygen tranefer must be considered from two angles: supply (oxygen transfer rate from air bubbles into liquid) and demand oxygen uptake rate by microbes).

4. The oxygen transfer from air bubbles to the fermentation medium is likely to be the rate-limiting step. In steady state, a mutual relationship among the specific rate of oxygen uptake (respiration), Q_{O_2}, the cell density, X, and the volumetric oxygen-transfer coefficient, $k_L a$, with respect to the dissolved oxygen concentration, $\bar{C}$, in the medium is:

$$\bar{C} = C^* - \frac{Q_{O_2} X}{k_L a} \qquad \text{[Eq. (6.11)]}$$

In turn, the specific growth rate of microbes, μ, is regulated by the value of $\bar{C}$; it is most likely a hyperbolic function of $\bar{C}$.

5. The various physical and chemical factors that affect the values of $k_L a$ are: air flow rate, agitation speed of impeller, temperature of liquid, organic substances, surface active agents, mycelial concentration, and type of sparger.

6. Equations for estimating the volumetric oxygen-transfer coefficient for Newtonian broths are well established [Eqs. (6.37) and (6.38) and the mechanisms of transfer are relatively well understood. For non-Newtonian broths, however, a satisfactory and generalized equation has yet to be developed.

NOMENCLATURE

a = interfacial area between liquid and gas per unit volume of liquid, m²/m³;

$$12\left(R\Big/\sqrt{\frac{6\,\bar{C}D}{\rho_{\mathrm{m}}Q_{\mathrm{O}_2}}}\right)^2$$

$\bar{a}$ = surface area of individual cell, m², cm²

C = concentration of dissolved gas (oxygen), kg mole/m³, g/l, ppm

$C_{\mathrm{crit.}}$ = critical concentration of dissolved oxygen, ppm

$\bar{C}$ = concentration of dissolved gas (oxygen) in bulk liquid, kg mole/m³, ppm

C^* = concentration of dissolved gas (oxygen) which is in equilibrium with partial pressure, $\bar{p}$, in bulk gas phase, kg mole/m³

C_{s} = concentration of surface active agent in liquid, ppm

D = molecular diffusivity of oxygen in water, cm²/sec, m²/hr; dilution rate, hr⁻¹

D_1 = impeller diameter, m

D_1' = disk diameter of turbine, m

D_{t} = tank diameter, m

d = orifice diameter, m; representative diameter of microbe, m, cm

d_{B} = bubble diameter, m, mm

f = function

f_{c} = correction factor

F = flow rate of air, m³/min, cm³/min

g_{c} = conversion factor, kg m/Kg sec²

H = Henry's constant, kg mole/m³ atm

H_1 = location of impeller from bottom of vessel, m

H_{L} = liquid depth, m

H_0 = gas (air) hold-up in liquid (water), % or fraction

K = consistency index (Fig. 6.4), g cm⁻¹ sec$^{n'-2}$

K_{m} = Michaelis constant, kg mole/m³

K_{v} = volumetric absorption coefficient, kg mole/m³ hr atm

k_{L} = mass (oxygen)-transfer coefficient of liquid film, m/hr, cm/sec

$k_{\mathrm{L}}a$ = volumetric oxygen-transfer coefficient, hr⁻¹

L = distance between impellers, m

N = rotation speed of impeller, min⁻¹

N_{a} = aeration number ($= F/n\,D_1^3$)

N_{b} = number of baffle plates

N_{P} = power number ($= Pg_{\mathrm{c}}/\rho\,n^3D_1^5$)

N_{Re} = Reynolds number ($= nD_1^2\rho/\mu,\ nD_1^2\rho/\mu_{\mathrm{a}},\ d_{\mathrm{B}}v_{\mathrm{B}}\rho/\mu$)

N_{Sc} = Schmidt number ($= \mu/\rho D$)

N_{Sh} = Sherwood number ($= k_{\mathrm{L}}d/D$)

n = rotation speed of impeller, sec⁻¹; number concentration of microbes, cm⁻³

n' = flow behavior index (Fig. 6.4)

n_{O_2} = oxygen flux, kg mole/m² hr, g mole/cm² sec

n_{O_2}' = volumetric rate of oxygen transfer, kg mole/m³ hr

n_{p} = number of blades

P = power consumption of liquid agitation in ungassed system, Kg m/sec, HP

P_g = power consumption of liquid agitation in gassed system, Kg m/sec, HP

P_v = power requirements per unit volume of liquid, HP/m³, Fig. 6.13

$\bar{p}$ = partial pressure of oxygen in gas phase in bulk, atm

p^* = partial pressure of oxygen which is in equilibrium with bulk concentration in liquid phase, atm

Q_{O_2} = specific rate of oxygen uptake (respiration) by microbe at $C=\bar{C}$, g mole O_2/g-cell hr, mg mole O_2/g-cell hr, g mole O_2/mg-cell hr

Q_{O_2}' = specific rate of respiration at $C=C$, g mole O_2/g-cell hr

Q_{O_2}'' = maximum value of specific respiration rate, g mole O_2/g-cell hr

$(Q_{O_2})_m$ = specific rate of respiration required for maintenance, g mole O_2/g-cell hr

$\bar{Q}_{O_2}$ = apparent value of Q_{O_2} for pellet

R = radius, m, cm

R_c = critical radius, m, cm

r = radial distance, m, cm

t = time, min, sec

V = liquid volume, m³

v = liquid velocity, Eq. (6.22); relative velocity between liquid and microbe, Eq. (6.6)

v_B = ascending velocity of bubbles, terminal velocity, m/sec, cm/sec

v_s = nominal (superficial) velocity of gas (air) based on cross-sectional area of vessel, m/hr

W_b = width of baffle plate, m

W_1 = impeller width, m

X = cell mass concentration, g/l, mg/cm³

x = r/R

y = $C/\bar{C}$

subscript

min = minimum value

Greek letters

α, α' = empirical exponents

α'' = empirical coefficient

β = empirical constant

β' = specific value of μ

β'' = empirical exponent; specific value of Q_{O_2}

γ = empirical exponent; proportionality constant

γ' = specific value of $\bar{C}$

δ'' = empirical exponent

μ = liquid viscosity, kg/m sec, g/cm sec

μ_a = apparent viscosity of liquid, kg/m sec

ν = specific rate of product formation, mg/mg cell·hr

ρ = liquid density, kg/m³, g/cm³

ρ_m = mycelial density, g/cm³

σ = surface tension, dyne/cm

τ = shear stress, Kg/m², G/cm², Eq.(6.28)

REFERENCES

1. Aiba, S. (1960). *Biochemical Engineering* p. 42. Nikkan-Kogyo Pub. Co. Tokyo.
2. Aiba, S., and Yamada, T. (1961). "Oxygen absorption in bubble aeration. Part 1." *J. Gen. Appl. Microbiol.* **7**, 100.
3. Aiba, S., and Toda, K. (1963). "The effect of surface active agent on oxygen absorption in bubble aeration I." *J. Gen. Appl. Microbiol.* **9**, 443.
4. Aiba, S., Hara, M., and Someya, J. (1963). "Mass transfer in fermentation; oxygen absorption in gluconic acid fermentation." *J. Ferm. Tech. (Japan)* **41**, 74.
5. Aiba, S., and Toda, K. (1964). "Effect of surface active agent on oxygen absorption in bubble aeration II." *J. Gen. Appl. Microbiol.* **10**, 157.
6. Aiba, S., and Kobayashi, K. (1971). "Comments on Oxygen Transfer within a Mold Pellet." *Biotech. & Bioeng.* **13**, 583.
7. Bégin-Heick, N., and Blum, J. J. (1967). "Oxygen Toxicity in *Astasia*." *Biochem. J.* **105**, 813.
8. Brierley, M. R., and Steel, R. (1959). "Agitation-aeration in submerged fermentation. Part 2. Effect of solid disperse phase on oxygen absorption in a fermentor." *Appl. Microbiol.* **7**, 57.
9. Calderbank, P. H. (1958). "Physical rate processes in industrial fermentation. Part 1. The interfacial area in gas-liquid contacting with mechanical agitation." *Trans. Instn. Chem. Engrs.* **36**, 443.
10. Calderbank, P. H., Moo-Young, M. B. (1959), and (1961). "The prediction of power consumption in the agitation of non-Newtonian fluids." *Trans. Instn. Chem. Engrs.* **37**, 26. "The power characteristics of agitators for the mixing of Newtonian and non-Newtonian fluids." *Trans. Instn. Chem. Engrs.* **39**, 337.
11. Calderbank, P. H., Johnson, D. S. L., and Loudon, J. (1970). "Mechanics and mass transfer of single bubbles in free rise through some Newtonian and non-Newtonian liquids." *Chem. Eng. Sci.* **25**, 235.
12. Carver, C. E., Jr. (1955). "Absorption of oxygen in bubble aeration." *Biological Treatment of Sewage and Industrial Wastes* (Eds.) Eckenfelder, W. W., and McCabe, J. **1**, 149. Reinhold, New York.
13. Cooper, C. M., Fernstrom, G. A., and Miller, S. A. (1944). "Performance of agitated gas-liquid contactors." *Ind. Eng. Chem.* **36**, 504.
14. Chem. Engrs' Handbook (1968). p. 1089. Maruzen Book Co., Tokyo.
15. Eckenfelder, W. W., Jr. (1956). "Process design of aeration system for biological waste treatment." *Chem. Eng. Progress* **52**, 286.
16. Eckenfelder, W. W., Jr. (1959). "Absorption of oxygen from air bubbles in water." *J. Sanitary Eng. Division, Proc. A.S.C.E.* **85**, No. SA 4.
17. Eckenfelder, W. W., Jr., and Barnhart, E. L. (1961). "The effect of organic substances on the transfer of oxygen from air bubbles in water." *A.I.Ch.E. Journal* **7**, 631.
18. Finn, R. K. (1954). "Agitation-aeration in the laboratory and in industry." *Bact. Review* **18**, 254.
19. Fukuda, H., Sumino, Y., and Kanzaki, T. (1968). "Scale-up of fermentors. Part II. Modified equation for power measurement." *J. Ferm. Tech. (Japan)* **46**, 838.
20. Mancy, K. H., and Okun, D. A. (1963). "Effect of surface active agents on the rate of oxygen transfer." *Advances in Biological Waste Treatment* (Eds.) Eckenfelder, W. W., and McCabe, J. p. 111. Pergamon Press, Oxford.

21. Metzner, A. B., and Taylor, J. S. (1960). "Flow patterns in agitated vessels." *A.I.Ch.E. Journal* **6**, 109.
22. Metzner, A. B., Feehs, R. H., Ramos, H. P., Otto, R. E., and Tuthill, J. D. (1961). "Agitation of viscous Newtonian and non-Newtonian fluids." *A.I.Ch.E. Journal* **7**, 3.
23. Michel B. J., and Miller, S. A. (1962). "Power requirements of gas-liquid agitated systems." *A.I.Ch.E. Journal* **8**, 262.
24. Miller, D. N. (1971). "Scale-up of agitated vessels mass transfer from suspended solute particles." *Ind. Eng. Chem. Process Des. Develop.* **10**, 365.
25. McKeown, J. J., and Okun, D. A. (1963). "Effects of surface active agents on oxygen bubble characteristics." *Advances in Biological Waste Treatment* (Eds.). Eckenfelder W. W., Jr., and McCabe, J. p. 113. Pergamon Press, Oxford.
26. Nagai, S. and Aiba, S. (1972). "Reassessment of maintenance and energy uncoupling in the growth of *Azotobacter vinelandii*." *J. Gen. Microbiol.* **73**, 531.
27. Nishizawa, Y., Nagai, S., and Aiba, S. (1971). "Effect of dissolved oxygen on electron transport system of *Azotobacter vinelandii* in glucose-limited and oxygen-limited chemostat cultures." *J. Gen. Appl. Microbiol.* **17**, 131.
28. O'Connor, D. J. (1955). D. Sc. Thesis, N.Y. Univ.
29. Ohyama, Y., and Endoh, K. (1955). "Power characteristics of gas-liquid contacting mixers." *Chem. Eng. (Japan)* **19**, 2.
30. Peebles, F. N., and Garber, H. J. (1953). "Studies on the motion of gas bubbles in liquids." *Chem. Eng. Progress* **49**, 88.
31. Richards, J. W. (1961). "Studies in aeration and agitation." *Progress in Industrial Microbiology* **3**, 143.
32. Rushton, J. H., Costich, E. W., and Everett, H. J. (1950). "Power characteristics of mixing impellers. Part 2." *Chem. Eng. Progress* **46**, 467.
33. Rushton, J. H. (1951). "The use of pilot-plant mixing data." *Chem. Eng. Progress* **47**, 485.
34. Steinberger, R. L., and Treybal, R. E. (1960). "Mass transfer from a solid soluble sphere to a flowing liquid stream." *A.I.Ch.E. Journal* **6**, 227.
35. Schwartzberg, H. G., and Treybal, R. E. (1968). "Fluid and particle motion in turbulent stirred tanks." *I & EC Fundamentals* **7**, 6.
36. Taguchi, H., and Kimura, T. (1970). "Studies on geometric parameters in fermentor design Part I. Effect of impeller spacing on power consumption and volumetric oxygen transfer coefficient." *J. Ferm. Tech. (Japan)* **48**, 117.
37. Terui, G. (1963). "Panel discussion on oxygen transfer problems in the fermentation industry." p. 66. *Proc. 5th Symposium, Inst. Appl. Microbiol. Univ. of Tokyo.*
38. Timson, W. J., and Dunn, C. G. (1960). "Mechanism of gas absorption from bubbles under shear." *Ind. Eng. Chem.* **52**, 799.
39. Van Krevelen, D. W., and Hoftijzer, P. J. (1950). "Studies of gas-bubble formation. Calculation of interfacial area in bubble contactors." *Chem. Eng. Progress* **46**, 29.
40. Yano, T., Kodama, T., and Yamada, K. (1961). "Fundamental studies on the aerobic fermentation. Part III. Oxygen transfer within a mold pellet." *Agr. Biol. Chem. (Japan)* **25**, 580.
41. Yoshida, F., Ikeda, A., Imakawa, S., and Miura, Y. (1960). "Oxygen absorption rates in stirred gas-liquid contactors." *Ind. Eng. Chem.* **52**, 435.
42. Zieminski, S. A., Goodwin, C. C., and Hill, R. L. (1960). "The effect of some organic substances on oxygen absorption in bubble aeration." *Tappi* **43**, 1029.

Chapter 7

Scale-up

Scale-up is the study of the problem associated with transferring data obtained in laboratory and pilot-plant equipment to industrial production. Equipment for the fermentation industry's fermentors, heat exchangers, crystallizers, separators, and so forth, can be designed and operated properly only if scale-up technology is fully appreciated. This problem is not peculiar to biochemical engineering, but is shared by many branches of chemical engineering.

In this chapter, only those problems associated with the scale-up of fermentors will be discussed. To demonstrate an idea of the scaling-up, the following simple example will be presented.

Suppose the conditions of aeration which give the maximum productivity in a specific fermentation have been established using a bench-scale fermentor, and it is then proposed to transfer the fermentation to a large fermentor having the same geometrical design. The problem is to estimate the proper aeration rate in the large vessel, assuming for simplicity no mechanical agitation. Since the physical properties of the broth under consideration are the same in geometrically similar fermentors, Eq. (6.30) in Chapter 6 can be used. Then,

$$\frac{(k_L a)_1}{(k_L a)_2} = \frac{(F/V)_1}{(F/V)_2}\left(\frac{H_{L1}}{H_{L2}}\right)\left(\frac{d_{B2}}{d_{B1}}\right)^{3/2}\left(\frac{v_{B2}}{v_{B1}}\right)^{1/2} \tag{7.1}$$

where subscripts 1 and 2 relate to small- and large-scale equipment, respectively.

It may be assumed that $d_{B2} \doteq d_{B1}$ and $v_{B2} \doteq v_{B1}$ in Eq. (7.1), because the size of bubbles is considered not to differ appreciably with the fermentor size. Supposing that the aeration rate with the large fermentor be determined by equating the left-hand side of Eq. (7.1) to unity,

$$\frac{(F/V)_1}{(F/V)_2} = \frac{H_{L2}}{H_{L1}} \tag{7.2}$$

Equation (7.2) signifies, for example, that the aeration rate, $(F/V)_2$, with the large fermentor will be $(F/V)_2 = 5^{-1} = 0.2$ vvm, provided the value of $(F/V)_1 = 1.0$ vvm and the scale-up ratio, $H_{L2}/H_{L1} = 5$ (volumetric scale-up $\times$ 125).

However, the effect of oxygen transfer from the bubbling liquid surface and that associated with the transient state in the bubbles' emergence at a sparger cannot necessarily be disregarded when the bench-scale fermentor is used. If this effect is appreciated, it is suggested that the term, H_{L1}/H_{L2}, in the right-hand side of Eq. (7.1) be replaced by $(H_{L1}/H_{L2})^{2/3}$; i.e., the right-hand side of Eq. (7.2) will be modified as $(H_{L1}/H_{L2})^{2/3}$. If this is acceptable, the aeration rate, $(F/V)_2$, with the large fermentor will be $(F/V)_2 = 5^{-2/3} = 0.34$ vvm instead of $(F/V)_2 = 0.2$ vvm.

It is clear that problems of scale-up in a fermentor are associated with the behavior of liquid in the fermentor and the metabolic reactions of the organisms. Although studies of chemical engineering have already provided some insight into the behavior of liquids in stirred vessels, knowledge of the behavior of organisms in fermentation vessels is less well understood, and further investigations of both the biological and technological aspects of the system are required.

7.1. BASES OF SCALE-UP

7.1.1. Physical concept

Physical properties of the broth in geometrically similar and fully baffled fermentors, i.e., medium composition, temperature, pH, dissolved oxygen concentration, and so forth are assumed the same and the microbes are also assumed well dispersed in fully turbulent systems. Items relevant to the liquid behavior in agitated fermentor vessels are:

Power requirements of agitation, P, or power requirements of agitation, P_g, in a gassed system (*cf.* Section 6.2.2., Chapter 6), rotation speed of the impeller, n, and pumping rate of the impeller, F.

For turbulent liquid motion,

$$P \propto n^3 D_i^5 \qquad (cf. \text{ Fig. 6.5, Chapter 6}) \tag{7.3}$$

$$P_g/P = f(N_a)$$

provided:

$$N_a = \text{aeration number } (= F/nD_i^3) \; [cf. \text{ Eq. (6.25)}]$$

In addition,

$$V \propto D_i^3 \tag{7.4}$$

$$F \propto nD_i^3 \qquad \qquad 13 \tag{7.5}$$

Physical terms which are useful for scaling-up are:

a. Power consumed per unit volume of liquid, P/V,

$$P/V \propto n^3 D_i^2 \tag{7.6}$$

b. Liquid circulation rate inside the vessel, F/V,

$$F/V \propto n \tag{7.7}$$

c. Impeller tip velocity, v, representing liquid shear rate (see reference 21 in Chapter 6),

$$v \propto nD_i \tag{7.8}$$

d. The modified Reynolds number, $nD_i^2\rho/\mu$, where ρ=liquid density and μ= liquid viscosity,

$$nD_i^2\rho/\mu \propto nD_i^2 \tag{7.9}$$

Table 7.1 points out that all the physical properties for small- and large-scale equipment cannot be equated simultaneously (scale-up ratio, $D_{i2}/D_{i1}=5$, volumetric scale-up ratio $\times$ 125).[14] Supposing that the property, P/V, be equated, for instance, some figures in the table will be derived below.

TABLE 7.1

RELATIONSHIPS BETWEEN PROPERTIES FOR SCALE-UP.[14]

Property	Small scale 80 l		Large scale 10,000 l		
P	1.0	125	3125	25	0.2
P/V	1.0	1.0	25	0.2	0.0016
n	1.0	0.34	1.0	0.2	0.04
D_i	1.0	5.0	5.0	5.0	5.0
F	1.0	42.5	125	25	5.0
F/V	1.0	0.34	1.0	0.2	0.04
nD_i	1.0	1.7	5.0	1.0	0.2
$nD_i^2\rho/\mu$	1.0	8.5	25	5.0	1.0

From Eq. (7.6),

$$n_2^3 D_{i2}^2 = n_1^3 D_{i1}^2 \tag{7.6$'$}$$

Then,

$$n_2/n_1 = (D_{i1}/D_{i2})^{2/3} = 5^{-2/3} = 0.34$$

From Eq. (7.3),

$$P_2/P_1 = n_2^3 D_{i2}^5/n_1^3 D_{i1}^5 \tag{7.3$'$}$$

Then, from Eqs. (7.3)$'$ and (7.6)$'$,

$$P_2/P_1 = (D_{i2}/D_{i1})^3 = 5^3 = 125$$

For nD_1,

$$n_2 D_{12}/n_1 D_{11} = 0.34 \times 5 = 1.7 \qquad \text{(see Table 7.1)}$$

If the scale-up is attempted on the basis of equal power per unit volume of liquid, it is apparent that tip velocity of the impeller increases, whereas the value of F/V or shear rate of liquid decreases. With decrease of F/V, the mixing time is expected to increase.[13]

If a fermentation can be scaled-up successfully on the basis of equal power per unit volume of liquid, the fermentation concerned is presumably one which is less sensitive to the inevitable increase in tip velocity of the impeller and in the mixing time (see Figs. 7.3 and 7.4).

The proper choice of physical properties for scale-up changes from case to case. If the heat-transfer coefficient, h, of a liquid film with a coil in an agitated reactor is of interest, a plausible correlation is obtained using the values of h, P/V, and F/V.[13] The dimensional analysis of mass (oxygen)-transfer coefficient, k_L, shown in Eq. (6.32) in Chapter 6 can be applied to cases other than mass transfer.

If the Sherwood number, $k_L D_1/D$, is replaced by the Nusselt number, hD_1/k, where $h=$film coefficient of heat transfer, and $k=$thermal conductivity, and the Schmidt number, $\mu/\rho D$, by the Prandtl number, $c_p \mu/k$, where $c_p=$ specific heat, the analogous equation thus obtained can be used to calculate the liquid-mixing performance of an impeller in terms of heat transfer.

The following discussion assumes that the aeration conditions in the fermentors are such that the aeration number, N_a, is the same, or at least not significantly different, from one geometrically similar system to the next. As seen in Fig. 6.6 in Chapter 6, the power requirements of an impeller in gassed systems, which most fermentors are, can be represented in geometrically similar systems by P (the power consumed in ungassed systems).

In order to achieve the same value of k_L in Eq. (6.32), the following relation must hold (assuming the physical properties of the liquid are constant):

$$n_1/n_2 = (D_{12}/D_{11})^{\frac{2\alpha-1}{\alpha}} \qquad (7.10)$$

From Eq. (7.6),

$$\frac{P_2/V_2}{P_1/V_1} = (n_2/n_1)^3 (D_{12}/D_{11})^2 \qquad (7.11)$$

Cancelling out the term, n_1/n_2, from Eqs. (7.10) and (7.11),

$$\frac{P_2/V_2}{P_1/V_1} = (D_{12}/D_{11})^{2-3\frac{2\alpha-1}{\alpha}} \qquad (7.12)$$

In Fig. 7.1, the left-hand side of Eq. (7.12) is plotted against α, parameters being

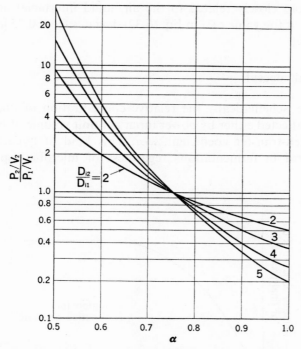

Fig. 7.1. P/V vs. α in scale-up;[17] parameters are scale-up ratio, D_{i2}/D_{i1}

the scale-up ratio, D_{i2}/D_{i1}. It is apparent from the figure that the "equal power per unit volume" concept can be applied effectively to the scale-up, irrespective of D_{i2}/D_{i1}, when the value of α is equal to 0.75.

If the value of $\alpha=0.5$ [see Eq. (6.33), Chapter 6] can be accepted for a particular fermentation broth in terms of oxygen transfer, Fig. 7.1 suggests that the same value of k_L as that for small-scale equipment cannot be expected for large-scale fermentors, when the value of P_2/V_2 is equated to P_1/V_1. The value of k_{L2} is about 58% of k_{L1}, provided $D_{i2}/D_{i1}=5$, for instance.

On the other hand, the interfacial area, a, between air bubbles and liquid per unit volume of liquid remains unaffected in the two fermentors, if $v_{s2} \doteq v_{s1}$, $v_{B2} \doteq v_{B1}$ and the value of P_g/V is equal in both cases, where the values of surface tension, σ, and the density, ρ, of liquid are constant [see Eq. (6.34)].

In microbial fermentations, power consumed per unit volume, P/V, volumetric oxygen-transfer coefficient, k_La, and the mean liquid velocity at a particular point in the vessel are the three criteria commonly used for scale-up. The last criterion is incorporated into the aeration number, N_a, in gassed systems [see Eq. (6.25)].

Referring to Eqs. (6.34) and (6.36), it might be inferred that the implications of using P/V (or P_g/V) and k_La as the criteria of scale-up of fermentors do not differ appreciably, since both criteria are indirectly and/or directly connected with the rate of oxygen transfer from bubbles into the liquid; but the property, P/V (or

P_g/V), seems more closely related to the air/liquid interfacial area, a, per unit volume of liquid, if the value of α in Eq. (6.32) deviates from 0.75 [*cf.* Fig. 7.1, Eqs. (6.34), and (6.35)].

7.1.2. Biological concept

Figure 7.2 shows schematically the relative concentration of final product in a fermentation as affected either by power input per unit volume of broth, P/V or by volumetric oxygen-transfer coefficient, k_La. This kind of hyperbolic pattern is generally observed in fermentations regardless of microbial species—bacteria, yeast, or fungi.

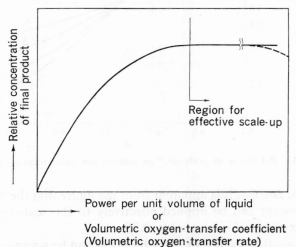

Fig. 7.2. Microbial performance as affected by operation variables (schematic presentation).

For the solid curve in Fig. 7.2, it is better to select an operation variable where the concentration of product levels off, whereas for the broken curve, selection at the maximum concentration of product is advisable. Other factors such as energy requirements, convenience of operation, etc., must be taken into account before the optimum value of P/V or k_La for scale-up can be determined from the plateau of the curve.

At this juncture, it is worthwhile emphasizing that the behavior of biological material is important in scale-up. Obviously some factor(s) closely related to the dissolved oxygen concentration in liquid must be influencing the yield when the concentration of final product decreases with decrease of P/V or k_La. In addition, the broken curve in Fig. 7.2 suggests the adverse effect of some inhibitor (and/or repressor) on the yield of microbial product (*cf.* Section 6.1.3., Chapter 6).

Even assuming that microbial metabolism is a factor in scale-up, very little information about the effect of scale on microbial metabolism is available. In order to illustrate this point, one simple example of an ethanol fermentation (respiration-deficient mutant of *Saccharomyces cerevisiae*) will be cited.

The kinetic equations established using a chemostat and anaerobic culture are:[1]

$$\frac{dX}{dt} = \frac{\mu_0}{1 + \frac{p}{K_p}} \cdot \frac{S}{K_s + S} X \tag{7.13}$$

$$\frac{dp}{dt} = \frac{\nu_0}{1 + \frac{p}{K_p'}} \cdot \frac{S}{K_s' + S} X \tag{7.14}$$

$$-\frac{dS}{dt} = \frac{1}{Y_{X/S}} \cdot \frac{dX}{dt} = \frac{1}{Y_{p/S}} \cdot \frac{dp}{dt} \quad \text{[see Eqs. (5.50) to (5.52), Chapter 5]} \tag{7.15}$$

where

X = cell mass concentration
p = product (ethanol) concentration
S = substrate (glucose) concentration
K_s, K_s' = saturation constants
K_p, K_p' = empirical constants
$Y_{X/S}$ = yield factor ($=\Delta X / -\Delta S$)
$Y_{p/S}$ = yield factor ($=\Delta p / -\Delta S$)
μ_0 = specific growth rate at $p=0$
ν_0 = specific rate of ethanol production at $p=0$
t = time

A fact to be emphasized is that the metabolic patterns as represented by Eqs. (7.13) to (7.15) are unchanged irrespective of the cultural procedure—Warburg respirometer, shaken flask, or bench-scale fermentor.[2] The only difference is in the values of empirical constants, K_p, K_p', K_s, etc., depending on the equipment (and/or scale-up). For example, the values of K_p and K_p' obtained in the different environments are as follows:

TABLE 7.2

VALUES OF K_p AND K_p'.[2]

Culture	K_p (g/l)	K_p' (g/l)
Batch (shaken flask for K_p, Warburg manometer for K_p')	16.0	71.5
Continuous (bench-scale fermentor)	55.0	12.5

Table 7.2 suggests that the effect of ethanol on the specific growth rate and the specific rate of ethanol production, in other words, on the microbial activity of the

yeast cells, depends apparently on the equipment. Equations (7.13) to (7.15) also depict the aerobic cultivation of yeast cells in batch in a bench fermentor, provided the many constants involved in the original equations can be reasonably and separately fixed (not a curve-fitting process; see reference 7 in Chapter 5).

This example of biological factors influencing scale-up indicates clearly the necessity of studying metabolic patterns. There is a need for further experimentation and discussion on the values of empirical constants used in the kinetic equations for a particular fermentation in order that the equations can be properly related to operation variables such as P/V, and/or $k_L a$ in scaling-up.

Here, in connection with Eq. (7.14), another idea to maximize the value of

$$\int_0^t (dp/dt)dt = \int_0^t \nu X \, dt \quad \text{(where } \nu = \text{specific rate of ethanol production)}$$

is of interest in scale-up. To expand the idea, a means of controlling the metabolic activity of a microorganism by controlling the oxygen uptake rate (demand) is also of significance in scale-up; this is achieved by controlling simultaneously both the rate of oxygen supply and the rate of supply of another independent nutrient such as sugar (cf. Section 6.1.3., Chapter 6).

It should be emphasized that even if the problems of scale-up are successfully solved from an engineering point of view, a comparison of behavior of microorganisms in small and large fermentors remains to be studied. It is claimed that the scale-up of bacterial or yeast fermentations presents less difficulty than that of mycelial fermentations primarily because it is easier to disperse the bacterial or yeast cells uniformly throughout the fermentor.

The study of the kinetics of mycelial fermentations, which usually exhibit non-Newtonian characteristics, requires information on mass (oxygen) transfer and that the mycelia be uniformly distributed throughout. Because of the importance of mycelial fermentations, the basic properties of non-Newtonian broths will be referred to in this chapter.

7.2. EXAMPLES OF SCALE-UP

7.2.1. Power per unit volume of liquid

Figures 7.3 and 7.4 are examples of fermentations where product concentration has been correlated with the power input for different-sized fermentors. The penicillin titre at the 108-th hr of fermentation is plotted against power input, HP/m³, in Fig. 7.3; the volumetric ratio of scale-up was 1:10, while the values of v_s were roughly doubled in each fermentor tested. Within the range of these data, it appears that scale-up was successful provided the power input exceeded 1.5 HP/m³. It is also noted from Fig. 7.3 that the yield of penicillin fell sharply with a power input of less than 1.0 HP/m³.

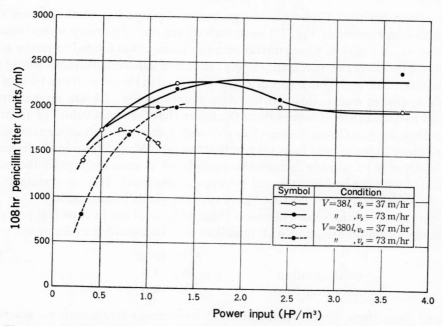

Fig. 7.3.   Effect of different values of power input on the yield of penicillin with different fermentation conditions.[6]

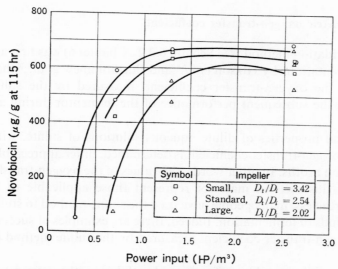

Fig. 7.4.   Effect of different values of power input on the yield of novobiocin in fermentors fitted with different-sized impellers.[12]

In principle, it is known that the performance of impellers can be assessed accurately, irrespective of the geometrical similarity or dissimilarity of the system, if the power imposed by the impeller on unit volume of agitated liquid is selected as the basis of comparison. Therefore, the geometrical properties of the system from which the data points of Fig. 7.3 were derived are not of primary importance.

However, in Fig. 7.4, where novobiocin production was related to power input, the above fact may not necessarily apply; standard flat-blade turbines whose sizes ranged from 2.02 to 3.42 in terms of D_t/D_i were used in this case; it can be seen that the three curves drawn through the data points in Fig. 7.4 are approximately parallel to each other. It is seen from the figure that for equal values of HP/m^3 the larger impeller gave lower yields of novobiocin compared with other sizes of impeller. These experimental facts suggest that the type of mycelium in the fermentation broth may be another factor to be considered in scale-up, besides the factors associated with the hydrodynamical behavior of the broth. It is interesting to note also the values of $HP/m^3 > 1.5 \ HP/m^3$ gave the best yields of antibiotic.

From reports of the values of power input in general use in chemical processes, the degree of liquid agitation with impellers may be classified as follows:[19]

	P/V HP/m^3
mild agitation	0.3 – 1.0
strong agitation	1.0 – 4.0

Judging from these figures, penicillin and novobiocin fermentations, where the titres of antibiotic are unaffected by values of HP/m^3, belong to the category of strongly agitated fermentations.

7.2.2. Volumetric oxygen-transfer coefficient

The sulfite-oxidation method (see Section 6.3.2., Chapter 6) was first applied to the study of fermentations by Hixson et al.[8] and Bartholomew et al.;[3] they correlated the value of the oxygen-transfer coefficient, measured by the sulfite-oxidation method, with the subsequent performance of the fermentor during an actual fermentation.

The physical properties of dilute aqueous solutions of sulfite, with which the value of the oxygen-transfer coefficient is determined, differ appreciably from those of fermentation broths. In other words, the values of the oxygen-transfer coefficient measured by this method may not represent those applicable to fermentation broths. Nevertheless, many people have used the sulfite method to study equipment for various kinds of fermentation. Indeed, there are examples of successful scale-up based on oxygen-transfer coefficient measured by the sulfite method (see Figs. 7.5 to 7.9).

To secure a really sound basis for control and design, the oxygen-transfer coefficient has recently been assessed by a number of people directly with the fermentation broth;[7,9,20] this derives from the widespread use of the membrane-covered oxygen sensor in the fermentation industry (cf. Section 12.3.3., Chapter 12).

However, so far as the correlation between microbial activity (and/or the final product concentration) and the oxygen-transfer coefficient is viewed, especially from the point of view of scale-up, the value of the oxygen-transfer coefficient might well be on a relative basis and even the estimation of the coefficient with the empirical equation presented by Cooper *et al.* (see Eqs. (6.37) and (6.38), Chapter 6) may become a basis of scale-up (see Figs. 7.7 and 7.8).[10,11]

Some typical examples are discussed briefly; readers who are interested in actual applications other than those exemplified here are suggested to refer to the respective original papers.[7,9,11,20]

Strohm *et al.* measured the yield of baker's yeast in vessels with different sulfite oxidation values; these results are shown in Fig. 7.5.[18] Data points in the figure are scattered considerably, but the yield appears to be fairly well correlated with the sulfite oxidation value. This good correlation was a little unexpected, because the experiments were conducted in both geometrically similar and dissimilar vessels, with and without mechanical agitation. The results reported in Fig. 7.5 suggest that the type of fermentor and the method of aeration did not affect the yield of baker's yeast appreciably if the sulfite oxidation value was greater than 150 mmole $O_2/l/hr$.

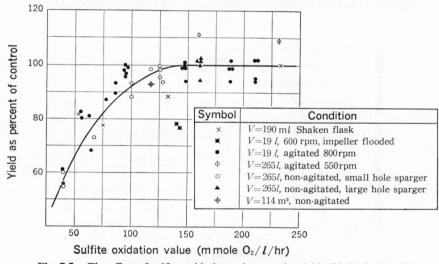

Symbol	Condition
×	$V=190$ ml Shaken flask
✷	$V=19$ l, 600 rpm, impeller flooded
•	$V=19$ l, agitated 800 rpm
φ	$V=265$ l, agitated 550 rpm
○	$V=265$ l, non-agitated, small hole sparger
▲	$V=265$ l, non-agitated, large hole sparger
⊕	$V=114$ m³, non-agitated

Fig. 7.5.   The effect of sulfite oxidation values on the yield of baker's yeast.[18]

The production of ustilagic acid provides another example of a good correlation between the yield of product and the sulfite oxidation value, as shown in Fig. 7.6.[16] All data in the figure, except for that with + symbol, were obtained using 5-l fermentors; the yield of ustilagic acid from a medium containing 7.5% glucose monohydrate was determined in fermentations using different degrees of aeration and agitation. The yield was plotted against the corresponding value of sulfite oxidation. Here, the yield increased with increase in sulfite oxidation value, until it approached 125 mmole $O_2/l/hr$; above this value, there was no improvement in yield. A similar trend was seen in Fig. 7.5 for the production of baker's yeast.

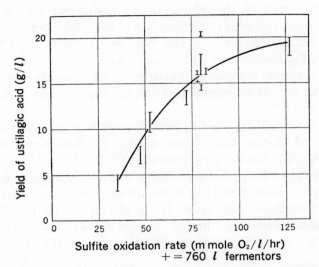

Fig. 7.6.   The effect of sulfite oxidation value on the yield of ustilagic acid.[16]

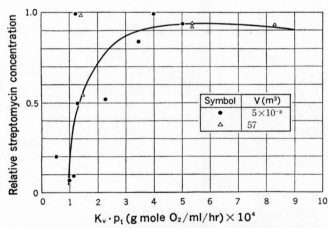

Fig. 7.7.   Comparison of the yields of streptomycin at different mass-transfer coefficients of oxygen applying in vessels of different size.[10]

In Figs. 7.7 to 7.9, the abscissae represent values of $K_v p_t$, which is the product of the volumetric oxygen-transfer coefficient and the total pressure. The principal reason for this representation is to correct for the driving force expressed by the pressure term, because, in large-scale fermentors, the pressure increases greatly; this results in inconsistencies when comparing the results of large-scale fermentors with those from bench-scale experiments.

In Figs. 7.7 and 7.8, the values of K_v for production equipment were calculated from the empirical equation presented by Cooper *et al.* (see Eq. (6.37), Chapter 6).

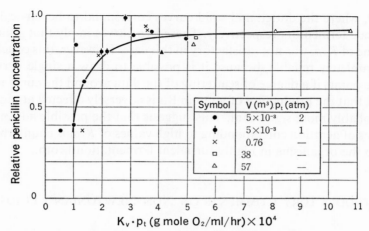

Fig. 7.8. Comparison of the yields of penicillin at different mass-transfer coefficients of oxygen applying in vessels of different size.[10]

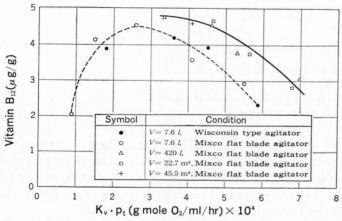

Fig. 7.9. Yield of vitamin B_{12} in different sized fermentors with different values of mass-transfer coefficient.[4]

Although the calculation of $K_v p_t$ from Eq. (6.37) is only an approximation, the calculated values correlate well with the final titers of streptomycin (Fig. 7.7) and penicillin (Fig. 7.8) obtained experimentally.[11] These figures clearly indicate that above certain minimum values of $K_v p_t$, the yields of antibiotic leveled off, as was found in Figs. 7.5 and 7.6. The data suggest that the penicillin and streptomycin fermentations can be scaled-up from Figs. 7.7 and 7.8 if the aeration rate and agitator speed are fixed such that the values of $K_v p_t$ calculated from Cooper's equation are beyond the critical value of about (5 to 6) $\times 10^{-4}$ g mole O_2/ml/hr.

In the scale-up of a bacterial fermentation for the production of vitamin B_{12}, some peculiar phenomena were observed by Bartholomew et al.[4] These are shown in Fig. 7.9, where the dotted and solid curves correspond to bench-scale and production-scale fermentors respectively. These two curves imply that scale-up based on the value of $K_v p_t$ was not satisfactory in this case. However, this conclusion may not be completely reliable, because it is possible to draw a single curve through these same points if a degree of scattering of the data around the curve is accepted.

The fact that the production of vitamin B_{12} is adversely affected at high values of $K_v p_t$ is probably more significant. This suggests that the possible advantage of the higher rates of aeration corresponding to high values of $K_v p_t$ is counterbalanced by damage to the organisms in highly turbulent fermentation broth.

7.3. INTRODUCTORY COMMENTS ON NON-NEWTONIAN FLUIDS

7.3.1. Definition

The viscosity of Newtonian fluids is defined as follows:

$$\tau g_c = \mu \frac{dv}{dr} \tag{7.16}$$

where

$$\mu = \text{liquid viscosity}$$
$$\tau = \text{shear stress}$$
$$dv/dr = \text{velocity gradient} = \text{shear rate}$$
$$g_c = \text{conversion factor}$$

The viscosity of μ in a Newtonian fluid is independent of shear rate, dv/dr, and is usually dependent on the temperature of the fluid. Therefore, in a plot of τ against dv/dr as shown in Fig. 7.10, the flow characteristics of a Newtonian fluid are represented by a straight line passing through the origin of the figure; the tangent, δ_1, is equal to the fluid viscosity, μ.

On the other hand, there are materials that have other flow characteristics; fluids with flow curves exhibiting shapes other than the straight line passing through the origin are called non-Newtonian. Such characteristic curves have been designated as Bingham plastic, plastic, pseudoplastic, and dilatant (see Fig. 7.10). The flow curve for a Bingham plastic is expressed as follows:

$$(\tau - \tau_y)g_c = \eta\left(\frac{dv}{dr}\right) \tag{7.17}$$

where

$$\tau_y = \text{yield value of stress, which means that the fluid will not “flow” unless a stress larger than } \tau_y \text{ is imposed upon it}$$
$$\eta = \text{coefficient of rigidity or plastic viscosity, the value of which is independent of } dv/dr$$

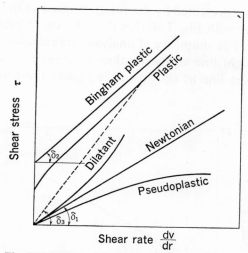

Fig. 7.10.   Shear stress vs. shear rate. Flow curves
of non-Newtonian and Newtonian fluids.

If the value of apparent viscosity, μ_a, is defined by the tangent of the angle subtended between the abscissa and an imaginary straight line connecting any point on the non-Newtonian flow curves with the origin of Fig. 7.10, it can be seen from the figure that the value of μ_a is dependent on dv/dr. For Bingham plastic, plastic, and pseudoplastic fluids, the values of μ_a will decrease with increase of dv/dr; the situation is reversed in a dilatant fluid.

In some non-Newtonian fluids, the flow curves may be approximated by:

$$\tau g_c = K \left(\frac{dv}{dr}\right)^{n'} \tag{7.18}$$

where

$\quad K$ = consistency index
$\quad n'$ = flow behavior index

Flow curves like those illustrated in Fig. 7.10 are usually determined with one of the rotational types of viscometer.

Another type of non-Newtonian flow, not shown in Fig. 7.10, is called thixotropic flow. Such fluids exhibit a hysteresis loop (the curve of viscosity obtained when dv/dr is gradually increased differs from that obtained by the reverse procedure).

Generally, mycelial fermentation broths exhibit non-Newtonian characteristics, but only a few rheological studies of these broths have been published so far.[5,15]

7.3.2.   Non-Newtonian characteristics of fermentation broths

The rheological character of *Streptomyces griseus* broth is shown in Fig. 7.11, in which the viscometer reading (corresponding to shear stress) is plotted against the shear rate, which was represented by the rotation speed of the viscometer.[15] It can be seen by comparison with Fig. 7.10 that this flow curve belongs to the category of pseudoplastic flow. For simplicity of analysis and calculation, the curve will be assumed to be a straight line at lower values of shear rate, extrapolating to the ordinate with a broken line as shown in the figure; the broth thus simulates a Bingham plastic.

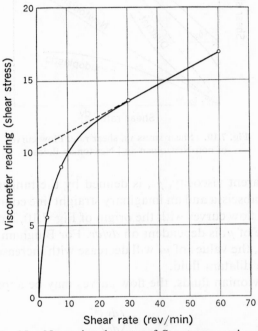

Fig. 7.11. Non-Newtonian character of *Streptomyces griseus* broth.[15]

Assuming that the *Streptomyces griseus* broth behaves like a Bingham plastic, if the values of η and τ_y are determined in broth sampled at intervals during the fermentation and plotted against time, t, the results are as shown in Figs. 7.12 and 7.13 respectively. It is seen from the figures that the values of both η and τ_y are closely dependent on the age of the mycelium. The clear dependence of η and τ_y on the stage of the fermentation may permit such curves to be used to gage the progress of the fermentation, in other words, as a means of controlling the fermentation.

Although there are examples (see Figs. 7.3, 7.7, etc.) where the scale-up of mycelial fermentations has been accomplished successfully using as a basis either the power input per unit volume or the overall oxygen-transfer coefficient, this does not preclude further study of the scale-up of non-Newtonian fermentation broths. On the contrary, much more information about properties like oxygen-transfer rates in bubble aeration, bubble motion, and mixing time in non-Newtonian fer-

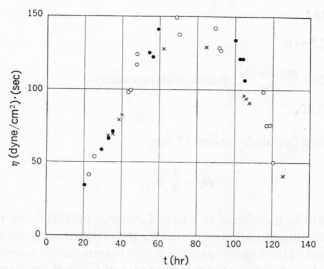

Fig. 7.12. Variation of plastic viscosity with the age of the broth (*Streptomyces griseus*).[15]

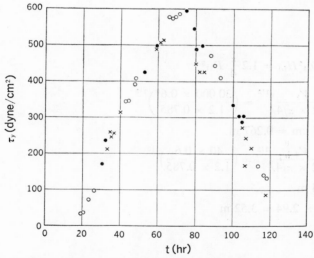

Fig. 7.13. Variation of yield stress with the age of the broth (*Streptomyces griseus*).[15]

mentation broths are needed to provide a better understanding of scale-up as a whole.

7.4. DESIGN EXAMPLE OF SCALE-UP

The optimum conditions for a bacterial fermentation were established with a bench-scale fermentor, 30 *l*, as follows:

$$V_1 = 18\ l$$

$$\left(\frac{F}{V}\right)_1 = 2.2\ \text{vvm}$$

$$K_v p = 200\ \frac{\text{mmole O}_2}{l\ \text{hr}} \quad \text{(sulfite oxidation value)}$$

$$H_L = 1.2 D_t$$

Impeller: standard flat-blade turbine (2 sets)

$$D_1 = \frac{1}{3} D_t$$

Production is to be transferred to a large fermentor (40,000 l) and it is desired to estimate the operating conditions required for the production fermentor, especially the rotation speed of the impeller and the power requirements, assuming the physical properties of the broth are the same as those of water at room temperatures in the scale-up calculation.

Solution

$$V_1 = \frac{\pi}{4} D_{t1}{}^2 H_{L1} = 1.2 \frac{\pi}{4} D_{t1}{}^3$$

$$D_{t1} = \left(\frac{V_1}{1.2 \times \pi/4}\right)^{1/3} = \left(\frac{30{,}000 \times 0.6^*}{1.2 \times 0.785}\right)^{1/3}$$

$$= 26.7\ \text{cm} = 0.267\ \text{m}$$

$$D_{t2} = \left(\frac{V_2}{1.2 \times \pi/4}\right)^{1/3} = \left(\frac{40 \times 0.6}{1.2 \times 0.785}\right)^{1/3}$$

$$= 2.94\ \text{m}$$

$$H_{L2} = 1.2 \times 2.94 = 3.52\ \text{m}$$

Assuming that

$$\left(\frac{F}{V}\right)_2 = \left(\frac{F}{V}\right)_1 \left(\frac{H_{L2}}{H_{L1}}\right)^{-2/3} \quad [cf.\ \text{modification of Eq. (7.2)}]$$

$$= \left(\frac{F}{V}\right)_1 \left(\frac{D_{t2}}{D_{t1}}\right)^{-2/3} = 2.20 \times \left(\frac{2.94}{0.267}\right)^{-2/3}$$

$$= 0.44\ \text{vvm}$$

$$(v_s)_2 = \frac{F_2}{(\pi/4)D_{t2}{}^2} = \frac{0.44\ V_2 \times 60}{(\pi/4)D_{t2}{}^2} = \frac{0.44 \times 24 \times 60}{0.785 \times 2.94^2}$$

$$= 94\ \text{m/hr}$$

* Actual working volume is assumed to be 60% of the nominal volume.

$$(K_v p)_1 = 200 \; \frac{\text{mmole } O_2}{l \; hr}$$

$$= 200 \; \frac{10^{-6} \text{ kg mole } O_2}{10^{-3} \text{ m}^3 \text{ hr}}$$

$$= 0.2 \; \frac{\text{kg mole } O_2}{\text{m}^3 \text{ hr}}$$

$$= (K_v p)_2$$

Oxygen-transfer efficiency for the production fermentor

$$= \frac{\text{Oxygen transferred}}{\text{Oxygen supplied}} = \frac{K_v p \times V_2}{V_2 \times 0.44 \times 60 \times 0.21/22.4}$$

$$= \frac{0.20}{0.44 \times 60 \times 0.21/22.4}$$

$$= 0.807 \text{ (specific value based on the sulfite oxidation)}$$

$$p_{2\text{in}} = (0.21) \left(p_t + \frac{3.52}{10.3} \right) = (0.21) \left(1.3 + \frac{3.52}{10.3} \right)$$

$$= 0.345 \text{ atm}$$

provided:

$$p_t = \text{internal pressure of production fermentor}$$

$$= 1.3 \text{ atm (assumed)}$$

$$p_{2\text{out}} = (0.21)(1 - 0.807)p_t = 0.21 \times 0.193 \times 1.3$$

$$= 0.0526 \text{ atm}$$

$$(p)_2 = \frac{p_{2\text{in}} - p_{2\text{out}}}{\ln (p_{2\text{in}}/p_{2\text{out}})} = \frac{0.345 - 0.0526}{2.3 \log (0.345/0.0526)}$$

$$= 0.155 \text{ atm}$$

$$(K_v)_2 = \frac{0.20}{0.155} = 1.29 \; \frac{\text{kg mole } O_2}{\text{m}^3 \text{ hr atm}}$$

Assuming that Eq. (6.37) can be approximated to the case where $H_L = 1.2 \; D_t$, and assuming also that the coefficient, 0.0635, of the equation be halved when the standard flat-blade turbine is concerned,

$$1.29 = 0.0318 \left(\frac{P_g}{V} \right)_2^{0.95} (v_s)_2^{0.67}$$

$$\left(\frac{P_g}{V} \right)_2 = \left(\frac{1.29}{0.0318 \times 94^{0.67}} \right)^{1/0.95}$$

$$= 2$$

$$(P_g)_2 = 2 \times 24 = 48 \text{ HP}$$

Assuming that

$$\left(\frac{Pg_c}{n^3 D_1^5 \rho}\right)_2 = 6 \times 2 \qquad (cf.\ \text{Fig. 6.5 and Section 6.2.2.1., Chapter 6}),$$

$$(P)_2 = \frac{12 \times n^3 \times (2.94/3)^5 \times 10^3}{9.81} = 1.10 \times 10^3 \times n^3$$

$$= 14.5 \times n^3 \text{ HP}$$

$$\left(\frac{P_g}{P}\right)_2 = \frac{48}{14.5 \times n^3} = 3.30 \times n^{-3} \qquad\qquad (7.19)$$

On the other hand, from Eq. (6.25) in Chapter 6,

$$(N_a)_2 = \frac{F_2}{60 \times n \times (2.94/3)^3} = \frac{0.44 \times 24}{60 \times n \times (2.94/3)^3}$$

$$= 0.187 \times n^{-1} \qquad\qquad (7.20)$$

From Eqs. (7.19), (7.20), and curve A in Fig. 6.6, Chapter 6, the value of n (rotation speed of the impeller in a large vessel) is found by a trial-and-error method as follows:

$$n = 1.8 \text{ rps}$$

From Eq. (7.20),

$$N_a = 1.87 \times 10^{-1} \times \frac{1}{1.8} = 0.104$$

From Eq. (7.19),

$$\left(\frac{P_g}{P}\right)_2 = 3.30 \times \frac{1}{1.8^3} = 0.57$$

From Fig. 6.6,

$$\left(\frac{P_g}{P}\right)_2 = 0.57$$

$$N = 1.8 \times 60 = 108 \text{ rpm}$$

$$(P)_2 = \frac{48}{0.57} = 84 \text{ HP}$$

$$\left(\frac{P_g}{V}\right)_2 = \frac{48}{24} = 2 \qquad \text{HP/m}^3 > 0.1 \text{ HP/m}^3$$

$$\left(\frac{n D_1^2 \rho}{\mu}\right)_2 = \frac{1.8 \times 0.98^2 \times 10^3}{10^{-3}} \div 10^6$$

Therefore, the use of both Eq. (6.37) and power number=6 is permissible. However, the application of Eq. (7.2) (its modification) and Fig. 6.6 to this problem is only an approximation, since the former deals with an estimation of aeration rate in a system without mechanical agitation, while the latter is relevant to a fermentor with only one impeller. Despite the approximations adopted, these are not considered to introduce grave errors in calculation.

7.5. SUMMARY

1. Constant volumetric oxygen-transfer coefficient is one basis of scale-up; although many correlations exist, one of the most satisfactory relationships between volumetric oxygen-transfer coefficient and operating variables is:

$$\frac{(k_L a)_1}{(k_L a)_2} = \frac{(F/V)_1}{(F/V)_2} \left(\frac{H_{L1}}{H_{L2}}\right) \left(\frac{d_{B2}}{d_{B1}}\right)^{3/2} \left(\frac{v_{B2}}{v_{B1}}\right)^{1/2} \qquad \text{[Eq. (7.1)]}$$

2. Constant power per unit volume of liquid is another criterion of scale-up;

$$\frac{P}{V} \propto n^3 D_1^2 \qquad \text{[Eq. (7.6)]}$$

This concept is easily applied in an aerobic cultivation, and power requirements during sterilization of fermentor vessels.

3. In addition to scale-up on the basis of oxygen transfer rate, a simultaneous control of both the supply rates of oxygen and another independent nutrient such as sugar to effect the rate of oxygen demand of the organism is significant in scaling-up.

4. Biological properties, especially various constants involved in kinetic equations for fermentations depend on scale-up, though the metabolic patterns remain unchanged.

5. Typical examples of scale-up of fermentors for bacteria, yeast and mycelial fermentations are described.

6. Further study on oxygen transfer in bubble aeration and bubble motion in non-Newtonian fermentation broths is needed for the scaling-up of these fermentations.

NOMENCLATURE

a = interfacial area between air bubbles and liquid per unit volume of liquid, m^2/m^3
c_p = specific heat, kcal/kg °C
D = molecular diffusivity, m^2/sec
D_1 = impeller diameter, m

D_t = vessel diameter, m

d_B = bubble size, m

F = aeration rate, m³/sec, m³/min; pumping rate of liquid by impeller, m³/sec, m³/min

g_c = conversion factor, kg m/Kg sec²

H_L = liquid depth

h = heat-transfer coefficient kcal/m² hr °C

K_v = volumetric oxygen-transfer coefficient, kg mole O_2/m³ hr atm

K_La = volumetric oxygen-transfer coefficient, hr⁻¹

K = consistency index, g cm⁻¹ sec$^{n'-2}$

K_p, K_p' = empirical constants

K_s, K_s' = saturation constants

k = thermal conductivity, kcal/m hr °C

k_L = mass-transfer coefficient, m/hr

N = rotation speed of impeller, min⁻¹

N_a = aeration number (= F/nD_i^3)

N_P = power number (= $Pg_c/n^3D_i^5\rho$)

N_{Pr} = Prandtl number (= $c_p\mu/k$)

N_{Re} = Reynolds number (= $nD_i^2\rho/\mu$)

N_{Sh} = Sherwood number (= k_LD_i/D)

N_{Nu} = Nusselt number (= hD_i/k)

n = rotation speed of impeller, sec⁻¹

n' = flow behavior index

P = power consumed for agitation in ungassed system, Kg m/sec, HP

P_g = power consumed for agitation in gassed system, Kg m/sec, HP

p = partial pressure of oxygen, atm; product (ethanol) concentration, g/l

p_t = internal pressure of fermentor, atm; total pressure of incoming air, atm

S = substrate (glucose) concentration, g/l

t = time

V = liquid volume, m³

v_s = nominal superficial velocity of air, m/hr

dv/dr = shear rate, sec⁻¹

X = cell mass concentration, g/l

$Y_{p/s}$ = yield factor (= $\Delta p/-\Delta S$)

$Y_{X/s}$ = yield factor (= $\Delta X/-\Delta S$)

subscripts

1, 2 = small and large vessels, respectively

in = inlet

out = outlet

Greek letters

α = experimental exponent

δ = angle subtended between τ and dv/dr (Fig. 7.10)

η = plastic viscosity, (dyne/cm²) · sec (Fig. 7.13), g/cm sec

μ = liquid viscosity, kg/m sec

μ_0 = specific growth rate at $p=0$, hr⁻¹

μ_a = apparent viscosity, kg/m sec
ν = specific rate of ethanol production, hr^{-1}
ν_0 = specific rate of ethanol production at $p=0$, hr^{-1}
ρ = liquid density, kg/m^3
τ = shear stress, Kg/cm^2, G/cm^2
τ_y = yield value of shear stress, $dyne/cm^2$ (Fig. 7.13), Kg/cm^2, G/cm^2

REFERENCES

1. Aiba, S., Shoda, M., and Nagatani, M. (1968). "Kinetics of product inhibition in alcohol fermentation."*Biotech. & Bioeng.* **10**, 845.
2. Aiba, S., and Shoda, M. (1969). "Reassessment of the product inhibition in alcohol fermentation." *J. Ferm. Tech. (Japan)* **47**, 790.
3. Bartholomew, W. H., Karow, E. O., Sfat, M. R., and Wilhelm, R. H. (1950). "Oxygen transfer and agitation in submerged fermentation." *Ind. Eng. Chem.* **42**, 1801.
4. Bartholomew, W. H. (1960). "Scale-up of submerged fermentation." *Adv. Appl. Microbiol.* (Ed.) Umbreit, W. W. **2**, 289. Academic Press, New York.
5. Deindoerfer, F. H., and West, J. M. (1960). "Rheological examination of some fermentation broths." *J. Biochem. Microbiol. Tech. & Eng.* **2**, 165.
6. Gaden, E. L., Jr. (1961). "Aeration and agitation in fermentation." *Sci. Repts. Instituto Superore di Sanita* **1**, 161.
7. Hirose, Y., Sonoda, H., Kinoshita, K., and Okada, H. (1966). "Studies on oxygen transfer in submerged fermentations. Part V. The effect of aeration on glutamic acid fermentation (1)." *Agr. Biol. Chem. (Japan)* **30**, 585.
8. Hixson, A. W., and Gaden, E. L., Jr. (1950). "Oxygen transfer in submerged fermentation." *Ind. Eng. Chem.* **42**, 1792.
9. Kanzaki, T., Sumino, Y., and Fukuda, H. (1968). "Oxygen transfer in L-glutamic acid fermentation by oleic acid-requiring organism." *J. Ferm. Tech. (Japan)* **46**, 1031.
10. Karow, E. O., Bartholomew, W. H., and Sfat, M. R. (1953). "Oxygen transfer and agitation in submerged fermentations." *J. Agr. Food. Chem.* **1**, 302.
11. Kodama, A., Chikaike, T., Yamaguchi, J., and Aizawa, M. (1968). "Studies on the leucomycin fermentation." *J. Ferm. Tech. (Japan)* **46**, 225.
12. Maxon, W. D. (1959). "Aeration-agitation studies on the novobiocin fermentation." *J. Biochem. Microbiol. Tech. & Eng.* **1**, 311.
13. Nakajima, S. (1957). "Practical designing of an agitator." *Chem. Eng. (Japan)* **21**, 96.
14. Oldshue, S. Y. (1966). "Fermentation mixing scale-up technique." *Biotech. & Bioeng.* **8**, 3.
15. Richards, J. W. (1961). "Studies in aeration and agitation." *Progress in Industrial Microbiology* **3**, 143.
16. Roxburgh, J. M., Spencer, J. F. T., and Salans, H. R. (1954). "Factors affecting the production of ustilagic acid by *Ustilago zeae.*" *J. Agr. Food Chem.* **2**, 1121.
17. Rushton, J. H. (1951). "The use of pilot-plant mixing data." *Chem. Eng. Progress* **47**, 485.
18. Strohm, J., Dale, H. F., and Peppler, H. J. (1959). "Polarographic measurement of dissolved oxygen in yeast fermentation." *Appl. Microbiol.* **7**, 235.
19. Society of Chem. Eng. (Japan) Subcommittee of liquid mixing survey (1956). "Survey of agitators used in the chemical industry in Japan." *Chem. Eng. (Japan)* **20**, 634.
20. Taguchi, H., Imanaka, T., Teramoto, S., Takatsu, M., and Sato, M. (1968). "Scale-up of glucamylase fermentation by *Endomyces* sp." *J. Ferm. Tech. (Japan)* **46**, 823.

Chapter 8

Translation of Laboratory Culture Results to Plant Operation

In addition to the problem of scaling up a new fermentation process, there exists another problem with well-established fermentations of translating culture data obtainable with bench-scale or shaken flasks to plant operation for process improvement. Process improvement presents a continuing problem. As far as a product is economically viable, a manufacturer has to undertake a program to maintain its viability. Figure 8.1 illustrates the improvement history of a typical antibiotic fermentation. It should be noted that many different kinds of improvements contributed to the viability of the process. Laboratory experiments are carried out to search for better strains, for medium and physical environment modifications, and for equipment changes that will yield substantial process improvements. For the most part, the improvements sought are those of increased product concentrations

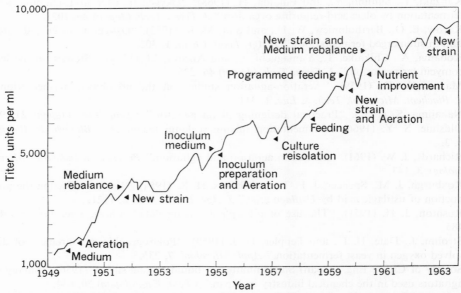

Fig. 8.1. Influence of fermentation development studies on a typical antibiotic. (*Courtesy* Bristol laboratories, Syracuse, N.Y., U.S.A.)

at a fixed time. Other factors such as improved conversion efficiencies, increased production rates, decreased recovery costs, etc., can all result in increased economic yield.

Once a plant is built, the conditions of agitation, aeration, mass (oxygen) transfer, and heat transfer are more or less set. Therefore, it has been suggested that the problem of translating process improvements is not one of scale-up, but rather scale-down. Those environmental conditions achievable in plant-scale equipment should be scaled down to the pilot plant and screen size of equipment (shaken flask) to insure that improvement studies are carried out under conditions that can be duplicated.

A probable prerequisite for scale-up and/or scale-down is to make the environmental conditions identical between plant- and bench-scale equipment, but the identity does not always guarantee the similar behavior of microorganisms. This situation remains very difficult to overcome. In view of the fact that microbial metabolism abounds with uncertain factors which are beyond the reach of simple physical environmental analyses, as has been illustrated in Section 7.1.2., Chapter 7, some practical ideas to assist scale-down will be discussed in this chapter.

8.1. SCALE-DOWN

By scale-down it is meant here that a microbial production rate of a specific metabolite in a full-scale plant is reproduced by some means in pilot-plant vessels. Identities in both systems with respect to time-dependence of sugar consumption rate, pH, dissolved oxygen concentration, etc., must be underlying the identity of product accumulation pattern. Referring to the criteria on scale-up in the previous chapter (*cf.* Table 7.1), one may choose again either power input per unit volume of liquid, shear rate of impeller or volumetric oxygen-transfer coefficient, etc. in scale-down. However, a proper selection cannot be made *a priori*, because a direct correlation between microbial activity and environmental conditions, which are represented by the above scale-up (and/or scale-down) criteria, still seems prohibitive.

8.1.1. Use of pilot plant

Figure 8.2 (upper diagram) presents schematically an idea for scaling down a production fermentor to pilot-plant vessels. Though aeration and agitation in the production vessel are fixed, the pilot vessels are assumed to be equipped with variable-speed motors to permit independently a wide range of variation in the rotation speed of the impellers. Geometrical similarity is also assumed between the plant and the pilot vessels. Here, the agitation intensity, keeping aeration rate constant among the 4 pilot vessels in Fig. 8.2, will be changed from one vessel to another.

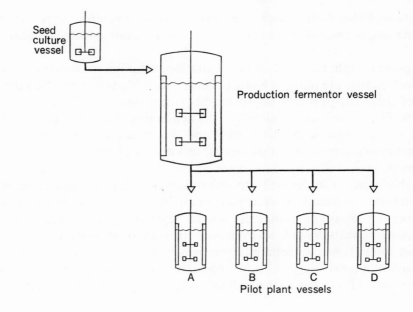

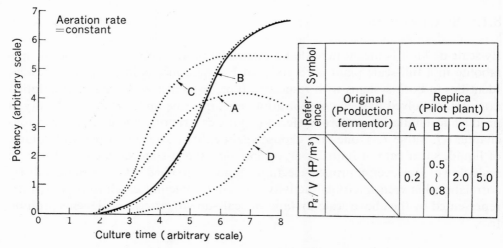

Fig. 8.2. Schematic diagram illustrating scale-down.

In order to secure the exact identity of fermentation media, sterilizing conditions and inoculum size, the media in the production vessel are transferred partly into the pilot vessels right after the inoculation into the large vessel. Then, the fermentation in both systems is started simultaneously. Control devices such as pH, temperature, anti-foam, etc. are installed on the pilot vessels and are operated if required.

The lower diagram of Fig. 8.2 shows, also schematically, the results of these scale-down experiments. Suppose that a dotted curve (curve B) agreed incidentally after a series of runs with the solid curve from the production fermentor. In this illustration, the power input per unit volume of liquid, P_g/V, is changed from vessel to vessel, and an appropriate value to allow overall reproduction of the production pattern is found; P_g/V exceeding or not attaining the appropriate value fails to reproduce the original pattern. Numerical values of P_g/V illustrated in the lower diagram have no particular significance. Now, in this example, the operation condition for vessel B are assumed successful for scaling down the production plant. Needless to say, parameters other than P_g/V may be employed depending on the fermentation characteristics to be scaled down.

Once the scale-down procedure has been established, it is preferable that the condition for vessel B be used exclusively in the following translation of laboratory culture results back to plant operation (see Section 8.2.). Although the concept of scale-up in this context suggests the use of shaken flasks, bench-scale (or jar) fermentors, followed by pilot plants and finally industrial plant, the above procedure skips the use of bench-scale fermentors; in fact, the size of pilot plant used to reproduce the production plant performance might well be as large as possible, presumably more than 3,000 l in nominal volume. As one extreme, some of the production plants themselves are frequently converted as very good "substitutes" for the pilot plant.

This is a good indication that the microbial performance is sophisticated, being far from the usual concept of environmental similarity in the hydrodynamic sense of the word; controversial as this argument may sound, neither the scale-up bases discussed in the previous chapter nor a highly instrumented bench- or pilot-scale fermentor such as those which will be referred to later on in Chapter 12 are invalidated. On the contrary, the bench-scale (or jar) fermentors are fully utilized to prepare test materials for animal or hospital use, while the highly instrumented pilot plant permits a rapid analysis of physical and metabolic data, the accumulation of which may lead eventually to a more rational scale-up and/or scale-down.

8.1.2. Use of oxygen uptake rate

Another idea of scale-down (or scale-up) originates from the work of Shu,[6] who attempted to reproduce a pattern of microbial uptake rate of oxygen by controlling the agitation speed of the impeller in the pilot plant via signals from the curve of oxygen uptake rate pre-marked on a recorder chart. Essentially, this reproduction stems from that the impeller speed increased automatically when the oxygen uptake rate is lower than the predetermined level, as detected from an oxygen analyzer installed on the exit line, whereas the impeller speed decreases when the deviation from the predetermined mark is indicative of a higher oxygen uptake rate.[1]

An example of this scale-down is shown in Fig. 8.3. This example deals with a scale-down from tank to tank, the latter being of 400 l capacity, while the volume of the former, presumably on an industrial scale is not clear from the paper.[3] The microorganism they used was *Streptomyces aureofaciens* which accumulated

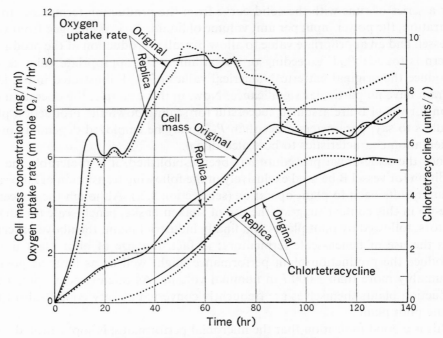

Fig. 8.3. Example of scale-down (cont'd).

chlortetracycline in the broth, composed initially of corn starch, corn steep liquor, lard oil and salts. According to the authors, the original deep tank fermentor was operated at a constant impeller speed of 150 rpm, while the replica vessel was manipulated automatically in a range from 175 to 250 rpm.[3]

The irregular pattern of oxygen uptake rate, as seen from Fig. 8.3, might have resulted from an interaction between different phases of metabolic activity and dissolved oxygen available to the cells. It is interesting to note from the figure that the patterns of chlortetracycline accumulation and oxygen uptake rate could be closely followed, although the cell mass concentration in the replica at the end of fermentation was about 18% higher than the original. It should be remarked that this idea of scale-down (or scale-up) pays particular attention to microbial physiological activity, disregarding geometrical similarity between the original and the replica. Though examples of this sort are not common yet, this idea may find application to various aerobic fermentations other than this specific actinomycete fermentation.[4]

8.2. TRANSLATION OF LABORATORY CULTURE DATA TO PLANT OPERATION

In collaboration with a routine program of production in a fermentation plant, there is always a requirement to enhance the product potency in the broth. Though a remodelling of plant equipment now in operation may be an alternate to the enhancement of potency in the product, it is usually very difficult to accomplish this renovation in the midst of the production program. Another alternate to this is to improve the strain by breeding and/or to introduce a sophisticated means of feeding (*cf.* Chapter 6) which requires little modification of the plant facilities. Both methods, concerned with strain and medium, must be studied in laboratories. Herein lies the absolute necessity of translating laboratory culture data to plant operation. In this section, a practical concept for translating strain breeding to plant operation will be exemplified, minimizing detailed descriptions, because there may exist various means of translation from case to case other than the following.

Figure 8.4 shows a schematic diagram for translating laboratory culture data to plant operation. The objective of the first screening, coupled with various strain

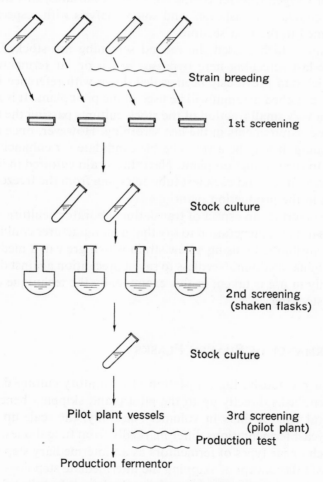

Fig. 8.4. Translation of laboratory culture data.

breedings (*cf.* Section 2.4.5., Chapter 2) is to search for better strains which could grow and metabolize the desired product in an enriched medium and in addition, could withstand poor oxygen supply, the capacity for which is assumed limited by the plant already built. Those strains which passed the first screening in plating and cup methods for instance, are stocked, followed by the second screening with lots of shaken flasks.

Obviously, the advantage of employing shaken flasks in this screening lies in enabling one to comprehensively change culture conditions from one flask to another, and/or to economize the time required for repeated runs. On the other hand, it is not practical to trace from one flask to another the patterns of pH, cell mass concentration and dissolved oxygen concentration with progress of culture time. Since it is important to know under what circumstances microbial activities in shaken flasks are most likely to be limited by dissolved oxygen deficiency or by lower values of oxygen-transfer coefficient of flask closures, this deserves consideration in the course of translation, and some analysis with respect to this point will be mentioned in the next section.

Specific strains which passed the second screening are stocked for the third screening. The last screening here bypasses bench or jar fermentors, using the pilot plant which had previously been scaled down with reference to the production plant. The enriched medium will be used in the pilot plant. It is not unusual to find that only a very small fraction of the stock cultures passing the second screening do fulfill the requirements in the last screening. However, once a strain passes the third screening, it may be a very eligible candidate for enhancing the product concentration in the production plant. Next, the strain cultured in the thick medium is freeze-dried *in situ* and each test tube emerging from the freeze-drying is used in every batch in the production plant.

This picture describes an aspect of translating laboratory culture results to plant operation. It is not an exaggeration to say that a manufacturer could quadruple the potency of an antibiotic by using a new strain which grew in a medium 2 times as thick as the original medium. Needless to say, a precaution against deterioration of strains currently in use is taken, in this example, by the technique of mono-spore isolation and stocking.

8.3. PERFORMANCE OF SHAKEN FLASKS

As pointed out previously, the translation of laboratory culture data starts from data for shaken flasks directly up to the pilot plant, skipping bench types of fermentors, several tens of liters in volume. Naturally, the scale-up and -down of bacterial and yeast fermentations differ markedly from fungal ones, and pay much tribute to bench or jar types of fermentors as an intermediary step of the process. In other words, this concept of skipping the intermediate step does not always apply to all cases of scale-up (and conversely, scale-down). In spite of this controversy, it may be true that more analysis is required of shaken flask experiments. The following discussion will be focused on some physical factors limiting the microbial growth under the specific environment. To begin with, the resistance of shaken flask closures to oxygen transfer will be considered.

8.3.1. Resistance of closure to oxygen transfer

Shaken flasks are usually provided with closures such as cotton plugs, filter paper, specific filters made of PVA, etc., to preclude airborne contaminants from coming through into the flasks. Suppose that the interior of an empty shaken flask is filled with carbon dioxide (atmospheric pressure) and that a certain kind of closure be fitted to the flask opening at the top. Right after the insertion of this closure, the flask is subjected to shaking. During the shaking the carbon dioxide will be replaced by oxygen from ambient air. Distribution of oxygen partial pressure inside the closure will eventually become independent of time after experiencing an unsteady state which inevitably accompanies the start of this shaking. Then, it is feasible to determine the transfer coefficient of oxygen through the closure by observing the increase of oxygen partial pressure inside the flask as detected with an oxygen sensor.

Designating p_g, p_0, V_g, and $(K_c)_{O_2}$ as partial pressures of oxygen inside the shaken flask and in ambient air, free volume of the shaken flask, and transfer coefficient of oxygen through the closure, respectively, the rate of change, dp_g/dt, will be shown by the following equation, assuming unidirectional diffusion of oxygen.

$$V_g \frac{dp_g}{dt} = -\frac{DA}{L}(p_g - p_0) \tag{8.1}$$

where

A = cross-sectional area of closure
L = axial length of closure (see Fig. 8.5)
D = apparent diffusivity of oxygen in closure

Setting $DA/L = (K_c)_{O_2}$, Eq. (8.1) is rearranged and integrated from $t=0$ to $t=t$ to give:

$$(K_c)_{O_2} = -\frac{2.303 \, V_g \log\left(\dfrac{p_0 - p_g}{p_0 - p_{gi}}\right)}{t} \tag{8.2}$$

provided:

$p_{gi} = p_g \quad$ at $\quad t = 0$

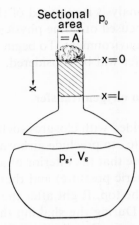

Fig. 8.5. Shaken flask with closure.

If a semi-logarithmic plot of $(p_0-p_g)/(p_0-p_{g1})$ against t yields a straight line when t becomes large, the slope of this line gives the value of $(K_c)_{O_2}/2.303V_g$.

If the above plot of log $\{(p_0-p_g)/(p_0-p_{g1})\}$ against t deviates from the straight line when t is small, the deviation should indicate the effect of the unsteady state in the distribution of oxygen partial pressure inside the closure.

Some consideration will be given below to correcting for the transfer coefficient of oxygen, $(K_c)_{O_2}'$, measured incidentally under unsteady state. A differential equation with respect to the partial pressure of oxygen, p, inside the closure as a function of location, x, and time, t, is given by:

$$\frac{\partial p}{\partial t} = D\frac{\partial^2 p}{\partial x^2}$$ (8.3)

Initial and boundary conditions are:

$$\left.\begin{array}{ll} p(x,\ 0) = p_c & \\ p(x,\ \infty) = p_0 & \\ p(0,\ t) = p_0 & (t>0) \\ p(L,\ t) = p_g & (t>0) \end{array}\right\}$$ (8.4)

Denoting N_{O_2} as molar flow rate of oxygen into the flask,

$$N_{O_2} = -\frac{DA}{RT}\left(\frac{\partial p}{\partial x}\right)_{x=L}$$ (8.5)

Then

$$p_g = p_{g1} + \left(\frac{RT}{V_g}\right)\int_0^t N_{O_2}\,dt$$

$$= p_{g1} - \frac{DA}{V_g} \int_0^t \left(\frac{\partial p}{\partial x} \right)_{x=L} dt \tag{8.6}$$

provided:

R = gas constant
T = absolute temperature

Setting,

$$\tilde{p} = \frac{p_0 - p}{p_0 - p_{g1}}, \qquad \tilde{x} = \frac{x}{L}$$

$$\tilde{t} = \frac{Dt}{L^2} \quad \text{and} \quad \nu = \frac{AL}{V_g} = \frac{V_c}{V_g}$$

where

V_c = volume occupied by closure

Equations (8.3) to (8.6) reduce to:

$$\frac{\partial \tilde{p}}{\partial \tilde{t}} = \frac{\partial^2 \tilde{p}}{\partial \tilde{x}^2} \tag{8.3'}$$

$$\left. \begin{array}{l} \tilde{p}(\tilde{x}, 0) = \tilde{p}_c \\ \tilde{p}(\tilde{x}, \infty) = 0 \\ \tilde{p}(0, t) = 0 \\ \tilde{p}(1, t) = \tilde{p}_g \end{array} \right\} \tag{8.4'}$$

$$\tilde{p}_g = 1 - \nu \int_0^t \left(\frac{\partial \tilde{p}}{\partial \tilde{x}} \right)_{\tilde{x}=1} d\tilde{t}$$

$$= \nu \int_{\tilde{t}}^\infty \left(\frac{\partial \tilde{p}}{\partial \tilde{x}} \right)_{\tilde{x}=1} d\tilde{t} \tag{8.6'}$$

Solving for $\tilde{p}$,[2]

$$\tilde{p} = \sum_{\tau=1}^\infty \left(\frac{2 \cos \delta_\tau}{\delta_\tau + \dfrac{\sin 2 \delta_\tau}{2}} \right) (\sin \delta_\tau \cdot \tilde{x}) \exp \left(-\delta_\tau^2 \tilde{t} \right) \tag{8.7}$$

$$\tilde{p}_g = \sum_{\tau=1}^\infty \left(\frac{\sin 2 \delta_\tau}{\delta_\tau + \dfrac{\sin 2 \delta_\tau}{2}} \right) \exp \left(-\delta_\tau^2 \tilde{t} \right) \tag{8.8}$$

provided:

the δ_τ ($\tau = 1, 2, 3, \ldots$) are the prositive roots, taken in order of magnitude, of the following equation

$$\tan \delta = \frac{\nu}{\delta} \tag{8.9}$$

Disregarding the second and higher orders in the infinite series of Eq. (8.8), and also omitting the subscript, τ, for δ in this case,

$$\tilde{p}_g \doteqdot \frac{\sin 2\delta}{\delta + \dfrac{\sin 2\delta}{2}} \exp\left(-\delta^2 \tilde{t}\right)$$

$$= f(\delta) \exp\left(-\delta^2 \tilde{t}\right) \tag{8.10}$$

provided:

$$f(\delta) = \frac{\sin 2\delta}{\delta + \dfrac{\sin 2\delta}{2}} \tag{8.11}$$

Taking logarithms of both sides of Eq. (8.10), converting nondimensional terms of $\tilde{p}_g$, $\tilde{t}$ to real values, and rearranging,

$$(K_c)_{o_2} = -\frac{2.303\, V_g \log\left(\dfrac{p_0 - p_g}{p_0 - p_{gi}}\right)}{t} \cdot \frac{\nu}{\delta^2} + \frac{\nu}{\delta^2} \cdot \frac{\ln f(\delta)}{t} \cdot V_g \tag{8.12}$$

In the above rearrangement,

$$\delta^2 \tilde{t} = \delta^2 \frac{D}{L^2} t$$

$$= \delta^2 \frac{DA}{L^2 A} t$$

$$= \delta^2 \frac{(K_c)_{o_2}}{\nu V_g} t$$

was used.

The first term on the right-hand side of Eq. (8.12), referring to Eq. (8.2), must be the apparent value, $(K_c)_{o_2}'$, of the transfer coefficient of the closure when the data for p_g are taken before the distribution of p within the closure reaches a steady state. Accordingly,

$$(K_c)_{o_2} = \frac{\nu}{\delta^2} (K_c)_{o_2}' + \frac{\nu}{\delta^2} \cdot \frac{\ln f(\delta)}{t} V_g \tag{8.13}$$

If a long time elapses before the data are taken, the second term on the right-hand side of Eq. (8.13) should vanish. Then,

$$(K_c)_{O_2} = \frac{\nu}{\delta^2} (K_c)_{O_2}' \tag{8.14}$$

Expanding $\tan \delta$ in Eq. (8.9),

$$
\begin{aligned}
\frac{\nu}{\delta^2} &= 1 + \frac{1}{3} \delta^2 + \frac{2}{15} \delta^4 + \frac{17}{315} \delta^6 + \cdots \\
&= 1 + \frac{1}{3} \nu \left(1 + \frac{1}{15} \delta^2 + \frac{2}{315} \delta^4 + \cdots \right) \\
&= 1 + \frac{1}{3} \nu \left\{ 1 + \frac{1}{15} \nu \left(1 - \frac{5}{21} \delta^2 + \cdots \right) \right\} \\
&\doteqdot 1 + \frac{1}{3} \nu + \frac{1}{45} \nu^2 - \frac{1}{189} \nu^3 \\
&= 1 + 0.3333 \nu + 0.022 \nu^2 - 0.0053 \nu^3
\end{aligned}
\tag{8.15}
$$

It is apparent from Eqs. (8.14) and (8.15) that the apparent value of $(K_c)_{O_2}'$ should be corrected for $(K_c)_{O_2}$ with respect to V_c/V_g even if the experiment is conducted such that the second term on the right-hand side of Eq. (8.13) can be disregarded.

The upper diagram of Fig. 8.6 shows several types of closure. In this example the value of $\nu (=V_c/V_g)$ was less than several percent, so the correction in Eqs. (8.14) and (8.15) is not required; but when test tubes are considered, this correction cannot always be disregarded.

Using the closures shown in Fig. 8.6, Hara determined $(K_c)_{O_2}$ values by replacing carbon dioxide in empty shaken flasks with oxygen penetrating via diffusion from ambient air into the flask. These flasks during observation at 29° to 30°C were fixed on a table which rotated at 210 rpm with an eccentricity of 3.5 cm. These data are taken on the abscissa in the lower diagram of Fig. 8.6.[2]

8.3.2. Evaporation

Evaporation of water through the closure during the period of shaking, which continues for several days in some instances, cannot be overlooked. Suppose that distilled water only is charged into the shaken flask (see Fig. 8.5) to an extent of about 10% of V_g in the figure, and then subjected to shaking. Partial pressure, p_g, of water vapor inside the flask is assumed to be always saturated at the temperature specified, while the partial pressure, p_0, of water vapor in ambient air is also assumed constant.

Introducing a transfer coefficient, $(K_c)_{H_2O}$, for water vapor through the closure, the rate of evaporation, N_{H_2O}, moles per unit time, can be formulated in a steady state as follows:

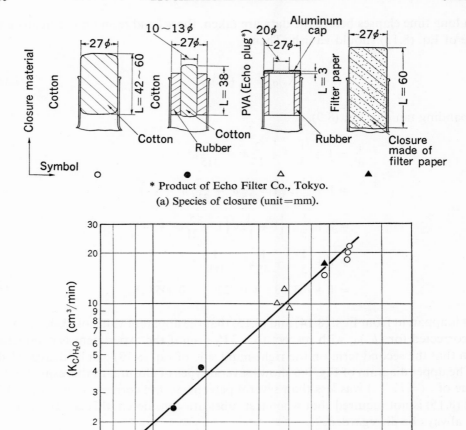

(a) Species of closure (unit=mm).

(b) Transfer coefficients with respect to water vapor and oxygen.

Fig. 8.6.   Oxygen transfer and water evaporation[2] (shaken flask).

$$N_{H_2O} = \frac{(K_c)_{H_2O}}{RT}(p_g - p_0) \qquad (8.16)$$

Hara also measured $(K_c)_{H_2O}$ values by measuring the rate of decrease in mass of water in shaken flasks in parallel with the previous experiment of assessing the values of $(K_c)_{O_2}$.[2] The result is shown as the ordinate in the lower diagram of Fig. 8.6; though a solid line in the figure has no particular significance, a fairly close correlation between $(K_c)_{H_2O}$ and $(K_c)_{O_2}$ is interesting.

To illustrate explicitly the water loss per day, the use of shaking flasks at 30°C in a room with a relative humidity of 29% at the same temperature and a cotton closures whose $(K_c)_{H_2O}$ values extend from 18 to 21 cm³/min, corresponds to a water loss of 0.6 to 0.7 g/day.

8.3.3. Surface aeration

Here again, partial pressures of oxygen in ambient air, in the free space in shaken flasks, and in the liquid (volume $= V_l$) in the flasks are designated as p_0, p_g, and p_g', respectively. These flasks, provided with appropriate closures, are assumed to be shaken. Due to microbial uptake of oxygen, a steady state in terms of p_0, p_g, and p_g' is also assumed. Further, the assumption is made that the unidirectional transfer (diffusion) of oxygen from ambient air into the broth is hardly affected by the counter-current diffusions of carbon dioxide and water vapor.[5]

The molar rate, N_{O_2}, of oxygen transfer in the steady state can be expressed by the following equation.

$$N_{O_2} = \frac{p_0 - p_g'}{\dfrac{RT}{(K_c)_{O_2}} + \dfrac{1}{H \cdot k_L a \cdot V_l}} \tag{8.17}$$

$$= \frac{p_0 - p_g'}{R_c + R_{O_2}} \tag{8.17'}$$

where

R_c = resistance to oxygen transfer due to the closure

$\quad = \dfrac{RT}{(K_c)_{O_2}}$

R_{O_2} = resistance to oxygen transfer in surface area

$\quad = \dfrac{1}{H \cdot k_L a \cdot V_l}$

H = Henry's constant

$k_L a$ = volumetric coefficient of oxygen transfer, commensurate with surface area

In the denominator of Eq. (8.17) or Eq. (8.17)' the resistance to oxygen transfer from the broth into microbes has to be added in general; however, as has been stated in Section 6.1.2., Chapter 6, oxygen uptake by microbes must be interpreted not in the light of physical transfer, but in terms of enzymatic reaction. Consequently, due to the difficulty of expressing the enzymatic reaction resistance in a fashion similar to those in Eq. (8.17), the last term, of this category, is omitted in this analysis.

Solomons presented some data on oxygen transfer with shaken flasks which are operated under various shaking conditions. Using data of oxygen-transfer rate, mmole O_2/l hr, taken by the sulfite oxidation method without the use of any closure at the top of each flask, $k_L a$ can be estimated, provided Henry's constant ($=1.17 \times 10^{-3}$ mole O_2/l atm for water at 30°C) or $C^*=7.5 \times 10^{-3}$ g/l at 30°C in atmospheric air is assumed. His data indicate that $k_L a$ values, thus converted, depend not only on the frequency and amplitude of shaking, but also on the volume of liquid (more exactly, volumetric ratio of liquid to gas in the flasks); values of $k_L a$ are estimated to be of the order of 17 to 450 hr^{-1} and $k_L a \cdot V_l$ values extend from (28 to 850) $\times 10^{-3}$ l/min.[7]

In connection with surface area in shaken flasks, it may be worthwhile to remind oneself of the effect of free surface area in the aeration of a deep fermentor. Clearly, the contribution of oxygen transfer from the free surface to the value of $k_L a$ should not be underrated when the liquid height decreases. This is indicated by the fact that the empirical correlation between $k_L a$ and operating variables incorporates a factor relevant to liquid height.[1] It has been suggested that the contribution of surface aeration to the $k_L a$ value in the conventional aerator is negligible when the liquid height exceeds about 3 to 5 m.[1]

Referring to Fig. 8.6, values of $(K_c)_{o_2} \approx 10$ cm^3/min are taken for the following comparison.

From Eq. (8.17)′

$$\left.\begin{aligned}
R_c &= \frac{RT}{(K_c)_{o_2}} \\
&= \frac{(82.05) \times (303)}{10} = 2.49 \times 10^3 \text{ atm min/mole} \\
R_{O_2} &= \frac{1}{H \cdot k_L a \cdot V_l} \\
&= \frac{1}{(1.17 \times 10^{-3})\{(28 \sim 850) \times 10^{-3}\}} \\
&= (1 \sim 31) \times 10^3 \text{ atm min/mole}
\end{aligned}\right\} \qquad (8.18)$$

It is noted from the above comparison that the resistance of the closure to oxygen transfer may become, in some cases, a limiting factor in transferring oxygen from ambient air into the broth in shaken flasks.

8.3.4. Dissolved oxygen

The discussion in this section is not necessarily for surface aeration, but applies also to a deep fermentor vessel with sparged aeration. However, in view of the fact that simultaneous measurement of dissolved oxygen concentration in the broth of all shaken flasks is not practical, the analysis here to permit the estimation of oxygen-transfer coefficient from data on concentrations of cell mass and limiting substrate (taken here as dissolved oxygen) is not considered to be wide of the mark.

Equations required for this discussion are as follows (*cf.* Sections 4.4.1. and 6.1.4., Chapters 4 and 6):

$$\frac{dX}{dt} = \mu_{max} \frac{\bar{C}}{K_s + \bar{C}} X \tag{8.19}$$

$$\frac{d\bar{C}}{dt} = k_L a (C^* - \bar{C}) - Q_{O_2}'' \frac{\bar{C}}{K_m + \bar{C}} X \tag{8.20}$$

where

$\bar{C}$ = dissolved oxygen concentration in bulk liquid
X = cell mass concentration
K_s = saturation constant
μ_{max} = maximum value of specific growth rate
$k_L a$ = volumetric coefficient of oxygen transfer in aeration
C^* = dissolved oxygen concentration in equilibrium with oxygen partial pressure in bulk air
t = time
Q_{O_2}'' = maximum value of specific respiration rate
K_m = Michaelis constant

For convenience of calculation the following dimensionless terms are introduced.

$x = X/X_0, \qquad y = \bar{C}/C^*$
$A = k_L a/\mu_{max}, \qquad B = Q_{O_2}'' X_0/C^* \mu_{max}$
$\alpha = K_s/C^*, \qquad \beta = K_m/C^*$
$\theta = t \cdot \mu_{max}$

provided:

$X_0 = X \qquad$ at $\qquad t = 0$

Equations (8.19) and (8.20) are then rearranged below.

$$\frac{dx}{d\theta} = \frac{xy}{\alpha + y} \tag{8.21}$$

$$\frac{dy}{d\theta} = A(1 - y) - B \frac{xy}{\beta + y} \tag{8.22}$$

If $y \gg \alpha$, and $y \gg \beta$, Eq. (8.22) reduces to:

$$\frac{dy}{d\theta} = A(1 - y) - Be^{\theta} \tag{8.23}$$

Then

$$y = e^{-\int A d\theta} \left\{ \int (A - Be^{\theta}) e^{\int A d\theta} \, d\theta + \text{const.} \right\}$$

From the initial condition

$$y = 0 \quad \text{and} \quad \theta = 0$$

the solution to Eq. (8.23) is:

$$y = 1 - \frac{B}{A+1} e^{\theta} + \left(\frac{B}{A+1} - 1\right) e^{-A\theta} \tag{8.24}$$

Assuming that

$$A \gg 1, \quad y = 1 - \frac{B}{A} e^{\theta} \tag{8.25}$$

For the region $dy/d\theta \doteq 0$ in Eq. (8.22), the equation can be rearranged as below.

$$xy = \frac{A}{B}(1-y)(\beta+y)$$

$$= \gamma(1-y)(\beta+y) \tag{8.26}$$

provided:

$$\gamma = \frac{A}{B}$$

From Eq. (8.26),

$$\frac{dx}{d\theta} = \gamma\left(-\frac{\beta}{y^2} - 1\right)\frac{dy}{d\theta} \tag{8.27}$$

From Eqs. (8.21), (8.26), and (8.27),

$$-\left(\frac{y^2 + \beta}{y^2}\right)\frac{dy}{d\theta} = \frac{(1-y)(\beta+y)}{\alpha+y} \tag{8.28}$$

Setting

$$\frac{y^2+\beta}{y^2}\frac{\alpha+y}{(1-y)(\beta+y)} = \frac{\gamma_1}{y} + \frac{\gamma_2}{y^2} + \frac{\gamma_3}{1-y} + \frac{\gamma_4}{\beta+y}$$

Coefficients of γ_1 to γ_4 are determined as:

$$\gamma_1 = (1+\alpha) - \frac{\alpha}{\beta}, \quad \gamma_2 = \alpha, \quad \gamma_3 = 1+\alpha, \quad \text{and} \quad \gamma_4 = \frac{\alpha}{\beta} - 1$$

Consequently, Eq. (8.28) is rearranged as shown below.

$$\left[-\left\{\frac{\alpha}{\beta} - (1+\alpha)\right\}\frac{1}{y} + \frac{\alpha}{y^2} + \frac{1+\alpha}{1-y} + \frac{(\alpha/\beta)-1}{\beta+y}\right]dy = -d\theta \tag{8.29}$$

Integrating the above equation and rearranging,

$$\left\{\frac{\alpha}{\beta}-(1+\alpha)\right\}\ln\frac{1}{y}-\frac{\alpha}{y}+(1+\alpha)\ln\frac{1}{1-y}-\left(\frac{\alpha}{\beta}+1\right)\ln\frac{1}{\beta+y}=\text{const.}-\theta \quad (8.30)$$

Equations (8.25) and (8.30) are analytical solutions to Eqs. (8.21) and (8.22) for large and/or small values of y. For intermediate ranges of y, Eqs. (8.21) and (8.22) are solved numerically by the Runge-Kutta-Gill method (HITAC 5020, Computer Center, University of Tokyo); an example of the solutions is shown in Fig. 8.7, provided:

$$\alpha = K_s/C^* = 0.027,$$
$$\beta = K_m/C^* = 0.0053$$
$$B = \frac{Q_{O_2}''X_0}{\mu_{max}C^*} = 40; \quad \text{parameters are } A\left(=\frac{k_La}{\mu_{max}}\right)$$

Since K_m and K_s values for many microorganisms range as follows:

$$K_m = 10^{-5} \text{ to } 10^{-7} \text{ mole } O_2/l \quad \text{and}$$
$$K_s = 10^{-5} \text{ to } 10^{-6} \text{ mole } O_2/l \quad (cf. \text{ Tables 4.1 and 4.3}),$$

the median values ($K_m = 10^{-6}$ mole/l = 0.032 ppm, $K_s = 5 \times 10^{-6}$ mole/l = 0.160 ppm), and $C^* = 6$ ppm are taken in Fig. 8.7; the above values of α ($= K_s/C^* = 0.160/6 = 0.027$), and β ($= K_m/C^* = 0.032/6 = 0.0053$) originate from these assumptions. Whatever values α and β may take with respect to a specific microbe, Eq. (8.30) permits a quick estimation of y vs. θ when $dy/d\theta \doteq 0$.

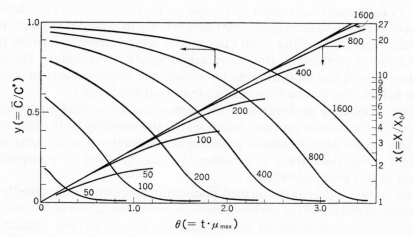

Fig. 8.7. Microbial growth pattern dependent on k_La and dissolved oxygen concentration; $\alpha = K_s/C^* = 0.027$; $\beta = K_m/C^* = 0.0053$; $B = (Q_{O_2}''/X_0)/(\mu_{max}C^*) = 40$; parameters: $A = k_La/\mu_{max}$.

Another relationship between y and θ for $A \gg 1$ and $dy/d\theta \neq 0$ (see Eq. (8.25)) is versatile, because the solid curves given for this region in Fig. 8.7 remain unchanged so long as the ratio of B to A taken in the figure remains unaltered; e.g., the curve for $A = 800$ in Fig. 8.7 ($B/A = 40/800$) can be used *in situ* for that of $A = 200$, if the value of B is assumed to be 1/4 of the value illustrated in the figure, keeping the ratio, $B/A = 1/20$, constant.

It is clearly noted from the figure that microbial growth deviates from the logarithmic phase at specific values of θ corresponding to the point where dissolved oxygen concentration begins to limit the growth. As a probable application in the transfer of laboratory culture data, Fig. 8.7 may present a clue to estimate the values of $k_L a$ required in a cultivation, provided the critical value of dissolved oxygen, below which the microbial activity deteriorates, is given, and if specific aeration is demanded, guaranteeing that the dissolved oxygen concentration will be above the critical value throughout the whole culture period.

8.4. SUMMARY

1. In addition to the problem of scaling up a new fermentation process, there exists the problem with a well established fermentation of translating improvements in the laboratory-scale process (improvements of strain and medium) to the plant scale.

2. Once a particular strain has been selected, the equipment used to determine optimum production conditions at the laboratory scale should duplicate, as nearly as possible, those achievable in plant equipment.

3. Translation of fermentation processes is usually done in three steps: i.e., laboratory (or bench) scale, pilot plant, and plant. For the most part, laboratory equipment involves shaken flasks, 500 to 1,000 ml in nominal volume. In pilot plant, large vessels exceeding 3,000 l in capacity are preferred to jar-type fermentors, and this is particularly true in an actinomycete fermentation. These larger pilot plant vessels are indispensable to confirm the improvements of strain and medium before they can be translated to industrial production.

4. Plant scheduling problems and the capabilities of existing equipment may constrain the translation process. Additionally, the handling of the inoculum for production vessels presents special problems not encountered on the laboratory or pilot-plant scale, since there is considerable time involved in transferring the inoculum at this production scale.

5. One or two highly instrumented pilot-plant fermentors undoubtedly give a more complete picture of the environmental history of the fermentation, but proper evaluations of the resistance of shaken-flask closures to oxygen transfer, water evaporation, and the effects of surface aeration and dissolved oxygen concentration in these flasks are, nonetheless, significant in the data translation.

NOMENCLATURE

$A = k_La/\mu_{max}$; sectional area of shaken flask closure, cm^2

$B = Q_{O_2}''X_0/(\mu_{max}C^*)$

$\bar{C}$ = concentration of dissolved oxygen in liquid, mg/l

C^* = concentration of dissolved oxygen in liquid in equilibrium with partial pressure of oxygen in air (bubbles) in bulk, mg/l

D = molecular diffusivity of oxygen in shaken flask closure, cm^2/sec, cm^2/min

f = function of δ

H = Henry's constant, mole/l atm

$(K_c)_{O_2}$ = oxygen-transfer coefficient of shaken flask closure (true value), cm^3/min

$(K_c)_{O_2}'$ = oxygen-transfer coefficient of shaken flask closure, assuming steady state in partial pressure distribution of oxygen in the closure (apparent value), cm^3/min

$(K_c)_{H_2O}$ = transfer coefficient of water vapor through shaken flask closure, cm^3/min

K_m = Michaelis constant, mg/l

K_s = saturation constant, mg/l

k_La = volumetric oxygen-transfer coefficient, min^{-1}, hr^{-1}

L = thickness of shaken flask closure, mm, cm

N_{O_2} = oxygen transfer rate through closure, mole/min

N_{H_2O} = transfer rate of water vapor through closure, mole/min

P_g = power consumed for agitation in gassed system, Kg m/sec, HP

p, p_c = oxygen partial pressure inside closure of shaken flask, atm

p_g = oxygen partial pressure or water vapor pressure in free space in the shaken flask, atm

p_{g1} = initial value of oxygen partial pressure in free space in shaken flask, atm

p_0 = oxygen partial pressure or water vapor pressure in ambient air, atm

$\tilde{p} = \dfrac{p_0 - p}{p_0 - p_{g1}}$

$\tilde{p}_g = \dfrac{p_0 - p_g}{p_0 - p_{g1}}$

Q_{O_2} = specific rate of respiration, mmole O_2/mg hr, μl O_2/mg hr

Q_{O_2}'' = maximum value of Q_{O_2}

R = gas constant, 82.05 ml atm/mole °K

$R_c = RT/(K_c)_{O_2}$, resistance of closure to oxygen transfer, atm min/mole

$R_{O_2} = \dfrac{1}{H \cdot k_La \cdot V_l}$, resistance of oxygen transfer in surface aeration, atm min/mole

T = absolute temperature, °K

t = time, min, hr

$\bar{t} = \dfrac{Dt}{L^2}$

V = liquid volume, m^3, cm^3

$V_c = AL$, closure volume, cm^3

$V_g = $ gas volume in shaken flask, cm^3

$V_l = $ liquid volume in shaken flask, cm^3

$X = $ cell mass concentration in liquid, mg/ml

$X_0 = X$ at $t=0$

$x = X/X_0$; coordinate

$y = \bar{C}/C^*$

Greek letters

$\alpha = K_s/C^*$

$\beta = K_m/C^*$

$\gamma = A/B$

$\gamma_1, \gamma_2, \gamma_3, \gamma_4, = $ coefficients

$\delta_\tau = $ positive roots ($\tau = 1, 2, 3, \ldots$), taken in order of magnitude, of transcendental equation

$\theta = t \cdot \mu_{max}$

$\mu = $ specific growth rate of cells, hr^{-1}

$\mu_{max} = $ maximum value of μ, hr^{-1}

$\nu = AL/V_g = V_c/V_g$

$\tau = $ subscript for δ

References

1. Eckenfelder, W. W., Jr. (1959). "Absorption of oxygen from air bubbles in water." *J. Sanitary Eng. Division, Proc. A.S.C.E.* **85**, *No. SA 4*.

2. Hara, M. (1972). Private communication.

3. Jensen, A. L., Schultz, J. S., and Shu, P. (1966). "Scale-up of antibiotic fermentations by control of oxygen utilization." *Biotech. & Bioeng.* **8**, 525.

4. Kanzaki, T., Sumino, Y., and Fukuda, H. (1968). "Oxygen transfer in L-glutamic acid fermentation by oleic acid-requiring organism. (I). Flask-to-tank relations." *J. Ferm. Tech.* (*Japan*) **46**, 1031.

5. Schultz, J. S. (1964). "Cotton closure as an aeration barrier in shaken flask fermentations." *Appl. Microbiol.* **12**, 305.

6. Shu, P. (1956). "Control of oxygen uptake in deep tank fermentations." *Ind. Eng. Chem.* **48**, 2204.

7. Solomons, G. L. (1969). *Materials and methods in fermentation* p.2. Academic Press, London and New York.

Chapter 9

Media Sterilization

Microorganisms can be removed from fluids by mechanical methods, for example, by filtration, centrifugation, flotation, or electrostatically. They may also be destroyed by heat, chemical agents, or electromagnetic waves. Although cells may be disrupted and killed by mechanical abrasion on a small scale, this method is not satisfactory industrially. Similarly, X-rays, β-rays, ultra-violet light, and sonic irradiations, while useful in the laboratory, are not applicable to the sterilization of large volumes of fluids. Gamma rays on the other hand may prove useful, particularly in the food industry.

Antibacterial agents have an important place in the fermentation industry, particularly for the production of a pure-water supply but have little application for the sterilization of fermentation media. Therefore, a discussion of antibacterial chemicals is beyond the scope of this book. This chapter will be confined to a discussion of the application of moist heat to fermentation media. Despite the fact that heat sterilization of media is the most common method, little attention has been paid until recently to the engineering aspects of heat sterilization.[8]

Interest in continuous methods of sterilizing media is increasing, but for the successful operation of a continuous sterilizer, foaming of the medium must be carefully controlled and the viscosity of the media must be relatively low.[12] The advantages of continuous sterilization of media are as follows:

a. Increase of productivity since the short period of exposure to heat minimizes damage to media constituents,
b. Better control of quality,
c. Leveling of the demand for process steam, and
d. Suitability for automatic control.

At present, most media in the fermentation industry are sterilized by batch methods. Over-exposure of the medium to heat is inherent in batch sterilization processes. Procedures which minimize damage to the medium will be outlined later, but before discussing the design and operation of equipment for sterilizing media, the concept of thermal death of microorganisms will be introduced. This is important since the rational design of sterilizers must be based on knowledge of the kinetics of the death of microorganisms.

In this connection, the deterministic model which has been widely applied in the study of microbial death kinetics will be compared with the probabilistic model. The sterilization of media is necessarily concerned with the progressive reduction of microbes throughout the sterilization cycle, and hence the probabilistic approach is appropriate for the design and operation of equipment.

Although it is known that metallic ions, amino acids, and the pH of the medium all affect the resistance of microorganisms to heat, detailed discussions of these topics are also beyond the scope of this book (see reference 5).

9.1. THERMAL DEATH OF MICROORGANISMS

9.1.1. Theory

The destruction of microorganisms by heat implies loss of viability, not destruction in the physical sense. The destruction of organisms by heat at a specific temperature follows a monomolecular rate of reaction as shown in Eq. (9.1).

$$\frac{dN}{dt} = - kN \tag{9.1}$$

where

k = reaction rate constant, min^{-1} (function of temperature; see Section 9.1.2.)
N = number of viable organisms
t = time

Microbiologists sometimes prefer the term decimal reduction time, D, meaning the time of exposure to heat during which the original number of viable microbes is reduced by one-tenth.[6] Integrating Eq. (9.1) under the condition of $N=N_0$ at $t = 0$,

$$N = N_0 e^{-kt} \tag{9.2}$$

From the above definition of D,

$$\frac{N}{N_0} = \frac{1}{10} = e^{-kD}, \qquad D = 2.303/k \tag{9.3}$$

It is well known that spores are much more resistant to heat than are vegetative cells; in other words, k values for vegetative cells are much greater than those for spores. Although some workers consider that the dipicolinic acid present in spores may be responsible for their increased resistance to heat,[5] further study is necessary before definite conclusions can be drawn as to the mechanism of their increased resistance to heat.

Figures 9.1 and 9.2 show typical death rate data for bacterial spores and vege-

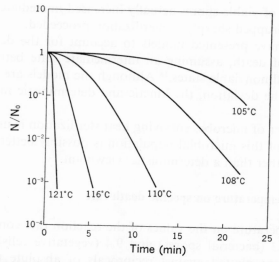

Fig. 9.1. Typical death rate data for spores of *Bacillus stearothermophilus* Fs 7954 in distilled water, where N = number of viable spores at any time, N_0 = original number of viable spores.

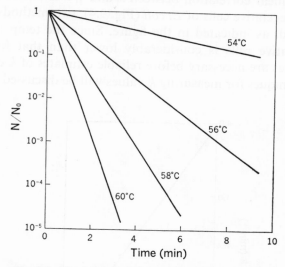

Fig. 9.2. Typical death rate data for *E. coli* in buffer, where N=number of viable cells at any time and N_0= original number of viable cells.

tative cells respectively; the parameters are different temperatures. For simplicity, data points are excluded from these figures. It is apparent from Figs. 9.1 and 9.2 that the resistance of bacterial spores to heat is much more marked than that of vegetative cells. It is clear from Fig. 9.1 that the logarithmic rate of death, as stated in Eq. (9.1), does not always hold for bacterial spores, particularly during the short period immediately following exposure to heat. There are cases reported where the number of viable spores actually increased immediately after exposure to heat and then dropped sharply as sterilization proceeded.

Some workers have presented models to account for the deviation from the logarithmic rate of death, assuming an intermediate state between the original viable and the final nonviable states.[13] Although the models are partially successful in explaining the deviation, the kinetic and deterministic models will not be elaborated here.

Since the number of microbes surviving heat sterilization is usually less than 1, the rate of death of this microbial population is possibly better considered from a probabilistic, rather than a deterministic, viewpoint.

9.1.2. Effect of temperature on specific death rate

The effect of temperature on the values of the reaction rate constant, k, is exemplified in Figs. 9.3 (bacterial spores) and 9.4 (vegetative cells). In both figures the values of k are plotted against reciprocals of absolute temperature, $1/T$. Although the data points are scattered considerably in the case of vegetative cells (Fig. 9.4), a linear correlation between k and $1/T$ is shown in these figures.

Regarding the vegetative cells of E. coli (Fig. 9.4), two methods of determining k values were used, as indicated in the figure. Since the temperature range for inactivating vegetative cells is considerably lower than that for spores, some elaborate techniques are necessary before reliable estimates of k can be obtained. Experimental techniques for measuring k values will be discussed in the following section.

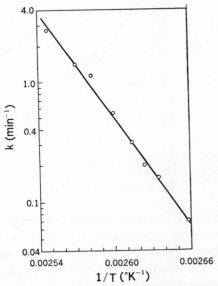

Fig. 9.3. Correlation of isothermal death rate data for *Bacillus stearothermophilus* Fs 7954, where $k =$ reaction rate constant and $T =$ absolute temperature. Value of E (activation energy) = 68.7 kcal/g mole.

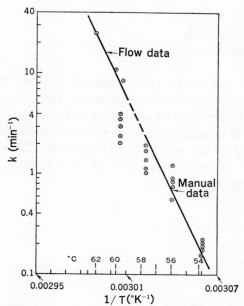

Fig. 9.4. Correlation of isothermal death rate data with temperature for *E. coli*, where $k=$reaction rate constant and $T=$absolute temperature. Data points at lower temperatures were measured with apparatus like that described in Fig. 9.6, while at higher temperatures apparatus like that described in Fig. 9.7 was used. Value of E(activation energy)$=127$ kcal/g mole.

The effect of temperature on the specific reaction rate, k, may be expressed by the Arrhenius equation as follows:

$$k = \alpha'e^{-E/RT} \qquad (9.4)$$

where

$\alpha' =$ empirical constant
$T =$ absolute temperature
$E =$ activation energy
$R =$ gas constant

On the basis of Eyring's theory of absolute reaction rate,

$$k = gT \exp(-\Delta H^*/RT) \exp(\Delta S^*/R) \qquad \text{[see Eq. (4.28), Chapter 4]} \qquad (9.5)$$

where

$g =$ factor including Boltzmann constant and Planck's constant
$\Delta H^* =$ heat of reaction of activation
$\Delta S^* =$ entropy change of activation

In 1921, Bigelow published the Q_{10} theory[6] which held that

$$D = \alpha'' \exp(-\beta''T') \qquad (9.6)$$

where

  α'', β'' = empirical constants
  T' = temperature

Thus, the value of the reaction rate constant, k, may be calculated by Eqs. (9.4) or (9.5), the value of the decimal reduction time, D, may be calculated by Eq. (9.6), and the relation of D to k is given in Eq. (9.3). To determine whether these different bases of calculation give significantly different results, an organism that has D equal to 7.18 min at 115°C and equal to 2.27 min at 126.5°C will be used for comparison of the calculations; ΔS^* in Eq. (9.5) is assumed to be independent of temperature.

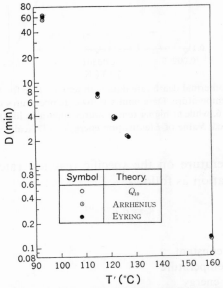

Fig. 9.5. Comparison of the Arrhenius, Eyring, and Q_{10} theories for the calculation of the decimal reduction time, D. Here D is plotted against temperature, T'.

In Fig. 9.5, the calculated values of D are plotted against temperature, T'; this shows the values of D calculated from k using the Arrhenius equation [Eq. (9.4)] or the Eyring theory [Eq. (9.5)] are not significantly different from each other. Equation (9.6), based on the Q_{10} theory, however, cannot be used to calculate D over a wide range of temperatures.

The value of E (activation energy) in Eq. (9.4) is considered to range from 50 to about 100 kcal/g mole with respect to vegetative cells and spores (cf. Figs. 9.3 and 9.4). These figures are considerably higher than those for enzymes or vitamins, which are in a range from 2 to 20 kcal/g mole.

9.1.3. Experimental determination of microbial death rate

The important points in determining microbial death rates are that:

a. The cell or spore suspension must be free of aggregates.
b. There must be no lag in heating or cooling the samples under examination.
c. The composition and pH values of the suspending medium must be such as to have the minimum inhibitory effect on the test organism in orde· to isolate the effect of temperature on the death rate of the organism.

It has been observed by many workers that clumps of spores or cells have increased resistance to heat compared with single cells (see Seection 9.2.2.). Accordingly, each suspension should be filtered to remove clumps and the uniformity checked microscopically.

The problem of lag in the time to heat and cool samples may be solved with the apparatus shown in Figs. 9.6 and 9.7.

In Fig. 9.6, the test suspension is sealed in a capillary and a number of capillaries hung over a cage on a "flipper" arm. The "flipper" allows the capillaries to be dipped into either a hot or a cold bath for a given time; the small volume in the capillary minimizes the time lag for the temperature to increase or decrease in the sample. This kind of apparatus can be operated manually and is suitable for experiments at lower temperatures (*cf.* Fig. 9.4). The number of viable organisms before and after exposure to elevated temperatures is usually determined biologically with a plate count.

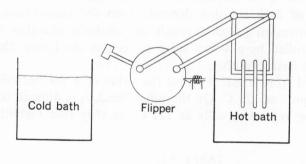

Cold bath   Flipper   Hot bath

Thermal death tube

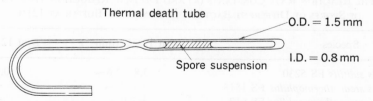

O.D. = 1.5 mm

Spore suspension   I.D. = 0.8 mm

Fig. 9.6. Manually operated apparatus for the determination of thermal death rates at relatively low temperatures.

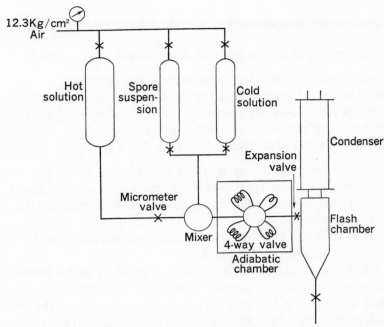

Fig. 9.7. Apparatus, operated by air pressure with a flash-cooling chamber, used for the determination of thermal death rates of organisms.

A more elaborate method to avoid the lag in change of temperature is exemplified in Fig. 9.7. The spore suspension is mixed with either a hot or a cold solution, the volume of each solution depending on the temperature desired. After mixing, the suspension passes through an adiabatic chamber where its retention time is controlled by a 4-way valve as seen in the figure; the sample is cooled instantly in a flash chamber.

It may be concluded from Table 9.1 that the value of k for bacterial spores is of the order of 1 min^{-1} at 121°C. On the other hand, it is difficult to give accurate values of k for vegetative cells at 121°C, as they lose viability almost

TABLE 9.1

THE REACTION RATE CONSTANTS (k) AND DECIMAL REDUCTION TIMES (D) OF SPORES OF DIFFERENT BACTERIA SUSPENDED IN BUFFER AT 121°C.

Species	k at 121°C (min^{-1})	D at 121°C (min)
Bacillus subtilis FS 5230	3.8 – 2.6~	0.6 – 0.9~
Bacillus stearothermophilus FS 1518	~0.77~	~3.0~
Bacillus stearothermophilus FS 617	~2.9~	~0.8~
Clostridium sporogenes PA 3679	~1.8~	~1.3~

instantaneously at this temperature; from Fig. 9.4, the value of k at 121°C for *E. coli* is of the order of 10^{13} min^{-1}.

Other values of k at 121°C for vegetative cells vary from 10 to about 10^{10} depending on the particular organism.

9.2. DETERMINISTIC VS. PROBABILISTIC APPROACH IN THE DESIGN OF STERILIZING EQUIPMENT

9.2.1. Life-span distribution of bacterial spores[1]

The life-span of spores is defined as the length of time for which spores exposed to a given temperature, remain viable. Each bacterial spore is assumed to have a specific life-span and a spore suspension to have some distribution of these.

If the ratio, N/N_0, of the number of viable spores, N, to the initial number of viable spores, N_0, is plotted against time (*cf.* Figs. 9.1 and 9.2) at a specific temperature, the death rate thus observable is an overall, or deterministic value, because the values of N which permit reliable determinations are of the order of 10 to 100.

Since the sterilization of media deals, in fact, with $N \leq 1$, the concept of the life-span of individual spores is significant for the probabilistic approach to the subject.

Suppose a spore population, N_0, is composed of spores, N_i in number and t_i in life-span at a specific temperature. A term, P_r, which is a probability associated with the distribution of life-span can then be introduced by the following equation.

$$P_r = \frac{N_0!}{\prod\limits_{i} N_i!} \cdot \frac{1}{\kappa}$$

(9.7)[10]

where

κ = normalization factor

It is apparent that

$$N_0 = \sum_{i=1}^{\infty} N_i$$

(9.8)

$$\bar{t} = \frac{1}{N_0} \sum_{i=1}^{\infty} N_i t_i$$

(9.9)

where

$\bar{t}$ = mean value of life-span

The maximum value of P_r, i.e., the most common life-span distribution, can be calculated from Eqs. (9.7), (9.8), and (9.9) by the Lagrange method,[15] provided that the value of N_0 is large. The result of this calculation shows:

$$N_i = \alpha' \exp(-\beta't_i) \tag{9.10}$$

where

α' and β' = empirical constants

If the values of N_i and t_i are continuous rather than discrete, Eq. (9.10) can be modified as follows:

$$-\frac{dN}{dt} = \alpha' \exp(-\beta't) \tag{9.11}$$

Equation (9.11) shows that an exponential distribution of life-span of individual spores is the most probable pattern.

From Eqs. (9.8) and (9.9), it follows that

$$N_0 = \int_0^\infty -\frac{dN}{dt}\, dt \tag{9.12}$$

and

$$\bar{t} = \frac{1}{N_0}\int_0^\infty -\frac{dN}{dt}\, t\, dt \tag{9.13}$$

Substituting Eq. (9.11) into Eqs. (9.12) and (9.13),

$$\alpha' = N_0/\bar{t} \tag{9.14}$$

$$\beta' = 1/\bar{t} \tag{9.15}$$

From Eqs. (9.11), (9.14), and (9.15),

$$\frac{-dN/dt}{N_0} = \frac{1}{\bar{t}}\exp\left(-\frac{1}{\bar{t}}t\right) \tag{9.16}$$

Integrating Eq. (9.16),

$$N = N_0 \exp\left(-\frac{1}{\bar{t}}t\right) \tag{9.17}$$

From Eqs. (9.2), and (9.17), it is evident that the reaction rate constant, k, min^{-1}, in the thermal death of spores can be interpreted as the reciprocal of the mean life-span of single spores, i.e.,

$$k = 1/\bar{t} \qquad (9.18)$$

9.2.2. Average life-span of bacterial spore-clump[3]

It can be seen from Eqs. (9.16) and (9.18) that the fraction, m, of spores whose life-span is less than t at a given temperature is shown by the following equation.

$$m = \int_0^t ke^{-kt}dt = 1 - e^{-kt} \qquad (9.19)$$

If one arbitrarily selects any single spore out of the population, the probability that the specific spore selected has a life of less than t may be expressed by:

$$p = 1 - e^{-kt} = 1 - \exp\left(-\frac{1}{\bar{t}}t\right) \qquad (9.20)$$

The probability, $P(t)$, that the life-span of single spores, n in total number is all less than t at the given temperature is:

$$P(t) = (1 - e^{-kt})^n \qquad (9.21)$$

Denoting $f(t)_n dt$ as the life-span distribution between t and $t+dt$ in a specific clump composed of n single spores,

$$P(t) = \int_0^t f(t)_n dt \qquad (9.22)$$

Then,

$$f(t)_n = \frac{dP(t)}{dt} \qquad (9.23)$$

Substituting Eq. (9.21) into Eq. (9.23),

$$f(t)_n = n(1 - e^{-kt})^{n-1}ke^{-kt} \qquad (9.24)$$

The mean value, $\bar{t}_n$, of life-span can be derived from Eq. (9.24) as follows:

$$\bar{t}_n = \int_0^\infty f(t)_n t dt$$

$$= \frac{1}{k}\sum_{r'=1}^n \frac{1}{r'} \qquad (9.25)$$

From Eq. (9.25),

$$\frac{\bar{t}_n}{\bar{t}_1} = \sum_{r'=1}^{n} \frac{1}{r'} \tag{9.26}$$

For large values of n $(n \geq 10)$, the right-hand side of the above equation can be approximated as follows:

$$\frac{\bar{t}_n}{\bar{t}_1} = \gamma + 2.303 \log n$$

$$= 0.577 + 2.303 \log n \tag{9.27}$$

where

γ = Euler's constant

An example of the calculation of life-span distribution and mean value of $\bar{t}_n$ using Eqs. (9.24) and (9.25) with a value of $k = 1$ min^{-1} is shown is Fig. 9.8.

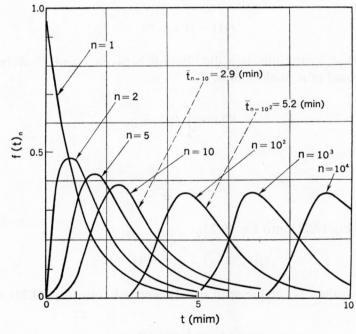

Fig. 9.8. $f(t)_n$ vs. t; $k = 1$ min^{-1}.[3]

It is interesting to observe from the figure that the mean value, $\bar{t}_n$, in life-span of the bacterial clump increases appreciably following the addition of single spores, n, to the clump.

9.2.3. Survival vs. sterilization charts[1]

Similar to Section 9.2.2., the probability that all the N_0 spores will be inactivated during the period of time, t, at a given temperature is:

$$P = (1 - e^{-kt})^{N_0} \tag{9.28}$$

Consequently, the probability that one spore, at least, survives during the heat exposure time, t:

$$1 - P = 1 - (1 - e^{-kt})^{N_0}$$

For $N_0 \gg 1$,

$$1 - P = 1 - e^{-N} \tag{9.29}$$

where

$$N = N_0 e^{-kt}$$

Figure 9.9 shows the relationship between $1 - P$ and kt, calculated by Eq. (9.29); parameters are the values of N_0. With increase of kt, it is clear from Eq. (9.29) that the probability of unsuccessful sterilization, $1 - P$, coincides with the number of viable spores, N, as expressed by the first-order reaction (cf. Eq. (9.2)). On the other hand, the correlation, $1 - P$ vs. kt, deviates from Eq. (9.2) (see broken lines in Fig. 9.9), as the term $1 - P$ approaches unity.

Generally, the death rate of bacterial spores at a given temperature is measured by plotting either N or N/N_0 against time (cf. Figs. 9.1 and 9.2). This expression of the data points may be compared with the survival chart by extrapolating (not shown) each broken line beyond $1 - P =$ unity in the figure; the survival chart exhibits a clear disagreement with the plot of $1 - P$ vs. kt. The latter plot is termed the sterilization chart.

Previous workers[9] used an extrapolation of the survival chart beyond $N \leq 1$ to estimate the time required for a specific sterilization, but the procedure should have included the probability concept as elaborated in this section.

Figure 9.9 is similar to the chart prepared by Rahn who calculated the theoretical curves regarding the thermal death of multicellular material rather than the single spores.[14]

For example, suppose the probability of unsuccessful sterilization, $1 - P$, be taken as 10^{-3} for $N_0 = 10^4$ at a given temperature; from Fig. 9.9, the value of kt is determined as 16; the time required for the sterilization can be estimated, provided that the value of k at the temperature is known.

If the value of N_0 per unit volume of the suspension is translated into the total number of viable spores, the value of kt for one contamination every 1,000

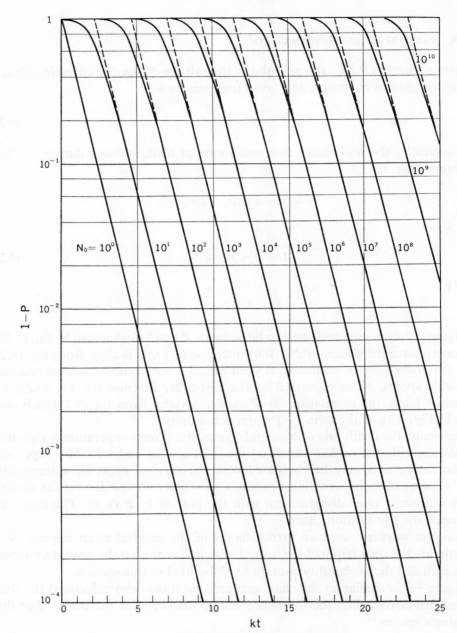

Fig. 9.9. Sterilization chart.[1]

sterilizations $(1-P=10^{-3})$ of 1,000 ml of medium with $N_0=10^4$/ml, will increase from $kt \doteq 16$ to $kt \doteq 23$ (see Fig. 9.9).

This implies that the sterilization time must be extended by about 44%, assuming the temperature of operation and the species of microbes are identical (k = constant). In other words, the latter way of evaluating N_0 (total number rather than concentration) is a more severe criterion for assessing the degree of sterilization.

The tdp (thermal death point) and tdt (thermal death time) tests have been used in the food industry as alternative techniques to the tdr (thermal death rate) used for testing sterility in the fermentation industry (*cf.* Section 9.1.3.). The tdp test determines the thermal death temperature (point) of a spore suspension which is exposed to heat for a definite period of time; the commonly used tdt test measures the length of time required for inactivating a spore suspension at a given temperature. Experimental data on either the tdp or the tdt tests may become more useful with reference to the sterilization curves as summarized in Fig. 9.9.[17]

Readers who are interested in either the experimental check of Fig. 9.9 or an application of the figure in assessing the sterilization "aftereffect," are recommended to consult the original papers.[1,4,16]

9.3. BATCH STERILIZATION OF MEDIA

9.3.1. Temperature-time profile and design calculation

Figure 9.10 shows different types of equipment for the batch sterilization of media, while Table 9.2 summarizes the design equations with which the temperature-time profile can be calculated for each type of equipment. Heat losses are neglected in these equations.

From Eqs. (9.1) and (9.4),

$$\frac{dN}{dt} = -kN = -\alpha' e^{-E/RT} N$$

Integrating,

$$\nabla_{\text{total}} = \ln \frac{N_0}{N} = \int_0^t k \, dt = \alpha' \int_0^t e^{-E/RT} dt \qquad (9.34)$$

where

∇_{total} = design criterion

It is apparent from Table 9.2 that the values of absolute temperatures, T, are functions of the time of exposure, t. In the design of batch sterilizers, the time, t, applying Eq. (9.34) must be determined. It is clear that some of the contaminating microbes will be inactivated during the time required to heat (t_1) and cool (t_3) the bulk of the medium.

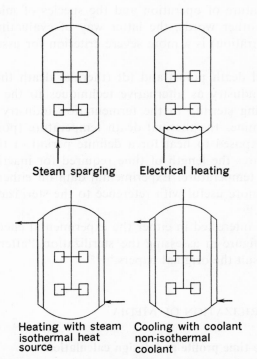

Steam sparging   Electrical heating

Heating with steam Cooling with coolant
isothermal heat   non-isothermal
source   coolant

Fig. 9.10. Types of equipment for batch sterilization of media.[9]

TABLE 9.2

TEMPERATURE-TIME PROFILE IN BATCH STERILIZATION.[9]

Type of heat transfer	Temperature-time profile	α, β, or γ
Steam sparging	$T = T_0\left(1 + \dfrac{\alpha t}{1 + \gamma t}\right)$ (Hyperbolic) (9.30)	$\alpha = \dfrac{hs}{Mc_p T_0}$ $\gamma = \dfrac{s}{M}$
Electrical heating	$T = T_0(1 + \alpha t)$ (Linear) (9.31)	$\alpha = \dfrac{q}{Mc_p T_0}$
Heating with steam	$T = T_H(1 + \beta e^{-\alpha t})$ (Exponential) (9.32)	$\alpha = \dfrac{UA}{Mc_p}$ $\beta = \dfrac{T_0 - T_H}{T_H}$
Cooling with coolant	$T = T_{co}(1 + \beta e^{-\alpha t})$ (Exponential) (9.33)	$\alpha = \left(\dfrac{wc_p{}'}{Mc_p}\right)\left\{1 - \exp\left(-\dfrac{UA}{wc_p{}'}\right)\right\}$ $\beta = \dfrac{T_0 - T_{co}}{T_{co}}$

Consequently, Eq. (9.34) can be subdivided as follows:

$$\nabla_{\text{total}} = \ln \frac{N_0}{N} = \nabla_{\text{heating}} + \nabla_{\text{holding}} + \nabla_{\text{cooling}} \tag{9.35}$$

$$\nabla_{\text{heating}} = \ln \frac{N_0}{N_1} = \int_0^{t_1} k \, dt = \alpha' \int_0^{t_1} e^{-E/RT} dt \tag{9.36}$$

$$\nabla_{\text{holding}} = \ln \frac{N_1}{N_2} = \int_0^{t_2} k \, dt = \alpha' \int_0^{t_2} e^{-E/RT} dt$$

$$= \alpha' e^{-E/RT} t_2 \tag{9.37}$$

$$\nabla_{\text{cooling}} = \ln \frac{N_2}{N} = \int_0^{t_3} k \, dt = \alpha' \int_0^{t_3} e^{-E/RT} dt \tag{9.38}$$

$$t = t_1 + t_2 + t_3 \tag{9.39}$$

where

N = sterility level (number of contaminating microbes after sterilization)
N_0 = contamination level (number of contaminating microbes before sterilization)
N_1 = number of contaminating microbes after heating period, t_1
N_2 = number of contaminating microbes after holding period, t_2

Supposing that the temperature-time profile is given as shown in Table 9.2, it is possible to calculate each term of ∇ for heating and cooling, but the analytical solution seems to be tedious. Accordingly, the graphical rather than the analytical solution to this subject will be discussed.

9.3.2. Example of calculation

A medium in a fermentor is sterilized batchwise at 120°C. The temperature-time profile observed with a recorder attached to the fermentor is as follows:

t (min)	0	10	30	36	43	50	55	58	63	70	102	120	140
T' (°C)		30	50	90	100	110	120	120	110	100	90	60	44	30

Assuming that the specific denaturation rate of contaminating bacterial spores,

$$k = 7.94 \times 10^{38} \exp\left(-\frac{68.7 \times 10^3}{RT}\right) \text{min}^{-1} \quad (cf. \text{ Fig. 9.3})$$

and that the initial number of the spores, $N_0 = 6 \times 10^{12}$, calculate the sterility level after the sterilization.

Solution

Figure 9.11 shows the temperature-time profile and the values of k as a function of t. A graphical integration yields:

$$\int_{t=34}^{t=64} k\,dt = 33.8$$

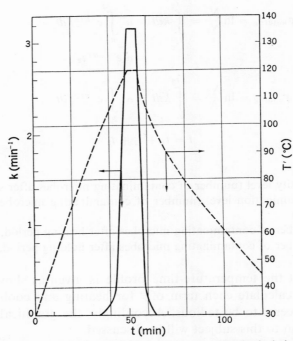

Fig. 9.11. k min^{-1} and T' °C vs. t min; example of calculation.

Regarding the periods before $t=34$ min and after $t=64$ min, the values of k are small enough to be neglected.

$$1 - P = N_0 e^{-kt}$$
$$= 6 \times 10^{12} \times e^{-33.8}$$
$$= 1.5 \times 10^{-2} \quad (cf.\ \text{Fig. 9.9})$$

9.4. Continuous Sterilization of Media

9.4.1. Equipment and temperature-time profile

Figure 9.12 shows equipment for continuous sterilization of fermentation media.

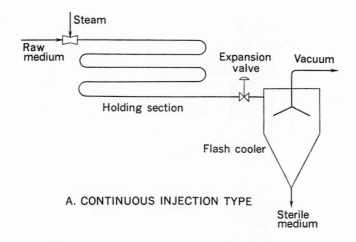

A. CONTINUOUS INJECTION TYPE

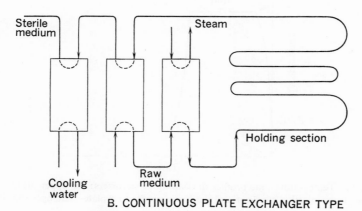

B. CONTINUOUS PLATE EXCHANGER TYPE

Fig. 9.12. Continuous sterilizers.

The upper part, A, of the figure shows an injection-type sterilizer; steam is injected directly into the raw medium; therefore the temperature rises almost instantly to the predetermined sterilizing temperature. The time for which the medium is held at this temperature is governed by the length of a pipe in the holding section. The sterilized medium is cooled instantly by passing through an expansion valve into a vacuum chamber, as shown in the figure.

An example of the temperature-time profile obtained with injection-type equipment is shown in the upper profile of Fig. 9.13. The numerical values given in the figure have no general significance; they will depend on the particular sterilization problem. Because of the short period of exposure to heat, it is possible to raise the temperature as high as 140°C without serious damage to the medium.

The lower part, B, of Fig. 9.12 shows a plate heat exchanger of the type fre-

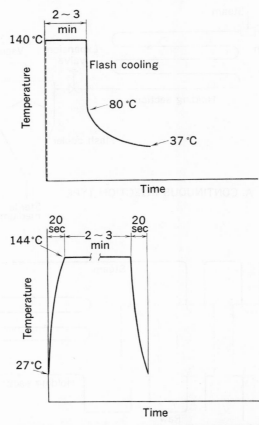

Fig. 9.13. Temperature-time profiles in continuous sterilizers; the upper profile applies to a steam injection sterilizer and the lower profile to a plate-exchanger sterilizer.

quently employed in the fermentation industry for the continuous sterilization of media.

Steam-heated plates raise the temperature of the raw medium, the medium is maintained at the elevated temperature for a certain period of time, and then cooled in another section of the plate exchanger, as shown in Fig. 9.12.

Although the time required to heat and cool the raw medium is much longer than with the steam injection type of continuous sterilization, the contribution of these periods to the sterilizing cycle will be much smaller (around 1 or 2%) than in the case with batch sterilization (*cf.* the lower profile in Figs. 9.13 and 9.11).

9.4.2. Residence time concept

To design and evaluate a continuous media sterilizer where the raw medium passes through round pipes, it is important to appreciate that not all portions of the medium spend the same length of time in the holding section of the sterilizer.

This is because, in both turbulent and viscous flow, the mean velocity, $\bar{u}$, of the fluid is a function of the radial distribution of fluid velocities occurring across the pipe. The mean velocity of a viscous-fluid flow through a pipe is one-half the maximum velocity found at the axis of the pipe (radial distribution of velocities is parabolic); in the turbulent condition, the mean velocity is 82% of the maximum value (radial distribution of velocities is governed by Karman-Prandtl's power law); these types of flow are illustrated in the two lower diagrams of Fig. 9.14.

With piston flow, the mean time in the holding section is exactly equal to the time of exposure to heat in all portions of the medium; this makes it easier to calculate the length of the holding section required to give a desired level of sterility. However, in practice it is difficult to realize ideal piston flow, although it would be most desirable since it avoids both over-cooking and under-cooking the medium.

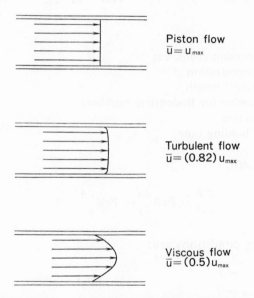

Fig. 9.14. Distribution of velocities in fluids exhibiting different types of flow inside round pipes; $\bar{u}=$mean velocity of the fluid.

It is difficult to predict the distribution of residence times in advance, so this must be determined empirically. This can be done by discharging a "marker" material and observing the pattern of its recovery at the outlet of the pipe; this pattern will, of course, be dependent on the type of flow in the pipe.

Before proceeding to the following discussion, it is important to make an underlying assumption clear. The assumption is that, as the length of pipe in the holding section becomes longer, the property of effective dispersion is more satisfactory than the distribution of velocities to describe the performance of sterilizers. This is particularly true for sterilizers where elbows or valves are present, to say nothing of the more complicated pathways existing in the plate-type heat exchanger.

The behavior of microbes suspended in a medium is considered to follow closely the mixing characteristics of the equipment with respect to effective dispersion, originating from the mixing of molecules.[2]

For simplicity, the residence time curves for a pipe will be considered. A differential equation is derived from the material balance as follows:

$$E_z \frac{\partial^2 n}{\partial x^2} - \bar{u} \frac{\partial n}{\partial x} = \frac{\partial n}{\partial t} \qquad (9.40)$$

Setting

$$\bar{n} = n/n_0, \qquad \chi = x/L, \qquad \text{PeB} = \bar{u}L/E_z, \qquad \phi = t/\bar{t}$$

where

$$E_z = \text{axial dispersion coefficient}$$
$$n_0 = \text{initial concentration}$$
$$L = \text{pipe (reactor) length}$$
$$\text{PeB} = \text{Péclet number (or Bodenstein number)}$$
$$x = \text{axial direction}$$
$$\bar{t} = \text{nominal holding time}$$

Rearranging Eq. (9.40),

$$\frac{\partial^2 \bar{n}}{\partial \chi^2} - \text{PeB} \frac{\partial \bar{n}}{\partial \chi} - \text{PeB} \frac{\partial \bar{n}}{\partial \phi} = 0 \qquad (9.41)$$

Initial and boundary conditions are:

$$\chi > 0, \qquad \bar{n}_{\phi=0} = 1$$
$$\chi < 0, \qquad \bar{n}_{\phi=0} = 0 \qquad \left.\right\} \text{ initial conditions}$$

$$\chi \to 0^+, \qquad \frac{\partial \bar{n}}{\partial \chi} - \text{PeB}\,\bar{n} = 0$$
$$\chi \to 1, \qquad \frac{\partial \bar{n}}{\partial \chi} = 0 \qquad \left.\right\} \text{ boundary conditions}$$

Solving for $\bar{n}$ [7,18]

$$\bar{n} = 32 \sum_{\tau=1}^{\infty} \frac{\text{PeB}\lambda_\tau (4\lambda_\tau^2 \cos 2\lambda_\tau \chi + \text{PeB} \sin 2\lambda_\tau \chi)}{(16\lambda_\tau^2 + 4\text{PeB} + \text{PeB}^2)(16\lambda_\tau^2 + \text{PeB}^2)} \exp\left(\frac{\text{PeB}}{2}\chi - \frac{\text{PeB}^2 + 16\lambda_\tau^2}{4\text{PeB}}\phi\right) \qquad (9.42)$$

provided: the λ_τ ($\tau=1, 2, 3, \ldots$) are the positive roots, taken in order of magnitude, of the transcendental equation

$$\tan 2\lambda = \frac{8\lambda\,\text{PeB}}{16\lambda^2 - \text{PeB}^2} \tag{9.43}$$

The exit concentration, $\bar{n}_{\chi=1}$, is given from Eq. (9.42) by setting $\chi=1$.

$$\bar{n}_{\chi=1} = R(\phi)$$

$$= 16\sum_{\tau=1}^{\infty}\frac{\lambda_\tau\sin 2\lambda_\tau}{(16\lambda_\tau^2 + 4\text{PeB} + \text{PeB}^2)}\exp\left(\frac{\text{PeB}}{2} - \frac{\text{PeB}^2 + 16\lambda_\tau^2}{4\text{PeB}}\phi\right) \tag{9.44}$$

where

$$E(\phi) = -\frac{dR(\phi)}{d\phi}$$

$$\int_0^\infty E(\phi)d\phi = 1.0$$

Numerical correlations between $R(\phi)$ and ϕ are plotted from Eq. (9.44) as shown in Fig. 9.15,[7] in which the values of PeB are taken as parameters.

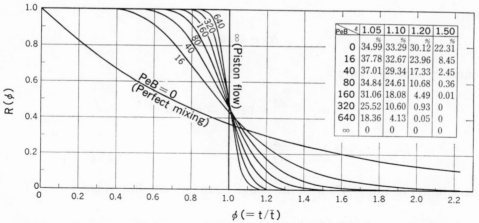

Fig. 9.15.   Effect of different types of flow (as shown by different PeB values and different ratios of the holding time, t, to the nominal holding time, $\bar{t}$), in continuous sterilization of media.

It is seen from the figure that the flow in the reactor approaches piston-type flow as the values of PeB are increased. Assuming that the medium is not overheated at all in the case of piston flow, the relative degree of overheating for any type of flow can be defined as

$$\int_{\xi}^{\infty} E(\phi)d\phi = R(\xi)$$ (9.45)

provided:

ξ = dimensionless residence time ($\xi > \phi = 1$)
$E(\phi)$ = residence time distribution function

Some values of $R(\xi)$ are listed in Fig. 9.15.[7]
Next, the inactivation of organisms by heat in continuous sterilizers will be considered. Since the rate of microbial death is expressed by Eq. (9.1), the material balance with a continuous sterilizer (straight pipe) at steady state is as follows:

$$E_z \frac{d^2n}{dx^2} - \bar{u}\frac{dn}{dx} - kn = 0$$ (9.46)

Changing the variables as shown previously in connection with Eq. (9.40) and rearranging Eq. (9.46),

$$\frac{d^2\bar{n}}{d\chi^2} - \text{PeB}\frac{d\bar{n}}{d\chi} - \text{PeB}N_r\bar{n} = 0$$ (9.47)

where

$$N_r = kL/\bar{u}$$

Boundary conditions are:

$$\chi \to 0^+, \qquad \frac{d\bar{n}}{d\chi} + \text{PeB}(1 - \bar{n}) = 0$$

$$\chi = 1, \qquad \frac{d\bar{n}}{d\chi} = 0$$

The solution of Eq. (9.47) with the above boundary conditions is:[19]

$$\bar{n}_{\chi=1} = (n/n_0)_{X=L}$$

$$= \frac{4\zeta\, e^{\text{PeB}/2}}{(1 + \zeta)^2\, e^{\text{PeB}/2} - (1 - \zeta)^2\, e^{\text{PeB}/2}}$$ (9.48)

provided:

$$\zeta = \sqrt{1 + \frac{4N_r}{\text{PeB}}}$$

A degree of sterility, N/N_0, is plotted against, $N_r(=kL/\bar{u})$ in Fig. 9.16 according to Eq. (9.48), parameters being PeB $(=\bar{u}L/E_z)$.

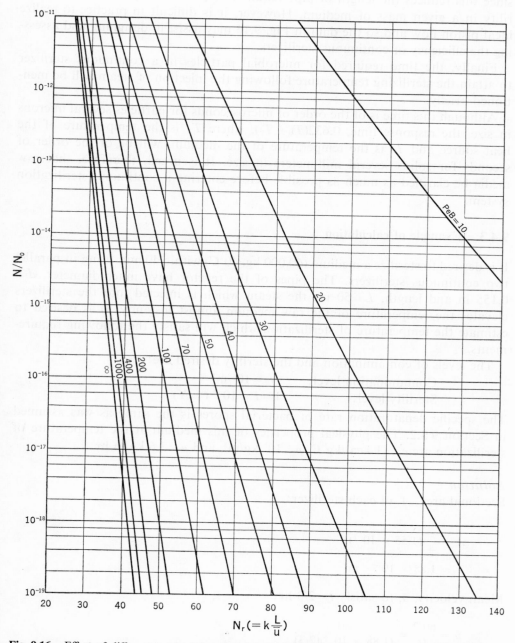

Fig. 9.16. Effect of different types of flow (as shown by different PeB values) on the destruction of organisms (N/N_0) at different rates of destruction (measured as $N_r=kL/\bar{u}$).

It is also evident from Fig. 9.16 that where $PeB=\infty$, in other words with ideal piston flow, conditions are most favorable for designing continuous sterilizers, since this reduces the length of pipe required to produce a given degree of sterility in a given mass of medium. However, it is difficult in practice to realize ideal piston flow and so the data in Fig. 9.16 may assist in designing and assessing the efficiency of continuous sterilizers.

Finally, the time required for microbial particles in a continuous sterilizer to attain the sterilizing temperature following the injection of steam will be mentioned briefly.

Although this time is of the order of microseconds for particles several microns in size, the response time, $0.632(T_H-T_0)$, where T_H is the temperature of the heat source and T_0 is the temperature of the medium, will be of the order of seconds for solids several millimeters in size. It is urged, therefore, that raw media be clarified as much as possible before entering a continuous sterilization system.

9.4.3. Example of calculation

It is desired to sterilize a medium (60,000 kg, 40°C) within 40 min using in parallel two continuous sterilizers. The pipes of the reactor have inner diameter, $d=$ 0.155 m and length, $L=50$ m; the steam which is injected into the sterilizers elevates the temperature of the raw medium almost instantly. It is desired to estimate the temperature of sterilization which will satisfy the following requirements.

The levels of contamination and the sterility desired are:

$\qquad$ Contamination level: $N_0=10^5/ml$
$\qquad$ Sterility level: $1-P=10^{-3}\ (=N)$

The specific denaturation rate of bacterial spores is the same as was assumed in Section 9.3.2. The physical properties of this medium at the temperature of sterilization are $c_p=1$ kcal/kg°C, $\rho=10^3$ kg/m³, and $\mu=3.6$ kg/m hr.

Solution
Sectional area, A, of each sterilizer:

$$A = \frac{\pi}{4}(1.55 \times 10^{-1})^2$$

$$= 1.88 \times 10^{-2}\ \ m^2$$

Mean velocity, $\bar{u}$, to satisfy $t=2/3$ hr:

$$\bar{u} = \frac{60/2}{At} = \frac{30}{(1.88 \times 10^{-2})(2/3)}$$

$$= 2.39 \times 10^3\ m/hr$$

$$N_{Re} = \frac{d\bar{u}\rho}{\mu} = \frac{(1.55 \times 10^{-1})(2.39 \times 10^3)(10^3)}{3.6}$$

$$\doteqdot 10^5$$

From the data which have been published by Levenspiel[11] regarding $E_z/\bar{u}d$ vs. N_{Re},

$$E_z/\bar{u}d \doteqdot 0.2 \qquad (N_{Re} = 10^5)$$

Then,

$$PeB = \frac{\bar{u}L}{E_z} = \frac{\bar{u}d}{E_z}\frac{L}{d}$$

$$= \frac{1}{0.2} \cdot \frac{50}{0.155} = 1{,}610$$

From Figs. 9.15 and 9.16, the condition of flow may be regarded approximately as piston-type flow.

The levels of contamination and the sterility assumed:

$$\frac{N}{N_0} = \frac{\dfrac{1}{60 \times 10^3 \times 10^3 \times 10^3}}{10^5} = 1.67 \times 10^{-16}$$

From Fig. 9.16,

$$N_r = kL/\bar{u}$$

$$= 36$$

$$k = N_r\bar{u}/L$$

$$= \frac{36 \times 2.39 \times 10^3}{50} = 1.72 \times 10^3 \quad hr^{-1}$$

$$= 28.7 \ min^{-1}$$

The relation between temperature, $T'°C$, and the value of the reaction rate constant, k, min^{-1}, is shown in Fig. 9.17; from this graph the temperature required to give the desired level of sterility can be shown as 129.5°C.

The nominal holding time, $\bar{t}$, of the raw medium within the sterilizer,

$$\bar{t} = L/\bar{u}$$

$$= \frac{50}{2.39 \times 10^3} = 2.09 \times 10^{-2} \ hr$$

$$= 1.25 \ min$$

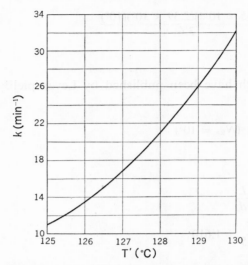

Fig. 9.17. Values of the reaction rate constant, k, at different sterilization temperatures, T', applying in a sample calculation.

If live steam (latent heat of vaporization=481.7 kcal/kg) at 9 Kg/cm² gage is injected into the sterilizer, the mass of steam, s, needed per hr:

$$481.7 \times s = \frac{60}{2} \times 10^3 \times (129.5 - 40)$$

$$s = \frac{30 \times 10^3 \times 89.5}{481.7} \doteq 5.6 \times 10^3 \text{ kg/hr}$$

The effect of the increase in the Reynolds number due to s on the value of $E_z/\bar{u}d$ is negligible.[11]

9.5. SUMMARY

1. The most probable distribution of the life-span of single spores is logarithmic at a given temperature; the key equation is:

$$-\frac{dN}{dt} = \alpha' e^{-\beta' t} \qquad \text{[Eq. (9.11)]}$$

2. The specific denaturation rate constant, k, min⁻¹, for spores is the reciprocal of the mean value of their life-span, i.e.,

$$k = 1/\bar{t} \qquad \text{[Eq. (9.18)]}$$

3. Once spores form clumps, they are more resistant to heat than single spores.

4. The relationship between the probabilistic and deterministic approaches, i.e., the distinction between the sterilization and survival rates have been presented. Though useful in designing pasteurizers, the survival rate concept has no basis, in fact, for designing sterilization processes where no spores survive. In this latter case, one must utilize the probabilistic approach (Fig. 9.9).

5. In the design of continuous systems for sterilizing fermentation media, one must employ the residence time distribution concept rather than a single residence time (Fig. 9.15).

NOMENCLATURE

A = heat transfer area, m²; sectional area of pipe, m²
c_p = specific heat of medium, kcal/kg °C
c_p' = specific heat of coolant, kcal/kg °C
D = decimal reduction time, min
d = inner diameter of pipe, m
E = activation energy, cal/g mole, kcal/g mole; residence time distribution function
E_z = axial dispersion coefficient, m²/sec
$f(t)_n$ = life-span distribution for n spores
g = constant, Eq. (9.5)
h = enthalpy of steam relative to raw medium, kcal/kg
ΔH^* = heat of reaction of activation
k = reaction rate constant; specific rate of denaturation, min⁻¹, sec⁻¹, hr⁻¹
L = length of reactor pipe, m
M = initial mass of medium in batch sterilization, kg
m = fraction of spores whose life-span is less than t
N = number of viable spores; number concentration of viable microbes
N_0 = initial number of viable spores (at $t = 0$); initial number concentration of viable microbes
N_r = $kL/\bar{u}$, dimensionless
N_{Re} = Reynolds number, ($= d\bar{u}\rho/\mu$)
n = concentration of dyestuff of electrolyte (simulation of microbial concentration); number of spores
n_0 = inlet concentration [Eq. (9.47)]; initial concentration [Eq. (9.40)]
$\bar{n}$ = n/n_0
PeB = Péclet number (or Bodenstein number), ($= \bar{u}L/E_z$)
P = probability of successful sterilization
P_r = probability of life-span distribution
p = probability of single spores whose life-span being less than t
q = rate of heat transfer, kcal/sec
R = gas constant, 1.98 cal/g mole °K
$R(\phi)$ = function of ϕ, residence time curve
r' = number, 1∼n
s = mass flow of steam, kg/sec, kg/min, kg/hr
ΔS^* = entropy change of activation

T = absolute temperature, °K

T_0 = initial temperature of medium, °K

T_H = temperature of heat source, °K

T_{co} = inlet temperature of coolant, °K

T' = temperature, °C

t = time; life-span

$\bar{t}$ = nominal holding time; mean value of life-span

U = overall heat transfer coefficient, kcal/m² hr °C

$\bar{u}$ = mean velocity of fluid through round pipe, m/hr, m/sec

u_{max} = maximum velocity of fluid through round pipe, m/hr, m/sec

w = mass flow rate of coolant, kg/sec, kg/min, kg/hr

x = axial direction

∇ = design criterion

Greek letters

$\alpha = \dfrac{hs}{Mc_pT_0}; \ \dfrac{q}{Mc_pT_0}; \ \dfrac{UA}{Mc}; \ \left(\dfrac{wc_p{}'}{Mc_p}\right)\left\{1-\exp\left(-\dfrac{UP}{wc_p{}'}\right)\right\}$

α' = empirical constant, Eqs. (9.4) and (9.10)

α'' = empirical constant, Eq. (9.6)

$\beta = \dfrac{T_0-T_H}{T_H}; \ \dfrac{T_0-T_{co}}{T_{co}}$

β' = empirical constant, Eq. (9.10)

β'' = empirical constant, Eq. (9.6)

$\gamma = s/M$; Euler's constant

$\zeta = \sqrt{\dfrac{1+4N_r}{PeB}}$

κ = normalization factor

μ = viscosity of medium, kg/m hr

ξ = dimensionless residence time, $(t/\bar{t})$, $(\xi > \phi = 1)$

ρ = density of medium, kg/m³

$\phi = t/\bar{t}$

$\chi = x/L$

References

1. Aiba, S., and Toda, K. (1965). "An analysis of bacterial spores thermal death rate." *J. Ferm. Tech. (Japan)* **43**, 527.
2. Aiba, S., and Sonoyama, T. (1965). "Residence-time distribution of a microbial suspension in a straight pipe." *J. Ferm. Tech. (Japan)* **43**, 534.
3. Aiba, S., and Toda, K. (1966). "Some analysis of thermal inactivation of a bacterial-spore clump." *J. Ferm. Tech. (Japan)* **44**, 301.
4. Aiba, S., and Toda, K. (1967). "Thermal death rate of bacterial spores." *Process Biochemistry* **2**, 35.
5. Amaha, M. (1953). "Heat resistance to Cameron's putrefactive anaerobe 3679 in phosphate buffer (*Clostridium sporogenes*)." *Food Research* **18**, 411.
6. Bigelow, W. E. (1921). "The logarithmic nature of thermal death time curves." *J. Infect. Diseases* **29**, 528.

7. Brenner, H. (1962). "The diffusion model of longitudinal mixing in beds of finite length. Numerical values." *Chem. Eng. Sci.* **17**, 229.

8. Burton, H. (1958). "An analysis of the performance of an ultra-high temperature milk sterilizing plant. Part 1. Introduction and physical measurements. Part 2. Calculation of the bactericidal effectiveness." *J. Dairy Research* **25**, 75; 324.

9. Deindoerfer, F. H., and Humphrey, A. E. (1959). "Analytical method for calculating heat sterilization times." *Appl. Microbiol.* **7**, 256.

10. Hirschfelder, J. O., Curtiss, C. F., and Bird, R. B. (1954). *Molecular Theory of Gases and Liquids* p. 97. John Wiley, New York, Chapman & Hall, London.

11. Levenspiel, O. (1958). "Longitudinal mixing of fluids flowing in circular pipes." *Ind. Eng. Chem.* **50**, 343.

12. Pfeifer, V. F., and Vojnovich, C. (1952). "Continuous sterilization of media in biochemical processes." *Ind. Eng. Chem.* **44**, 1940.

13. Prokop, A., and Humphrey, A. E. (1970). "Kinetics of disinfection." *Disinfection* (Ed.) Bernarde, M. A. p. 61. Marcel Deckler, New York.

14. Rahn, O. (1958). "Temperature and life." "Temperature, its measurement and control in science and industry." p. 409. Reinhold, New York.

15. Sokolnikoff, I. S., and Sokolnikoff, E. S. (1941). *Higher mathematics for engineers and physicists* p. 163. McGraw-Hill, New York.

16. Toda, K., and Aiba, S. (1967). "Studies on heat sterilization. Part 6. Effect of medium sterilization on yield of some fermentation products." *J. Ferm. Tech.* (*Japan*) **45**, 769.

17. Toda, K. (1968). "Studies on heat sterilization. Part 7. Thermal death time as assessed from the life-span distribution of microbes." *J. Ferm. Tech.* (*Japan*) **46**, 743.

18. Yagi, S., and Miyauchi, T. (1953). "On the residence time curves of continuous reactors." *Chem. Eng.* (*Japan*) **17**, 382.

19. Yagi, S., and Miyauchi, T. (1955). "Operational characteristics of continuous flow reactors in which the reactants are mixing." *Chem. Eng.* (*Japan*) **19**, 507.

Chapter 10

Air Sterilization

The problem of producing a large quantity of sterile air for aerobic fermentations is peculiar to biochemical engineering. On a laboratory scale, cotton plugs are satisfactory in test tubes or shaken flasks, and in pilot-plant fermentors, small fibrous filters present no serious problems. On an industrial scale, however, fibrous filters have certain disadvantages.

It should be stated at the outset that 100% efficiency of collection of airborne microorganisms cannot be expected in a filter bed packed with fibers or granular particles. The purpose of such air filters is to prolong the interval of time between the passage of one airborne organism and the next. The length of this interval of time required will be governed by the type of fermentation but should be sufficient to protect the fermentation at least for the first critical period of growth.

During the past two decades, valuable contributions have been made in both the theoretical and practical aspects of aerosol filtration. However, the removal of aerosols from industrial air differs in an important feature from the practice of sterilizing air for fermentations; the air to be sterilized is already clean by comparison with the aerosols commonly processed in other industries. In addition, an extremely high degree of particle removal is required for air sterilization (see Section 10.1.). Thus, the design and operation of air sterilization equipment requires a probabilistic rather than a deterministic approach.

This chapter will largely be concerned with the probabilistic approach to the design of fibrous filters for the fermentation industry, but in the final section, a plate-type filter made of polyvinyl alcohol (PVA) will be described; this filter is much smaller in size and has lower operating cost than the conventional fibrous filters.

10.1. SPECIES AND NUMBER OF AIRBORNE MICROBES

Table 10.1 lists species of bacteria and bacterial spores which are commonly detected in air;[5,20] in addition, yeast, fungi, and viruses are also present. The size of these organisms varies from several millimicrons to several hundred microns. Small organisms are, however, often adsorbed onto airborne dust and

TABLE 10.1

REPRESENTATIVE SPECIES OF AIRBORNE BACTERIA AND BACTERIAL SPORES.
(AIBA AND HUMPHREY)

Species	Width (μ)	Length (μ)
Aerobacter aerogenes	1.0 – 1.5	1.0 – 2.5
Bacillus cereus	1.3 – 2.0	8.1 – 25.8
Bacillus licheniformis	0.5 – 0.7	1.8 – 3.3
Bacillus megaterium	0.9 – 2.1	2.0 – 10.0
Bacillus mycoides	0.6 – 1.6	1.6 – 13.6
Bacillus subtilis	0.5 – 1.1	1.6 – 4.8
Micrococcus aureus	0.5 – 1.0	0.5 – 1.0
Proteus vulgaris	0.5 – 1.0	1.0 – 3.0
(Spores)		
Bacillus megaterium	0.6 – 1.2	0.9 – 1.7
Bacillus mycoides	0.8 – 1.2	0.8 – 1.8
Bacillus subtilis	0.5 – 1.0	0.9 – 1.8

can thus be easily removed from the air along with the larger dust particles
before attempting air sterilization proper. Therefore, the airborne particles
which have to be destroyed or collected during air sterilization are about the
size of small bacteria, namely, 0.5 to 1.0 μ (see Table 10.1). Incidentally, an
extensive study of the dust particles in both indoor and outdoor air shows that
the median size of dust particles is around 0.6 μ.[21]

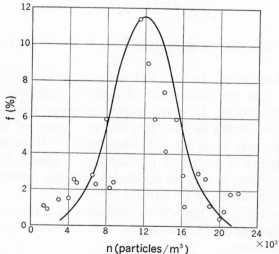

Fig. 10.1. Frequency of distribution of organisms in air in Tokyo.

The numbers of airborne microbes in indoor and outdoor air were determined at the Sanitary Engineering Station, Tokyo, in 471 samples of air taken over a period of one year.[28] In each experiment, 5 l of air were sucked into a sterile saline solution and shaken vigorously. The solution was then distributed over several Petri dishes containing nutrient agar. After incubation for 2 to 3 days at 30°C, the number of organisms per m³ of air, n, was calculated. This is shown in the abscissa of Fig. 10.1, where these data have been rearranged by Aiba to show the frequency of distribution of organisms per m³ of air.[7]

Now, suppose N is the total number of microbes which are found in a large volume of air. The probability, $P(n)$, with which a unit volume, m³, arbitrarily selected from the bulk air will contain n microbes can be expressed as follows:

$$P(n) = \binom{N}{n} p^n (1 - p)^{N-n} \tag{10.1}$$

where

p = probability with which each microbe out of N in the total will be found in the unit volume of air arbitrarily selected

Setting

$Np = \nu$ (expected value)

and assuming

$N \gg 1$, $p \ll 1$, and Np is of moderate magnitude,

Eq. (10.1) can be rearranged to give:

$$P(n) = \frac{\nu^n e^{-\nu}}{n!} \tag{10.2}$$

The curve in Fig. 10.1 was obtained from Eq. (10.2) with the value of $\nu = 12 \times 10^3$ particles/m³.

Although the data points in Fig. 10.1 are scattered considerably around the curve, it appears that the occurrence of airborne microbes is represented by Poisson's law (Eq. (10.2)). In other words, airborne microbes are randomly distributed. It is interesting to note that another report of the value of $\nu (=\bar{n})$ for air in London, was (3 to 9) $\times 10^3$ particles/m³.[18] Humphrey, on the other hand, suggests the use of $\bar{n} = 2 \times 10^3$ particles/m³, if reliable figures of $\bar{n}$ are not available.[20] At any rate, it may be concluded that the number of airborne microbes is of the order of 10^3 to 10^4 particles/m³.

The value of $\bar{n} = 10^3$ to 10^4 particles/m³ indicates that the microbial contamination of air is exceedingly light compared with the usual problem of cleaning dust-laden air in industry.

For example, the dust load of stack gas from an atomic reactor ranges from

0.4 to 1.4 mg/m³,[10] and if the radioactive particles are spherical with a diameter, $d_p = 0.2 - 0.7 \; \mu$ and density, $\rho = 1.0$ g/cm³, the value of $\bar{n}$ would be from 10^9 to 10^{10} particles/cm³. Industrially speaking, such figures of 0.4 to 1.4 mg/m³ fall in the category of very low loading, since a light loading is defined as from 3.5 to 25 mg/m³ according to Silverman.[30]

10.2. AIR STERILIZATION IN PRACTICE

10.2.1. Heat

Although bacterial spores are notoriously resistant to heat, they can be destroyed if temperatures are high enough. For instance, Decker *et al.* found that bacterial spores suspended in air could be killed at 218°C in 24 sec.[13]

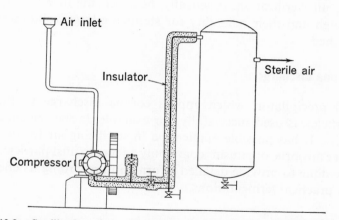

Fig. 10.2.   Sterilization of air with heat generated from the compression of air.[33]

Figure 10.2 is a simple process which Stark[33] designed for air sterilization. In this system, heat generated by air compression is used to kill airborne microbes; the outlet temperature of the air in a single-stage and adiabatic compressor operating at 3 Kg/cm² gage rises to 150°C and 220°C, starting from inlet temperatures of 20°C and 70°C, respectively. Stark carried out successful fermentations for the production of acetone and butanol, amylase and 2,3-butylene glycol using air sterilized by the heat generated during the compression of the air. (Inlet and outlet temperatures were 21°C and 187°C, respectively; outlet air pressure = 7 Kg/cm² gage; fermentor volume = 1 to 30 m³; air flow rate = 0.1 to 3 m³/min.)

The principle of utilizing heat generated by air compression is promising, especially for the production of large quantities of sterile air. However, before applying this idea to a practical fermentation, several other factors, including the location of the air compressor relative to the fermentor, the length of pipe connecting the compressor and the fermentor, and their sterility, must be carefully examined. Due to the complicated factors involved and to the difficulty of assessing them, it is usual to install, in addition to the air compressor, auxiliary equipment to filter the air before it enters the fermentor.

10.2.2.   Ultra-violet rays and other electromagnetic waves

Morgan *et al.* found ultra-violet rays of wavelength from 2,265 to 3,287 Å were the most effective for killing airborne microbes.[27] Sterile rooms in fermentation factories and operation rooms in hospitals are usually equipped with lamps emitting these rays, but it is questionable whether ultra-violet irradiation can be successfully applied to producing sterile air for fermentations.

Theoretically, sonic waves, high energy cathode rays, and gamma rays could be applied to air sterilization. Practically, however, the investment costs of such devices are high and their reliability for sterilizing large volumes of air has yet to be established.

10.2.3.   Corona discharge

The Cottrell precipitator, which applies corona discharge to the removal of airborne particles, is used successfully, for example, in the cement and sulphuric acid industries. It has possible application in sterilizing air for the fermentation industry, since airborne organisms commonly adhere to dust particles, but research has yet to be done to provide detailed information regarding its effectiveness and reliability in practical fermentations.

10.2.4.   Germicidal sprays

Carswell claimed a significant reduction in airborne bacteria by introducing a small amount of a germicide - for instance, phenol, ethylene oxide, or salts of heavy metals - into water which was sprayed and circulated within air-conditioning equipment.[12] Although this method may have application in cleaning air in rooms where aseptic operations are to be performed, it is likely that there will be considerable difficulty in removing all entrained vapors and mists from the treated air before it is used.

10.2.5.   Mechanical filtration

The fibrous air filters which are now widely used to produce sterile air for fermentations may be classed provisionally as mechanical filters; the mechanism of their operation is, however, too complicated to designate them simply as "mechanical."

Cotton fibers were used exclusively as the fibrous medium for air sterilization in the fermentation industry before World War II. With the production of antibiotics on a large scale, glass fibers have replaced cotton fibers; glass fibers give a lower drop in air pressure and are less liable to wetting or combustion. In Japan, filters packed with cylindrical pads woven of glass fibers 19μ in diameter are used industrially, while in the United States, the glass fibers are about 5μ in diameter.

However, in Japan, conventional air filters are being replaced by polyvinyl alcohol (PVA) plate-type filters. Though macroscopically PVA filters look rather differ from conventional filters, microscopically they have a fine, fibrous structure (see Fig. 10.15). In addition, the performance of PVA filters resembles that of a fibrous filter containing glass fibers (see Section 10.3.2. and Fig. 10.13). Accordingly, except for details of construction, much of the data relating to fibrous and glass fiber filters applies also to PVA filters.

Since airborne microbes are exposed to relatively high temperature during compression, the current practice of air sterilization is actually a combination of heat inactivation and mechanical filtration.

A few of the papers on the filtration of particles from air will now be mentioned. One of the earliest papers was by Kluyver et al., who measured the efficiency of collection of Bacillus cereus on cotton fibers.[23] Gaden and Humphrey determined the distribution of the spores of Bacillus subtilis in a filter consisting of a 3 mm layer of glass fibers and studied the effect of air velocity on the efficiency of collection.[18,19] Aiba et al. measured the distribution of Serratia marcescens within a 50 mm layer of glass fibers (see Section 10.3.1.).[1] In addition, Chen has published a review of the theoretical basis of aerosol filtration.[11] In connection with the removal of radioactive particles suspended in exhaust gas from atomic reactors, many theoretical as well as technological publications have appeared.[10]

10.3.   AIR STERILIZATION BY FIBROUS MEDIA

10.3.1.   Analysis of collection efficiency from a probabilistic approach[2]

Assumptions for this analysis are that:

a. A fibrous filter packed with glass fibers is composed of a number of grids, ξ, per unit length of the filter; the grid mesh depends primarily on the diameter of the glass fibers and their volume fraction in the packed bed.

b. A microbial particle has a probability, p, of collision with the glass fiber in passing through each grid; this probability is affected by the dynamics of the air stream through the packed bed, and by the ratio of particle to fiber diameters.

c. Microbial particles which experience up to m collisions with fibers in their passage through a filter of L bed thickness and ξL grids, can be re-entrained into the air stream and emerge from the filter.

Then, the number of bacterial particles retained within the filter will be given by the following equation.

$$\nu = \nu_0 \left\{ 1 - \sum_{m=0}^{m=m} \binom{\xi L}{m} p^m (1-p)^{\xi L - m} \right\}$$ (10.3)

where

ν_0 = total number of particles blown into the filter
ν = number of particles captured by the filter

The relationship between the probabilities of collision and of capture is not included in Eq. (10.3), nor is "collision" clearly differentiated from "re-entrainment." Even though Eq. (10.3) is poor in the strict sense of physical interpretation, it may be used to est'mate the longitudinal distribution of particles collected by the filter and to derive the interval between the emergence of successive particles from the filter.

In Eq. (10.3), when $m = 0$,

$$\nu = \nu_0 \{ 1 - (1-p)^{\xi L} \}$$

and

$$\ln \left(1 - \frac{\nu}{\nu_0} \right) = \xi L \ln (1-p)$$

$$= \xi L \left(-p - \frac{p^2}{2} - \frac{p^3}{3} - \cdots \right)$$ (10.4)

If, in the above equation, p is small, and only the first term on the right-hand side of Eq. (10.4) can be taken,

$$\nu \doteqdot \nu_0 (1 - e^{-\xi L p})$$

$$\frac{d\nu}{dL} \doteqdot \nu_0 \xi p e^{-\xi L p}$$ (10.5)

Equation (10.5) shows the "log penetration" law.

In Eq. (10.3), when $m = 1$,

$$\xi L - 1 \doteqdot \xi L \quad \text{and} \quad p \ll 1$$

$$\frac{d\nu}{dL} \doteqdot \nu_0 \xi^2 p^2 e^{-\xi L p} \left(\frac{L}{1!} \right)$$ (10.6)

In the case of fibers several microns in diameter, ξ is assumed to be of the order of 100 per cm of filter bed.

Likewise, when $m = 2, m = 3, \cdots$, assuming that $\xi^2 L^2 p \gg 1, \xi^3 L^3 p \gg 1, \cdots$ the following equations can be derived.

$$\frac{d\nu}{dL} \doteqdot \nu_0 \xi^3 p^3 e^{-\xi L p} \left(\frac{L^2}{2!}\right) \tag{10.7}$$

$$\frac{d\nu}{dL} \doteqdot \nu_0 \xi^4 p^4 e^{-\xi L p} \left(\frac{L^3}{3!}\right) \tag{10.8}$$

In general, when $m = m$

$$\frac{d\nu}{dL} = \nu_0' \xi p e^{-\xi L p} \frac{(\xi L p)^m}{m!}$$

$$= \nu_0' k e^{-kL} \frac{(kL)^m}{m!} \tag{10.9}$$

provided:

$$k = \xi p$$

$$\nu_0' \int_0^\infty k e^{-kL} \frac{(kL)^m}{m!} dL \equiv \nu_0$$

Designating the average number of particles blown into the filter per unit time, and the operation period as $\bar{\nu}_0'$ and t, respectively, Eq. (10.9) is rearranged as follows:

$$\nu_0 = \nu_0' \int_0^\infty k e^{-kL} \frac{(kL)^m}{m!} dL$$

$$= \nu_0' \kappa \int_0^\infty \frac{k}{\kappa} e^{-kL} \frac{(kL)^m}{m!} dL$$

$$= \bar{\nu}_0' t \int_0^\infty P(L) dL = \bar{\nu}_0' t \tag{10.10}$$

where

$$\kappa = \text{normalization factor}$$

$$\int_0^\infty P(L) dL = \int_0^\infty \frac{k}{\kappa} e^{-kL} \frac{(kL)^m}{m!} dL = 1$$

The number of particles retained in the filter, L in bed thickness, during the period of time t,

$$\nu = \bar{\nu}_0' t \int_0^L P(L) dL \tag{10.11}$$

The time interval, Δt, between one particle emerging from the exit of the filter and the next,

$$\Delta t = \frac{1}{\bar{v}_0'\left(1 - \int\limits_0^L P(L)dL\right)} \tag{10.12}$$

Since the longitudinal distribution of particles retained in the filter, $\int_0^L P(L)dL$, is independent of time (*cf.* Fig. 10.6), as are most operations for air sterilization in the fermentation industry, the value of Δt in Eq. (10.12) must be an average, $\overline{\Delta t}$, statistically speaking, of the time interval between the successive emergence of particles out of the filter. Actually, however, the value of Δt is accompanied by another sort of distribution in magnitude unless the statistical average is taken (*cf.* Figs. 10.8 and 10.9).

From Eq. (10.12), the overall collection efficiency, η, of the filter, in the deterministic sense, can be correlated with the probabilistic approach to the time interval of the particle's appearance from the filter exit by the following equation.

$$\eta = \frac{\bar{v}_0'\overline{\Delta t} - 1}{\bar{v}_0'\overline{\Delta t}} = \int\limits_0^L P(L)dL \tag{10.13}$$

It is, now, appropriate to discuss the implication of the overall collection efficiency, η; both theoretical and experimental analyses have been made from the deterministic approach.

10.3.2. Physical implication of single fiber collection efficiency

The mechanisms of collection of aerosol particles by fibrous media may be classified as follows:

a. Inertial impaction
b. Interception
c. Diffusion
d. Settling by gravitational force
e. Electrostatic force

With fibrous filters, mechanism *d* in the above list can be excluded, because the diameter of the particles to be collected is of the order of 1μ. Humphery measured the electrostatic charge of *Bacillus subtilis* spores in relation to mechanism *e* and found that about 70% of them had a positive charge, about 15% of them had a negative charge, while the remaining spores were neutral.[19] It would be expected that charged organisms would be collected more effectively than neutral ones, and this was indeed shown experimentally by Silverman,[31] and Lundgren et al.[24]

However, there is little quantitative data yet on the contribution of mechanism *e* to the overall collection efficiency of fibrous air sterilization filters. Therefore the discussions which follow will be confined to the remaining mechanisms *a*, *b*, and *c*.

For quantitative evaluation of these mechanisms it will be assumed that single cylindrical fibers are placed perpendicularly to the aerosol flow in an indefinite space, and that the air flow around the cylinders is laminar with no vortices (see Fig. 10.3). The second assumption is that all of the following analyses are 2-dimensional.

a. Inertial impaction

Assuming that the resistance to motion of a spherical particle (diameter=d_p and density=ρ_p) moving with a velocity, u, in a stream of air (velocity=v and viscosity=μ) is governed by Stokes' law, the equation of motion with regard to the particle will be shown by either of the following:

$$\left(\frac{\pi}{6}\right) d_p{}^3 \rho_p \frac{d\mathbf{u}}{dt} = -\frac{3\pi\mu d_p}{C}(\mathbf{u} - \mathbf{v}) \tag{10.14}$$

$$\frac{C\rho_p d_p{}^2}{18\mu}\frac{d\mathbf{u}}{dt} = -(\mathbf{u} - \mathbf{v}) \tag{10.15}$$

Using dimensionless terms with respect to the coordinates, velocity, and time, i.e.,

$$\tilde{x} = 2x/d_f, \quad \tilde{y} = 2y/d_f, \quad \tilde{v}_x = v_x/v_0, \quad \tilde{v}_y = v_y/v_0, \quad \tilde{t} = 2v_0 t/d_f,$$

the inertial parameter is introduced from Eq. (10.15).

$\phi = \dfrac{C\rho_p d_p{}^2 v_0}{18\mu d_f}$: inertial parameter

C = Cunningham's correction factor for slip flow

v_0 = upstream velocity of air

d_f = fiber diameter

It is noted from Fig. 10.3 that the flow pattern of the particles (see broken lines in the figure, calculable from Eq. (10.14) or Eq. (10.15)) deviate from those of air flow (also calculable theoretically), due to the inertia of the particles as they approach the cylindrical surface. In the figure, the width of the upstream air flow is denoted as *b*; particles that move in the streamlines of air beyond *b* will not touch the cylinder surface even after they deviate from the air stream-line near the cylinder. The theoretical value of the collection efficiency of single fibers due to the inertial effect of the particles, η_0', can be expressed by:

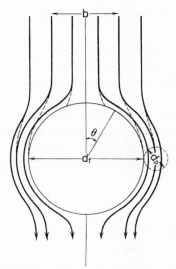

Fig. 10.3. Flow patterns around single cylindrical fibers. Solid lines indicate flow pattern of air and broken lines flow pattern of particles; d_f=diameter of fiber, d_p=diameter of particles, b=width of air stream.

$$\eta_0' = \frac{b}{d_f} \tag{10.16}$$

According to the calculation of Langmuir, the value of η_0' becomes zero when $\phi = 1/16$,[11] though other investigations have shown theoretically that other relationships exist between the values of η_0' and ϕ for cases in which various shapes other than the cylinder are involved.

Designating the critical air velocity as v_c, corresponding to the condition of $\phi = 1/16$,

$$v_c = (1.125) \frac{\mu d_f}{C \rho_p d_p^2} \tag{10.17}$$

Substituting the values of $\rho_p = 1.0$ g/cm³ and $\mu = 1.80 \times 10^{-4}$ g/cm sec (air at 20°C) into Eq. (10.17), Fig. 10.4 is obtained, the parameters being fiber diameter, d_f.[5] The value of C is determined from an empirical equation.[26]

Figure 10.4 is useful in determining quickly the approximate value of v_c, below which the inertial impaction of particles may be neglected. For a given particle the value of η_0' is expected to increase following increase of upstream velocity of air, v_0, provided an inertial mechanism of collection is predominant.

b. Interception
If particles had no mass, they would exactly follow the streamlines of air flow; and although bacteria have mass, it is so small that they do, in fact, closely fol-

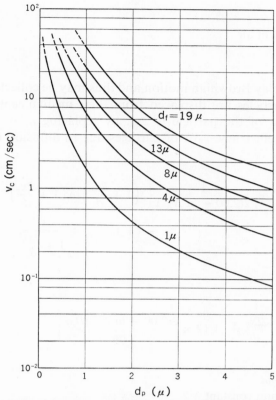

Fig. 10.4. Effect of fiber diameter, d_f, and particle diameter, d_p, on the critical velocity, v_c, for inertial impaction.

low the streamlines. When particles entrained in streamlines of air are collected by contact with the fibers (see Fig. 10.3) they are said to be intercepted. The streamline of air flow which is $d_p/2$ from the fiber surface at a location of $\theta = \pi/2$ (see Fig. 10.3) is a limiting condition for the deposition of entrained particles as they pass a cylindrical fiber. The collection efficiency, η_0'', of particles on single fibers due to interception is calculated using the limiting width of air streamlines, $d_p/2$ from the fiber surface mentioned above.

According to Langmuir, the value of η_0'' can be calculated with the following equation, in which $N_R = d_p/d_f$ and the upstream velocity of air, v_0, approximates the air velocity, v, around the cylinder:

$$\eta_0'' = \frac{1}{2(2.00 - \ln N_{Re})}\left\{2(1 + N_R)\ln(1 + N_R) - (1 + N_R) + \frac{1}{1 + N_R}\right\} \quad (10.18)^{11}$$

where

$$N_{\mathrm{Re}} = \frac{d_{\mathrm{f}} v \rho}{\mu}$$

$$\rho = \text{density of air}$$

c. Diffusion

Small particles display Brownian motion and thus may be collected on the surface of fibers as the particles are displaced from their median center of location. If the displacement of the particle is $2x_0$, by replacing d_{p} in Eq. (10.18) with $2x_0$, the collection efficiency, η_0''', of particles due to diffusion can be calculated from:[11]

$$\eta_0''' = \frac{1}{2(2.00 - \ln N_{\mathrm{Re}})} \left\{ 2\left(1 + \frac{2x_0}{d_{\mathrm{f}}}\right) \ln \left(1 + \frac{2x_0}{d_{\mathrm{f}}}\right) \right.$$
$$\left. - \left(1 + \frac{2x_0}{d_{\mathrm{f}}}\right) + \frac{1}{1 + \frac{2x_0}{d_{\mathrm{f}}}} \right\} \qquad (10.19)$$

Assuming $v_0 = v$,

$$\frac{2x_0}{d_{\mathrm{f}}} = \left\{ 1.12 \times \frac{2(2.00 - \ln N_{\mathrm{Re}}) D_{\mathrm{BM}}}{v d_{\mathrm{f}}} \right\}^{1/3} \qquad (10.20)[11]$$

where

$$D_{\mathrm{BM}} = C k T / 3 \pi \mu d_{\mathrm{p}}$$
$$k = \text{Boltzmann constant}$$
$$T = \text{absolute temperature}$$
$$D_{\mathrm{BM}} = \text{diffusivity of particles}$$

Collection efficiencies, η_0, with single fibers may be expressed by the following equation, assuming that the values of η_0', η_0'', and η_0''' are independent to each other:

$$\eta_0 = \eta_0' + \eta_0'' + \eta_0''' \qquad (10.21)$$

To simplify Eqs. (10.19) and (10.18), the following equation will be used; this holds nearly true over a range of values of N_{Re} from 10^{-4} to 10^{-1}:

$$\frac{1}{2.00 - \ln N_{\mathrm{Re}}} \propto N_{\mathrm{Re}}^{1/6} \qquad (10.22)$$

The terms, N_{R} and $2x_0/d_{\mathrm{f}}$, in the brackets of Eqs. (10.18) and (10.19) are then expanded, assuming that each term is small compared with unity and that the second and higher orders of each term can be neglected. Consequently, Eqs. (10.18) and (10.19) can be simplified as follows:

$$\eta_0'' \propto N_R{}^2 N_{Re}{}^{1/6} \tag{10.23}$$

$$\eta_0''' \propto N_{Sc}{}^{-2/3} N_{Re}{}^{-11/18} \tag{10.24}$$

where

$N_{Sc} = \mu/\rho D_{BM}$

Equations (10.23) and (10.24) are similar to the semi-theoretical analysis published by Friedlander.[17] It is noted from these equations that the geometrical ratio, N_R, and the diffusivity, D_{BM}, of particles are more significant for enhancing the values of η_0'' (due to interception) and η_0''' (due to diffusion), respectively.

The experience of practical sterilization of air by using a bed packed with glass fibers suggests sometimes that the collection efficiency, η_0', due to inertial impaction of particles can be disregarded in calculating the value of η_0 in Eq. (10.21). Thus, using Eqs. (10.23) and (10.24), Aiba attempted to rearrange the extensive data published on the efficiencies of collection of various sorts of particles with different fibrous materials.[5,11,19,25,29,32,34-36]

Aiba selected data on overall collection efficiencies, $\bar{\eta}$ [Eq. (10.25), where interception and diffusion probably predominate, see Eq. (10.27)]. From the value of $\bar{\eta}$, that of η_α (the collection efficiency of single fibers whose volume fraction in the fibrous bed is α) was calculated with Eq. (10.26).

$$\bar{\eta} = \frac{\nu}{\nu_0} \tag{10.25}$$

provided:

    ν_0 = number of organisms in the original aerosol
    ν = number of organisms in the aerosol retained by the filter

$$\eta_\alpha = \frac{\pi d_f (1 - \alpha)}{4L\alpha} \ln \frac{\nu_0}{\nu_0 - \nu}$$

$$= \frac{\pi d_f (1 - \alpha)}{4L\alpha} \ln \frac{1}{1 - \bar{\eta}} \tag{10.26}$$

where

    L = thickness of filter bed

Equation (10.26) indicates that the fraction of particles collected in any section within L is constant, the so-called "log penetration" relation. All of the data in these calculations were secured with relatively thin filter beds (less than 4 cm) and, under these conditions, Eq. (10.26) may be acceptable.

The following equation, which is an empirical relation presented by Chen,[11] was then used by Aiba to convert the values of η_α into η_0 starting from data on $\bar{\eta}$:

$$\eta_\alpha = \eta_0(1 + 4.5\alpha), \qquad 0 < \alpha < 0.10 \tag{10.27}$$

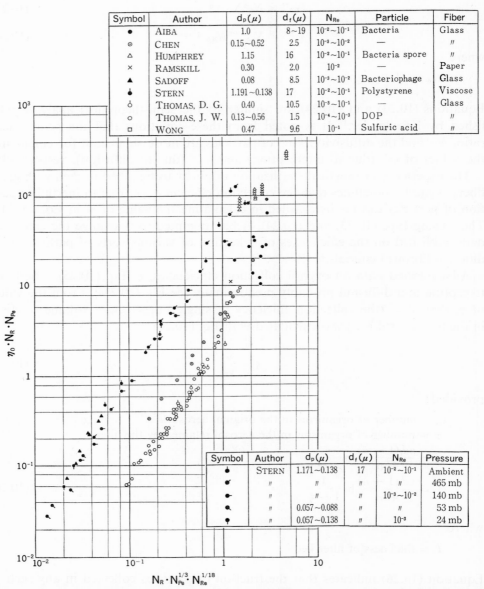

Fig. 10.5. Single fiber collection efficiencies. The ordinate and abscissa are $\eta_0 N_R N_{Pe}$ and $N_R N_{Pe}^{1/3}$ $N_{Re}^{1/18}$ respectively, where η_0 = single fiber collection efficiency, N_R = interception parameter, N_{Pe} = Péclet number, and N_{Re} = Reynolds number. For detailed information of these dimensionless terms, see the nomenclature at the end of this chapter.

Lastly, the values of $\eta_0 \cdot N_R \cdot N_{Pe}$ ($= \eta_0 \cdot N_R \cdot N_{Sc} \cdot N_{Re}$) and $N_R \cdot N_{Pe}^{1/3} \cdot N_{Re}^{1/18}$ were calculated and plotted as shown in Fig. 10.5. The data points in the figure are markedly scattered, but this may suggest that a curve with a slope of 3 can be drawn through points at higher values of the abscissa (>1) and with a slope of 1 at lower values of the abscissa ($\sim 10^{-1} \sim$), except for a series of data points presented by Stern et al.[32] If this is accepted, then the curve with a slope of 3 corresponds to situations where interception predominates and this is expressed by Eq. (10.23), while the other curve corresponds to situations where diffusion predominates and Eq. (10.24) applies.

In so far as the log penetration law holds true, Fig. 10.5 indicates a way of determining the thickness of the filter bed required under various operating conditions, i.e., various values of η, d_f, d_p, α, and v_0. The value of η_0 estimated from Fig. 10.5 for a specific set of operating conditions will lead to the determination of L by using Eqs. (10.26) and (10.27).

10.3.3. Experimental data of collection efficiency, deterministic vs. probabilistic

Figure 10.6 shows an example of the longitudinal distribution of bacterial cells retained within a fibrous bed; *Serratia marcescens* cells were labeled with [32]P

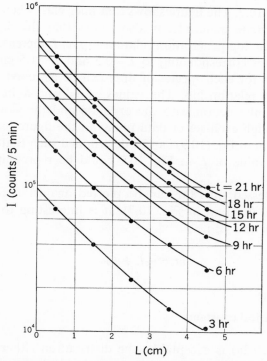

Fig. 10.6. Longitudinal distribution of bacterial cells captured in a fibrous bed; parameters, t, are operation periods.[3] Fiber diameter, $d_f = 8$ μ; volume fraction, $\alpha = 0.025$; superficial air velocity, $v_s = 5$ cm/sec; particle concentration of inlet air, $n \doteq 10^8$ particles/m³. Since radial distribution is averaged, the value of I could be assumed proportional to the total number of particles collected in each pad; also the same in the next figure.

and the filter bed, 10 cm in diameter and 5 cm in thickness, was subdivided into five pads, each 1 cm in thickness, to facilitate the measurement of β-rays emitted from the cells captured in each pad. The ordinate of the figure is the intensity, I, of β-rays in counts/5 min, which could be assumed proportional to the number of cells, ν, captured in each pad during a specific operation period, t.[1,3]

With curves constructed through each set of the data points in Fig. 10.6, $\frac{d\nu}{dL}$ can replace I as the ordinate; the figure shows that the distribution, $\int_0^L P(L)dL$, in Eq. (10.10) remains fairly constant. It is also noted from these curves that log penetration does not apply with bed thicknesses greater than 5 cm.

For ease of experimentation the numbers of cells in the original air stream were set at about $10^8/m^3$. In view of the very light loading ($\sim 10^4/m^3 \sim$) of the air sterilized in the fermentation industry (cf. Section 10.1.), the distribution of cells in a filter, $\int_0^L P(L)dL$, can be assumed virtually unchanged during the usual period of operation of a batch fermentation (10 to 200 hr).

For convenience, the value of $d\nu/dL$ in the first pad ($L = 0–1$ cm) along the air stream is taken as unity; the effect of fiber diameter on the distribution is exemplified in Fig. 10.7. The figure shows that finer glass fibers are more effective in collecting particulate materials; it also shows that the distribution deviates significantly from the log penetration relationship as L increases.

At bed thicknesses corresponding to $L=0.5$ and $L=1.5$ cm in Fig. 10.7, the distribution of cells for each species of fiber can be assumed to be governed by the log penetration relationship. The values of ν_0' and k in Eq. (10.9) can be assessed from the intersection with the ordinate and the slope, respectively, of a straight line through each set of data points in the figure [$m=0$ in Eq. (10.9)]. Then, the value of m in Eq. (10.9) can be determined such that the value of $d\nu/dL$ at the subsequent value of L in Fig. 10.7 could be represented exactly by Eq. (10.9).

Values of m assessed thus far can be plotted against L for each species of fiber (see the small diagram in Fig. 10.7). It is interesting to find a linear relationship between m and L; i.e.,

$$m = \alpha_0 L + \beta_0 \tag{10.28}$$

where

α_0, β_0 = empirical constants

As far as Eq. (10.28) is acceptable, the distribution either in Fig. 10.6 or in Fig. 10.7 can be extrapolated beyond the experimental range of L by using Eqs. (10.9) and (10.28);[2,4] this extrapolation becomes significant when the calculated values of $\bar{\eta}$ [see the right-hand side of Eq. (10.13)] are compared with the experimental data; see below.

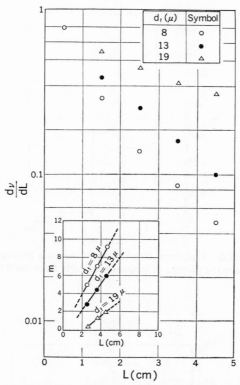

Fig. 10.7. Effect of fiber diameter on the longitudinal distribution of bacterial particles retained within the fibrous bed; volume fraction of fiber, $\alpha = 0.033$; superficial air velocity, $v_s = 5$ cm/sec; particle concentration of inlet air, $n \doteq 10^8$ particles/m^3.[2]

The time, Δt, between the emergence of bacterial cells from a fibrous filter was measured by impinging the air from the filter onto a Petri dish of nutrient agar, rotating at constant speed.[6]

Figure 10.8 is a histogram showing the frequency of the various values of the time interval, Δt, between the emergence of successive bacterial cells (*Serratia marcescens*) from the filter. The experimental conditions used to obtain the results shown in Fig. 10.8 are listed within the figure, where n' is the total number of colonies observed, $\bar{v}_0'$ the number of cells blown into the filter per second, and, v_s the nominal velocity of aerosol flow through the filter; for other symbols see the nomenclature at the end of this chapter.

It is seen from Fig. 10.8 that most of the cells pass the filter within short intervals of time; this trend was also observed with other experimental conditions. Since the passage of the bacterial cells through the filter is expected to be a random phenomenon, the mean value, $\overline{\Delta t}$, can therefore be determined accurately only after a significant number of bacteria have emerged from the filter.

To determine the number of bacteria needed to give a reliable value of $\overline{\Delta t}$,

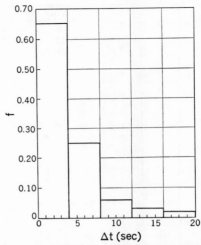

Fig. 10.8. Frequency distribution of the time interval, Δt, between the passage of successive bacteria from the filter.[6]

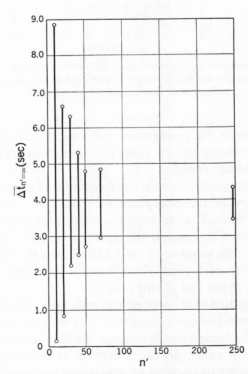

Fig. 10.9. Mean time interval $\overline{\Delta t}_{n'=\infty}$, between the emergence of successive bacteria from a filter exposed to an infinite population ($n' = \infty$).[6]

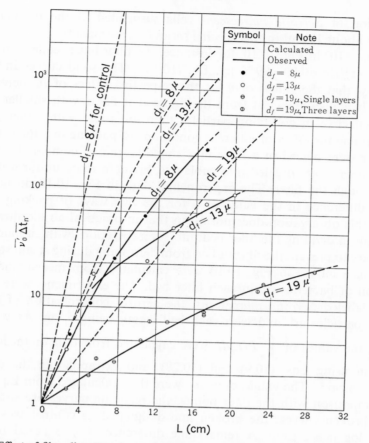

Fig. 10.10. Effect of fiber diameter, d_f, on the collection efficiency at different depths of filter bed, L. Superficial air velocity, $v_s = 5$ cm/sec; volume fraction of the filter, $\alpha = 0.033$; $\bar{\nu}_0'$ = number of particles entering filter per second; $\overline{\Delta t}_{n'}$ = mean time interval between the emergence of particles from the filter.[4]

the sample ($n' = 248$ in the example shown in Fig. 10.8) was divided into small groups, $n' = 10, 20, 30$, etc.

Using each subgroup or a combination of them, the range of $\overline{\Delta t}_{n'=\infty}$ (the mean of $\overline{\Delta t}$ in an infinite population of n') was estimated from the t-distribution with a level of significance of 5%. The result of this estimation is shown in Fig. 10.9; it is evident that the range of values of $\overline{\Delta t}_{n'=\infty}$ became smaller with increase of n', as could be expected. It was ascertained, from these results and many other experimental data, that the limits of confidence of $\overline{\Delta t}_{n'=\infty}$ were about $\pm 0.2 \Delta t_{n'}$ when n' was greater than 50 and the level of significance was 5%.

Figure 10.10 shows the effect of fiber diameter on the collection efficiency of the bacterial particles;[6] the values of $\bar{\nu}_0' \overline{\Delta t}_{n'}$, provided $\overline{\Delta t}_{n'}$ was measured with samples where $n' \geq 50$, are plotted against the bed length, L cm of the fibrous filter. To simulate the microbial concentration of air in practice, the number concentration of *Serratia marcescens* cells suspended in the original air was adjusted to about $10^4/m^3$ in this specific series of experiments.

For $\bar{v}_0' \varDelta t_{n'} = 10^2$ in Fig. 10.10, for instance, the filter has a collection efficiency of $\bar{\eta} = (10^2 - 1)/10^2 = 0.99 = 99\%$ [*cf.* Eq. (10.13)]. The solid curves in the figure clearly show that the values of $\bar{\eta}$ for the respective values of d_t increased as L increased. It is also apparent from the figure that fibers with smaller diameters are more effective in collecting particles than are those with larger diameters.

To determine the effect of different conditions of packing in a filter, fibers 19 μ in diameter were placed in a filter in a single layer and compared with a filter where the fibers were divided into sublayers, giving a more uniform flow of the aerosol through the filter. The experimental data in Fig. 10.10 do not indicate a marked difference in the value of $\bar{\eta}$ for the two ways of packing the filters. However, this observation does not mean that in practical air sterilization there is no benefit in dividing the filter bed into several sections with supporters; these are necessary to prevent the fibrous bed from deforming during steam sterilization.

The broken curves in Fig. 10.10 were calculated from measurements of the distribution of bacteria within each filter bed. The experiments were conducted using ^{32}P-labeled *Serratia marcescens*, the aerosol concentration of which was about 10^8 particles/m^3, and with each bed length being 5 cm. As was referred to earlier, the values of $\int_0^L P(L)dL$ were calculated from data on the longitudinal distribution using Eqs. (10.9) and (10.28), and extrapolating the distribution beyond $L = 5$ cm.[2,4] The values of $\bar{v}_0' \varDelta t_{n'}$ were then calculated from Eq. (10.13).[4]

For comparison with the data points, the resulting curves are shown in Fig. 10.10 as broken curves; the broken line designated as control was calculated assuming log penetration. A remarkable difference of $\bar{\eta}$ is noted between the calculated curve for $d_t = 8$ μ and the control, especially as the value of L is increased. For all fiber sizes, it should be noted that the curves for experimental determinations of $\bar{\eta}$ deviated considerably from those derived by calculation. This may be because the experimental curves were obtained with an aerosol concentration at 10^4 particles/m^3, while in the calculations much more concentrated aerosols (10^8 particles/m^3) were used. Although the collection efficiency ought to be independent of the aerosol concentration, the nonuniformity of passage of aerosol particles becomes more pronounced as the aerosol contamination is greatly reduced; this may account for the deviations between these curves. It seems probable, therefore, that the slopes of the broken curves for each filter size will decrease with increase in bed length, L, to approximate those of the corresponding solid curves, when the concentration of particles in the aerosol is decreased from 10^8 particles/m^3 to 10^4 particles/m^3.

10.3.4. Pressure drop of air flow

The drop in pressure as air flows through fibrous filters is related to the modified drag coefficient, C_{Dm} which was introduced by Kimura *et al.*[22] Experimental data plotted in Fig. 10.11 show the effect of changes in the Reynolds number, N_{Re} on the modified drag coefficient in filters packed with cotton or glass fibers.[7] When the fiber size is roughly the same, it is seen from Fig. 10.11 that cotton fibers exert more resistance to air flow than do glass fibers.

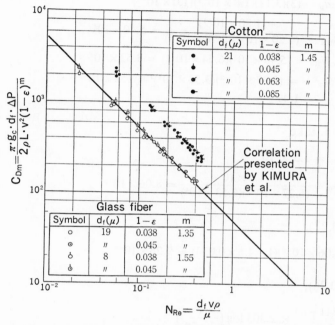

Fig. 10.11. Pressure drop in air flowing through cotton and glass fibers. The ordinate and abscissa are the modified drag coefficient, C_{Dm} and the Reynolds number, N_{Re}. The definitions of these two terms are also described in the figure; however, for detailed information regarding each term, see the nomenclature at the end of this chapter.

10.3.5. Design example of a filter for air sterilization

Air at 20°C will be assumed to contain 10^4 particles (1 μ in size)/m³. This dilute aerosol is to be filtered through a bed of glass fibers ($d_t = 19 \mu$ and $\alpha = 0.033$) at an air velocity, $v_s = 5$ cm/sec. Calculate the relation between bed length, L, and overall collection efficiency, η, of the filter. Calculate also the pressure drop, ΔP, in the air flowing through the filter.

Solution

1. *Calculation from the single fiber collection efficiency* (Fig. 10.5)
It is apparent from Fig. 10.4 that $v_s = 5$ cm/sec is far less than the velocity of v_c which should apply to this problem. Therefore, Fig. 10.5 may be used to estimate the value of η_0 (single fiber collection efficiency).

$$N_{\mathrm{Re}} = \frac{d_f \rho v}{\mu} = \frac{d_f \rho v_s}{\mu(1-\alpha)} = \frac{(19 \times 10^{-4})(1.20 \times 10^{-3})(5)}{(1.80 \times 10^{-4})(1-0.033)} = 6.54 \times 10^{-2}$$

$$N_{\mathrm{R}} = \frac{d_p}{d_f} = \frac{1.0 \times 10^{-4}}{19 \times 10^{-4}} = 5.27 \times 10^{-2}$$

$$D_{\mathrm{BM}} = \frac{CkT}{3\pi\mu d_p} = \frac{(1.16)(1.38 \times 10^{-16})(273+20)}{(3)(3.14)(1.8 \times 10^{-4})(1.0 \times 10^{-4})} = 2.78 \times 10^{-7} \text{ cm}^2/\text{sec}^*$$

$$N_{\mathrm{Sc}} = \frac{\mu}{\rho D_{\mathrm{BM}}} = \frac{(1.80 \times 10^{-4})}{(1.20 \times 10^{-3})(2.78 \times 10^{-7})} = 5.40 \times 10^{5}$$

$$N_{\mathrm{Pe}} = N_{\mathrm{Sc}} N_{\mathrm{Re}} = (5.40 \times 10^{5})(6.54 \times 10^{-2}) = 3.53 \times 10^{4}$$

$$N_{\mathrm{R}} N_{\mathrm{Pe}}{}^{1/3} N_{\mathrm{Re}}{}^{1/18} = (5.27 \times 10^{-2})(3.53 \times 10^{4})^{1/3}(6.54 \times 10^{-2})^{1/18}$$
$$= 1.49$$

From Fig. 10.5, take

$$\eta_0 N_{\mathrm{R}} N_{\mathrm{Pe}} = 1.5 \times 10$$

$$\therefore \quad \eta_0 = \frac{1.5 \times 10}{(6.27 \times 10^{-2})(3.53 \times 10^{4})} = 8.06 \times 10^{-3}$$

From Eq. (10.27),

$$\eta_\alpha = (8.06 \times 10^{-3})(1 + 4.5 \times 0.033) = 9.26 \times 10^{-3}$$

From Eq. (10.26),

$$L = \frac{\pi d_f (1-\alpha)}{4\eta_\alpha \alpha} \times 2.303 \log\left(\frac{1}{1-\bar{\eta}}\right)$$

$$= \frac{3.14 \times (19 \times 10^{-4})(1-0.033) \times 2.303}{4 \times (9.26 \times 10^{-3}) \times 0.033} \log\left(\frac{1}{1-\bar{\eta}}\right)$$

$$= 10.9 \times \log\left(\frac{1}{1-\bar{\eta}}\right) \tag{10.29}$$

2. *Empirical equation*—the following equation presented by Blasewitz et al.[10] will be used:

$$1 - \bar{\eta} = 10^{-r' L^{\alpha'} \rho_b{}^{\beta'} v_s{}^{r'}} \tag{10.30}$$

provided:

L = bed thickness, inch
ρ_b = packed density of glass fibers, lb/ft^3
v_s = superficial velocity of air, ft/min

Corresponding to the operating conditions of this problem,[10]

---

* The value of C was calculated from the equation presented by Ranz et al.[26]

$$\gamma' = 0.075$$
$$\alpha' = 0.9$$
$$\beta' = 1.0$$
$$\gamma'' = -0.4$$

Assuming that the true density of glass fibers is 2.5 g/cm³,

$$\rho_b = 0.033 \times 2.5 \text{ g/cm}^3 = 82.5 \text{ kg/m}^3 = 5.15 \text{ lb/ft}^3$$
$$v_s = 5 \text{ cm/sec} = 9.84 \text{ ft/min}$$

Substituting the above figures into Eq. (10.30),

$$\bar{\eta} = 1 - 10^{-\{(0.075)L^{0.9}(5.15)(9.84)^{-0.4}\}} \tag{10.31}$$

The results of calculations with Eqs. (10.29) and (10.31) are shown in Fig. 10.12, in which L is plotted against $\bar{\eta}$. The scale of the abscissa of the figure is derived from Eq. (10.13). The solid curve for a fiber, diameter, $d_t = 19 \, \mu$ in Fig. 10.10, can be extrapolated and will become curve B in this figure. The marked difference between curves A and B might be ascribed to the fact that the aerosol concentration, $n = 10^4$ particles/m³, in this example apparently decreased the value of $\bar{\eta}$ compared with the case of $n = 10^8$ particles/m³, from which Fig. 10.5 or Eq. (10.29) was derived.

3. *Pressure drop of air flow*
Corresponding to $N_{Re} = 6.54 \times 10^{-2}$,

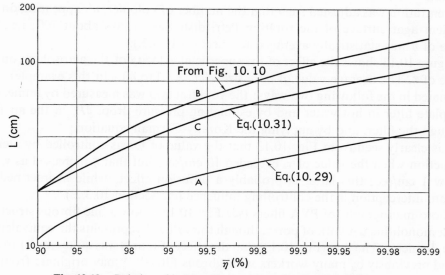

Fig. 10.12. Relation of bed length, L, to overall collection efficiency, $\bar{\eta}$

$$C_{\text{Dm}} = \frac{\pi g_c d_f \Delta P}{2\rho L v^2 (1-\varepsilon)^m} = 7.50 \times 10^2 \qquad \text{(Fig. 10.11)}$$

Taking the value of $m = 1.35$ from Fig. 10.11,

$$\frac{\Delta P}{L} = \frac{(7.50 \times 10^2) \times 2 \times 1.20 \left(\dfrac{0.05}{1-0.033}\right)^2 (1-0.967)^{1.35}}{3.14 \times 9.81 \times (19 \times 10^{-6})}$$

$$= 82 \ \text{Kg/m}^2/\text{m}$$

$$= 82 \ \text{mmH}_2\text{O/m}$$

In operating filters for air sterilization, it is suggested that the packed fibers be dried completely after sterilization with steam and that the air to be filtered be of low humidity to prevent the bed from becoming damp.

10.4. PVA Filter for Air Sterilization

A plate filter can be made by acetylating polyvinyl alcohol and by coating it with a heat-resistant resin, such as silicon resin to enable it to withstand repeated steam sterilizations. Such filters have been widely used to sterilize air in the fermentation industry, particularly in Japan.[8] The salient features of the filter (designated PVA filter) will be discussed.

Esumi et al. measured the overall collection efficiency, η of a PVA filter using an aerosol of *Bacillus cereus* spores ($d_p = 0.5$ to $1.0 \ \mu$) and a filter of thickness, $L = 0.5$ cm.[14] The methods used were similar to those of Aiba et al.,[6] but the aerosol concentration was adjusted such that the number of colonies, n', appearing on the nutrient agar surface of the rotating Petri dish was always about 100; i.e., the value of η was statistically averaged (cf. Section 10.3.3.).

Figure 10.13 shows the effect of the superficial velocity of the aerosol, v_s, on the value of η. The average value of pore size, d_e ($= 60$ to $80 \ \mu$ in this example), was estimated in the following two ways: the void fraction was measured by immersing the plate filter in hot water and by estimating pressure drop, ΔP, in the air flow through the filter, and by applying the Kozeny-Carman equation.

It is clearly seen from Fig. 10.13 that the value of η was controlled by inertial impaction when the value of v_s exceeded 10 cm/sec, but that η increased as v_s fell below 1 cm/sec; this latter is probably a diffusion effect, while in intermediate region, interception is the controlling influence (cf. Section 10.3.2.).

Photo-micrographs of PVA filters (see Fig. 10.15) show a fine fibrous structure, not a monolithic network of pores, though the value d_e represents the equivalent of pore diameter. The fact (Fig. 10.13) that the effect of v_s on η was quite similar to that found previously by many workers with fibrous filters[11,18] may originate from the fibrous nature of PVA plate filters.

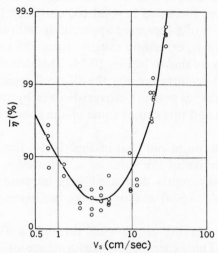

Fig. 10.13. Collection efficiency, $\bar{\eta}$, as affected by superficial velocity of air, v_s; PVA filter, equivalent pore size, $d_e = 60 - 80\ \mu$; thickness, $L = 0.5$ cm.[14]

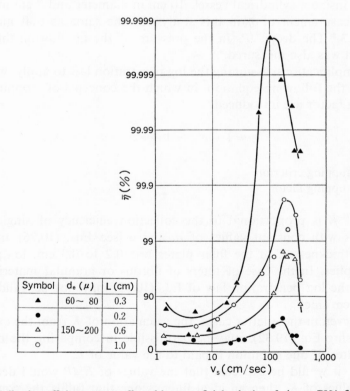

Symbol	$d_e\ (\mu)$	L (cm)
▲	$60 \sim 80$	0.3
●		0.2
△	$150 \sim 200$	0.6
○		1.0

Fig. 10.14. Collection efficiency, $\bar{\eta}$, as affected by superficial velocity of air, v_s; PVA filter (cont'd). "Re-entrainment."[15]

Figure 10.14 is another diagram to show the effect of v_s on $\bar{\eta}$, but in this experiment the values of v_s were increased beyond 100 cm/sec.[15] It is interesting to note in the figure that the value of $\bar{\eta}$ increased appreciably with increase in v_s; however, beyond a critical value of v_s, extending roughly from 150 to 200 cm/sec, the value of $\bar{\eta}$ deteriorated sharply as shown in Fig. 10.14. This deterioration of $\bar{\eta}$ is considered to reflect bacterial re-entrainment; the "force" which retains the bacterial spores captured by the fibrous network is overwhelmed by another "force" imposed from the aerosol stream and this causes some of the spores to be re-entrained into the stream.

It is also interesting to point out that in Fig. 10.14, the critical value of v_s was nearly the same, irrespective of the value of d_e. However, Esumi et al. observed that the critical value apparently decreased with increase in particle size, when they used yeast cells ($d_p = 5$ μ) and/or *Aspergillus oryzae* spores ($d_p = 7$ to 8 μ) as the test organisms.[16]

The drop in pressure in the air flowing through a PVA filter significantly affects the value of $\bar{\eta}$. The experimental performance of various filter media— granular active carbon, several types of fibers, metal and PVA filters—has been compared. Except for the plate-type filters ($L=0.2$ to 0.3 cm), the filter media were packed inside a cylindrical vessel, 10 cm in diameter and 7 cm in bed thickness. The measurement of $\bar{\eta}$ in this study was the same as that mentioned in Section 10.3.3.[6] The drop, ΔP, in the pressure of the air flowing through each filter medium was also measured.[9]

Certain simplifications, assuming the log penetration law to apply, will be made as shown by the following equation, in which the concepts of stopping criterion and stopping factor are introduced:

$$\bar{\eta} = 1 - e^{-S} = 1 - e^{-KL} \tag{10.32}$$

where

$\qquad S$ = stopping criterion
$\qquad K$ = stopping factor

The value of K is proportional to the collection efficiency of single fibers for fibrous filters with constant values of d_f and α [see Eqs. (10.26) and (10.27)].

Since the thickness L of the filter plates was 0.2 to 0.3 cm, log penetration probably applies. In the case of filters of fibrous or granular materials, on the other hand, the log penetration law of Eq. (10.32) may not be valid, since the bed was 0.7 cm thick.

However, even in these latter cases, the calculation of K from the experimental data of $\bar{\eta}$, using Eq. (10.32), may be permissible in comparing the relative efficiencies of filters if due attention is paid to the value of L.

In general, it would be expected that the values of $S/\Delta P$ would decrease with increasing values of K; then an ideal filter is one that fulfils the condition that both K and $S/\Delta P$ have high values. Accordingly, the values of $S/\Delta P$ were cal-

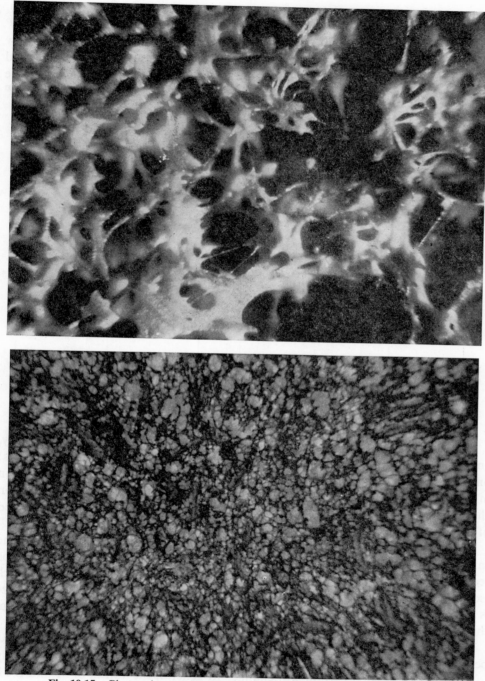

Fig. 10.15. Photo-micrographs of PVA filters. (magnification: 64 mm — 100 μ)
Upper: $d_e = 300\ \mu$ (for sterile room).
Lower: $d_e = 20\ \mu$ (for fermentor).

culated for each filter system from the measurements (η, ΔP) and Eq. (10.32).

The results of these comparisons are summarized in Fig. 10.16, in which $S/\Delta P$ (mmH$_2$O^{-1}) is plotted against K (m^{-1}). Rectangular regions in the figure show the extent of errors of mean square associated with respective values of K and $S/\Delta P$; these were determined from several replicated experiments, except in the case of the glass-fiber filter. In the case of glass fibers, the values of K and $S/\Delta P$ were calculated as exemplified in the preceding design examples.

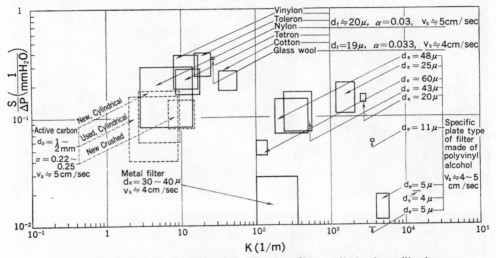

Fig. 10.16. Comparative performances of various filter media in air sterilization.

It is noted from the figure that all the fibrous materials performed similarly, and that the granular active carbon particles were not as effective as the fibers; one sample of cylindrical carbon granules had been used for air sterilization in a fermentation plant for about one year, and the other of crushed carbon showed, from a sieve analysis, that 95% by weight of the sample was composed of particles whose size was below $d_{95}=1.4$ mm.

If the value of K required in fermentation practice is considered to be in the range from 10^3 to 10^4, PVA filters, especially those of $d_e = 20$ to 30 μ, can be recommended for air sterilization using only a single layer 0.2 to 0.3 cm thick (see the values of K and $S/\Delta P$ in the figure). Filters whose effective pore size is 10 to 4 μ are less satisfactory due to the lower value of $S/\Delta P$. For this reason, the metallic filter is not a suitable substitute for the fibrous bed.

Furthermore, if the log penetration law does not apply to fibrous filters whose bed length exceeds several centimeters, the regions of Fig. 10.16 which apply to fibrous or activated carbon will change. In general, the values of K and $S/\Delta P$ would tend to shift towards the origin in Fig. 10.16 when measurements were made with filter beds of fibers or carbon thicker than 7 cm.

Although the fibrous network observed microscopically (Fig. 10.15) may partly account for the excellent performance of PVA filters, quantitative explanations have yet to be attempted. Since PVA filters, coated with heat-resistant resin, can be subjected to repeated sterilizations with steam, the installation of this type of filter in fermentation plants is promising both from the viewpoint of saving space and of overcoming various shortcomings, such as relatively high operation cost, troublesome renewal, etc. which apply to glass-fiber filters in the fermentation industry.

10.5. SUMMARY

1. Air to be sterilized for the fermentation industry should be originally clean and contain very low numbers of airborne microorganisms. The population of microorganisms in air has a Poisson distribution. A reasonable average number for a design basis is around 10^4 particles/m³.

2. Collection efficiencies of fibrous air sterilization filters should be assessed from a probabilistic approach.

3. The time interval between the successive emergence of airborne microbes from a fibrous filter, if statistically averaged, is related to the collection efficiency and to the longitudinal distribution of the organisms retained within the filter; the key equation is:

$$\eta = \frac{\bar{\nu}_0' \overline{\Delta t} - 1}{\bar{\nu}_0' \overline{\Delta t}} = \int_0^L P(L)dL \qquad \text{[Eq. (10.13)]}$$

4. Collection efficiencies are affected by the hydrodynamics of the air streams within the fibrous bed [Fig. 10.5 and Eqs. (10.14) to (10.20)].

5. Some of the salient features of the PVA filter (a plate filter made of polyvinyl alcohol) are that the pressure drop of air flow is lower and the collection efficiency is greater than a comparable fibrous filter (Fig. 10.16).

NOMENCLATURE

b = specific width of air stream, Fig. 10.3
C = Cunningham's correction factor for slip flow
C_{Dm} = modified drag coefficient
D_{BM} = diffusivity of particles, cm²/sec
d_f = fiber diameter, μ, cm
d_p = particle diameter, μ, cm
d_e = effective pore size, μ
d_{95} = diameter for particles whose size is below d_{95} comprise 95% by weight in sieve analysis

f = frequency

g_c = conversion factor, 9.81 kg m/Kg sec², 981 g cm/G sec²

I = intensity of β-rays, counts/5 min

K = stopping factor, m⁻¹, Eq. (10.32)

k = Bolzmann constant, 1.38×10^{-16} cm² g/sec² °K; in Eq. (10.9), $k = \xi p$

L = thickness of filter (bed), cm, m [inch in Eq. (10.30)]

m = empirical exponent, Fig. 10.11; number of particle collisions with fibrous grid, below the number of which particle can be re-entrained; a measure to show the deviation from log penetration, Eq. (10.3)

N = number of microbes in certain large volume of air

N_{Pe} = Péclet number ($= v d_f / D_{BM}$)

N_R = interception parameter ($= d_p / d_f$)

N_{Re} = Reynolds number ($= d_f v \rho / \mu$)

N_{Sc} = Schmidt number ($= \mu / \rho D_{BM}$)

n = particle concentration, particles/m³

n' = number of colonies

$P(L)$ = distribution (probability) function, Eq. (10.10)

$P(n)$ = probability, Eq. (10.1)

ΔP = pressure drop of air flow, mmH₂O, Kg/m², Kg/cm²

p = probability, Eq. (10.1), Eq. (10.3)

S = stopping criterion, Eq. (10.32)

T = absolute temperature, °K

t = time, hr, sec

Δt = time interval between particles, sec

u = particle velocity, Eq. (10.14)

v = air velocity [$= v_s/(1 - \alpha)$], Eq. (10.14), cm/sec, m/sec

v_0 = upstream air velocity, cm/sec

v_c = critical velocity of v, cm/sec, Eq. (10.17)

v_s = superficial air velocity, cm/sec, m/sec; ft/min in Eq. (10.30)

x_0 = effective radius of displacement of particles due to diffusion, Eq. (10.20)

x, y = coordinates

Bold face = vector

superscript

—: arithmetical mean; statistical average

~: nondimensional symbol

Greek letters

α = volume fraction

α' = empirical exponent, Eq. (10.30)

α_0 = empirical constant, Eq. (10.28)

β' = empirical exponent, Eq. (10.30)

β_0 = empirical constant, Eq. (10.28)

γ', γ'' = empirical exponents, Eq. (10.30)

ε = void fraction ($= 1 - \alpha$)

η_0 = collection efficiency of single fiber

η_0' = collection efficiency of single fiber due to inertial impaction
η_0'' = collection efficiency of single fiber due to interception
η_0''' = collection efficiency of single fiber due to diffusion
η_α = collection efficiency of single fiber whose volume fraction is α
$\bar{\eta}$ = overall collection efficiency of filter
θ = radian in Fig. 10.3
κ = normalization factor
μ = viscosity of air, g/cm sec
ρ = density of air, g/cm³, kg/m³
ρ_b = packed density of glass fiber, 1b/ft³, Eq. (10.30)
ρ_p = density of particle, g/cm³
ν = expected value ($= Np$); number of particles retained within filter
ν_0 = total number of particles blown into filter
$\bar{\nu}_0'$ = number of particles blown into filter per unit time, sec⁻¹
ξ = number of grids per unit length of filter bed
ϕ = inertial parameter ($= C\rho_p d_p^2 v_0/18\mu d_f$)

REFERENCES

1. Aiba, S., and Yamamoto, A. (1959). "Distribution of bacterial cells within fibrous air sterilization filters." *J. Biochem. Microbiol. Tech. & Eng.* **1**, 129.
2. Aiba, S., Kodama, T., and Sakamoto, H. (1959). "Distribution of bacterial cells within fibrous air sterilization filters." *J. Biochem. Microbiol. Tech. & Eng.* **1**, 325.
3. Aiba, S. (1960). *Biochemical Engineering* p. 16. Nikkan-Kogyo Pub. Co., Tokyo.
4. Aiba, S., Sakamoto, H., and Kodama, T. (1960). "Some analyses of the bacterial distribution within fibrous air sterilization filter." *J. Gen. Appl. Microbiol.* **6**, 15.
5. Aiba, S. (1961). "Design of air filters." *Chem. Tech. (Japan)* **13**, 43.
6. Aiba, S., Shimasaki, S., and Suzuki, S. (1961). "Experimental determination of the collection efficiencies of fibrous air sterilization filter." *J. Gen. Appl. Microbiol.* **7**, 192.
7. Aiba, S. (1962). "Design of fibrous air sterilization filters." *J. Gen. Appl. Microbiol.* **8**, 169.
8. Aiba, S., Nishikawa, S., and Ikeda, H. (1963). "A new type of air sterilization filter." *J. Gen. Appl. Microbiol.* **9**, 267.
9. Aiba, S., Nishikawa, S., and Niira, R. (1963). "Comparative studies on the performances of various sorts of materials as filter media in air sterilization." *Progress Rept. No. 26.* Biochem. Eng. Lab., Inst. Appl. Microbiol., Univ. of Tokyo.
10. Blasewitz, A. G., and Judson, B. F. (1955). "Filtration of radioactive aerosols by glass fibers." *Chem. Eng. Progress* 6–J.
11. Chen, C. Y. (1955). "Filtration of aerosols by fibrous media." *Chem. Review* **55**, 595.
12. Carswell, T. S., Doubly, J. A., and Nason, H. K. (1937). "Bacterial control in air conditioning." *Ind. Eng. Chem.* **29**, 85.
13. Decker, H. M., Citek, F. J., Harstad, J. B., Gross, N. H., and Piper, F. J. (1954). "Time-temperature studies of spore penetration through an electric air sterilizer." *Appl. Microbiol.* **2**, 33.
14. Esumi, S., and Ashida, K. (1965). "Experimental studies on "P.V.A." air-sterilization-filter. (I) Air flow pressure drop and collecting efficiency of air-borne bacteria." *J. Ferm. Tech. (Japan)* **43**, 547.
15. Esumi, S., and Ashida, K. (1966). "Experimental studies on "P.V.A." (II) Effect of pore size and thickness." *J. Ferm. Tech. (Japan)* **44**, 529.

16. Esumi, S., and Ashida, K. (1967). "Experimental studies on "P.V.A." (III) Efficiency and mechanism of collection as affected by particle size." *J. Ferm. Tech. (Japan)* **45**, 778.

17. Friedlander, S. K. (1958). "Theory of aerosol filtration." *Ind. Eng. Chem.* **50**, 1161.

18. Gaden, E. L., Jr., and Humphrey, A. E. (1956). "Fibrous filters for air sterilization." *Ind. Eng. Chem.* **48**, 2172.

19. Humphrey, A. E., and Gaden, E. L., Jr. (1955). "Air sterilization by fibrous media." *Ind. Eng. Chem.* **47**, 924.

20. Humphrey, A. E. (1960). Private communication.

21. Jacobs, M. B., Manohran, A., and Goldwater, L. J. (1962). "Comparison of dust counts of indoor and outdoor air." *Int. J. Air. Wat. Poll.* **6**, 205.

22. Kimura, N., and Iinoya, G. (1959). "Experimental studies on the pressure drop characteristics of fiber mats." *Chem. Eng. (Japan)* **23**, 792.

23. Kluyver, A. J., and Visser, J. (1950). "Some observations on filtration." *Antonie van Leeuwenhoek* **16**, 311.

24. Lundgren, D. A., and Whitby, K. T. (1965). "Effect of particle electrostatic charge on filtration by fibrous filters." *I & EC Process Des. Develop.* **4**, 345.

25. Ramskill, E. A., and Anderson, W. L. (1951). "The inertial mechanism in the mechanical filtration of aerosols." *J. Colloid Sci.* **6**, 416.

26. Ranz, W. E., and Wong, J. B. (1952). "Impaction of dust and smoke particles on surface and body collectors." *Ind. Eng. Chem.* **44**, 1371.

27. Robinson, F. W. (1939). "Ultra-violet air sanitation." *Ind. Eng. Chem.* **31**, 23.

28. Rept. Sanitary Eng. Station, Tokyo. (1925), (1927), (1928), (1929). "Hygienic examination of air in metropolitan areas." 342, 224, 57, 109.

29. Sadoff, H. L., and Almolf, J. W. (1956). "Testing of filters for phage removal." *Ind. Eng. Chem.* **48**, 2199.

30. Silverman, L. (1951). "Performance of industrial aerosol filters." *Chem. Eng. Progress* **47**, 462.

31. Silverman, L., Conners, E. W., Jr., and Anderson, D. M. (1955). "Mechanical electrostatic charging of fabrics for air filters." *Ind. Eng. Chem.* **47**, 952.

32. Stern, S. C., Zeller, H. W., and Schekman, A. I. (1960). "The aerosol efficiency and pressure drop of a fibrous filter at reduced pressures." *J. Colloid Sci.* **15**, 546.

33. Stark, W. H., and Pohler, G. M. (1950). "Sterile air for industrial fermentations." *Ind. Eng. Chem.* **42**, 1789.

34. Thomas, D. G. (1953). Ph.D. Thesis in Chem. Eng., Ohio State Univ.

35. Thomas, J. W., and Yoder, R. E. (1956). "Aerosol size for maximum penetration through fiberglas and sand filters." *A.M.A. Arch. Ind. Health* **13**, 545.

36. Wong, J. B., Ranz, W. E., and Johnstone, H. F. (1956). "Collection of aerosol particles." *J. Appl. Phys.* **27**, 161.

Chapter 11

Equipment Design and Asepsis

As in Chapter 7 for scale-up, the fermentor vessel is the principal subject here with regard to equipment design and asepsis. For a discussion on all kinds of auxiliary equipment needed for control of the fermentor, see Chapter 12. Before proceeding

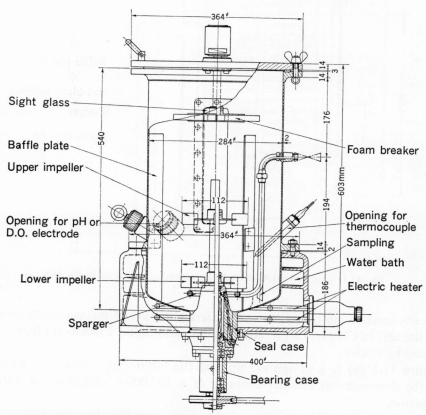

Fig. 11.1. Fermentor design examples. (a) A 30-*l* laboratory fermentor. (*Courtesy* Marubishi Laboratory Equipment Co., Ltd., Tokyo, Japan)

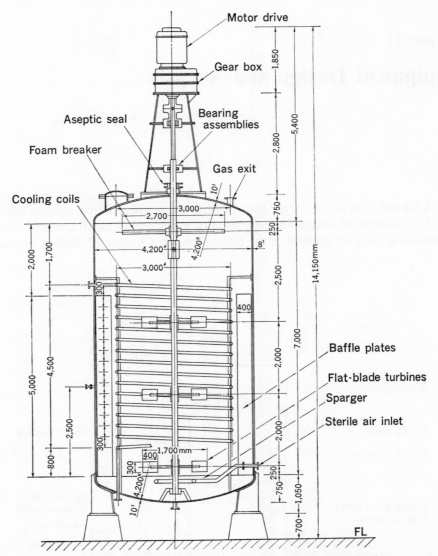

Fig. 11.1. Cont'd.   (b) A 100,000-*l* fermentor for antibiotic production.

to discuss some details of key parts of the fermentor vessel, it is necessary to supplement the text in Chapters 6 (aeration and agitation) and 7 (scale-up) from a more practical viewpoint.

Figure 11.1 (a) is a design example of the laboratory fermentor, while Fig. 11.1 (b) demonstrates another example of a 100,000-*l* fermentor for antibiotic production.

The agitator assembly consists of the shaft, impeller, bearing arrangements, aseptic seals and the motor drive. The shaft is designed so as to avoid whipping at high speeds and for easy disconnection from the drive to permit cleaning, if required. Though the flat-blade turbine is commonly used, other types of impeller, i.e., marine propellers, vaned discs, etc. (*cf.* Sections 6.2.2.1. and 6.2.2.2., and reference 6 in this chapter) can also be employed. In fact, when two or three impellers are installed on the same shaft, some urge that the vaned disc be located near the sparger and the marine propeller near the middle of the vessel.

Two types of sparging arrangements are used in the fermentation industry (*cf.* Section 6.4.5.). One is the single hole sparger, consisting of a pipe with a single hole located centrally below the impeller, while the other is the ring sparger. In Fig. 11.1, a circular tube, of less than the impeller diameter, is located below the impeller. Holes (2 to 3 mm in diameter) are drilled along the tube periphery at the bottom; the total area of the holes is suggested to be made equal to the cross-sectional area of the tube. In both types of spargers an optimal location, neither

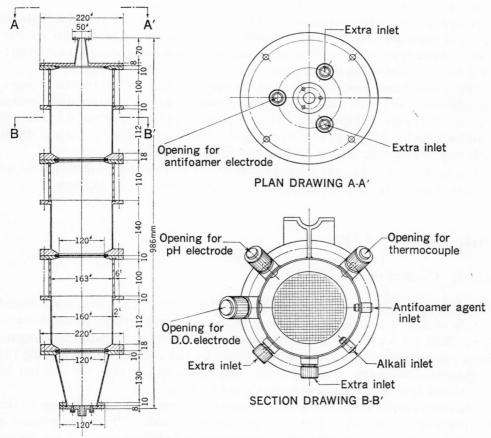

Fig. 11.2. Tower fermentor (laboratory scale).
(*Courtesy* Marubishi Laboratory Equipment Co., Ltd., Tokyo, Japan)

too close to the impeller, causing bubble coalescence and flooding nor so far away as to impair the break-up of bubbles by the impeller, must be considered.

The use of superficial air velocity, v_s [cf. Eqs. (6.37) and (6.38)], to represent aeration rate is a good guide to foaming and flooding problems. For example, the marine propeller and the paddle impeller flood when v_s exceeds 21 m/hr, whereas the flat-blade turbine can tolerate flooding up to $v_s \doteq 120$ m/hr when two sets of impellers are used on the same shaft (cf. Section 6.3.2.). The estimate of aeration rate [Eqs. (7.1) and (7.2)] in connection with this criterion will be of use in selecting the impeller type in scale-up.

In contrast to fermentors with mechanical agitation, air-lift and tower fermentors, neither of which are provided with mechanical agitators, will be mentioned briefly. In the former, a draft tube is installed at the center of the vessel to secure good circulation of liquid due to the draft effect of bubbles blown outside the tube. This has been utilized in some waste treatments, though usually open and sparged tanks (mostly made of concrete and very simple in construction) are used as "fermentor vessels" in the activated sludge process.

A design for a tower fermentor is illustrated in Fig. 11.2. The tall cylindrical vessel is subdivided into several compartments by perforated plates. Air bubbles ascend through the perforated plates, while the broth passes through the compartments concurrently with the air flow.[3]

The tower fermentor, characterized by larger values of volumetric oxygen-transfer coefficient with small power requirements, has the additional advantage that each compartment can function as an independent fermentor vessel and thus is suited for use in multi-stage continuous cultivation, if the perforated plate design is appropriate and back-mixing of the broth is minimized.

The tower fermentor, which has been used to cultivate microorganisms on an industrial scale, will find more applications in the fermentation industry.[2]

11.1. FERMENTOR DESIGN

11.1.1. Cardinal rules

Good design practice embodies several cardinal rules. The rules are directed mostly at aseptic practice, i.e., keeping the fermentation process contamination-free. There are additional design problems when dealing with pathogenic organisms, because the design must also be based on the necessity of containing the pathogens. The design of such facilities is a rather special case and will not be considered here. However, a warning should be issued.

Microorganisms are a rich source of protein, some of which is foreign to the human body. Large inhaled doses of any microorganism will usually produce a reaction in most humans. Therefore, care must be exercised to avoid any design procedure which would permit the escape of aerosol from an outlet of the fermentor. This means that safe practice demands either incineration, filtration or dilution (by dispersing from a high effluent stack) of the exit gas of the fermentor. It also means that whenever centrifuges are used (note: some fermentor inherently produces large quantities of aerosols), some consideration should be given to enclosing these units in a specially ventilated area.

With this warning note, here are the cardinal design rules.

a. There should be no direct connections between sterile and nonsterile parts of the system. Bacteria have been known to grow through closed valves.

b. Minimize flange connections. Due to equipment vibration and thermal expansion, flanged connections do move and contamination can work its way through many flanged seals.

c. Whenever possible, use all welded construction. Be sure to polish all welds so that no reservoirs exist for the accumulation of solid medium that could resist sterilization.

d. Avoid dead spaces, crevices, and the like. Solids can accumulate in these and provide an insulating environment for contaminants to escape sterilization.

e. Various parts of the system should be independently sterilizable.

f. Any connections to the vessel should be steam-sealed. For example, the valve on the sampling port should have live steam on the exit side whenever it is not in use.

g. Use valves which are easy to clean, maintain and sterilize. Examples are ball valves, diaphragm valves and globe valves with nonrising stems. (For items *f* and *g* above, *cf.* Section 11.3.).

h. Maintain a positive pressure in the fermentor so that leakage is always out. (With pathogenes this is not applicable.)

11.1.2. Materials of construction[5] and vessel size

Almost all fermentor vessels are constructed from glass or from stainless steel. Glass is limited to vessels less than 20 – 30 *l* in size. Glass vessels are of the following design: *a.* Glass jars with rounded or flat bottoms and a flanged top fitted with a plastic or stainless steel head, *b.* Glass cylinders fitted at both ends with head plates.

Fermentor vessels from 40 *l* on up have all parts in contact with the broth constructed of stainless steel. Sometimes in the name of economy the outside jacket may be made of ordinary steel. Sometimes stainless steel-clad vessels are used. In the long run, corrosion and maintenance problems caused by the use of different materials of construction hardly ever warrant the initial economy in capital costs.

For the most part the vessels have dished bottoms and tops. Plate tops can also be used in vessels less than 500 *l* in size, but this is dictated by the pressure code regulations.

Vessel size is dependent on the following considerations: *a.* Total projected fermentation capacity, b. The greater risk of costly losses when a large fermentor is contaminated, c. Fabrication capabilities of the constructor, i.e., erecting a vessel on site vs. shipping, d. Restriction or shipping size limitations. For example, when shipping by rail or by truck, bridge clearances, etc. may limit the size of vessel that can be delivered to a site.

In the days when the butanol-acetone and alcohol fermentations were economical, vessels of nearly 400 m^3 or larger were common. When antibiotic fermentations first became popular (late 1940's), stainless steel vessels (40 – 60 m^3) were ubiquitous. Today, as the technique of managing a fermentation aseptically improves, there is a tendency to have larger and larger fermentor vessels. Indeed, new construction frequently is based on vessels, 200 m^3 and larger.

However, there may well be a counter-trend emerging to this pattern. Recent practice of antibiotic and similar fermentations is to have batch extended with repeated draw-offs, semi-continuous and even continuous operations. Because of the increased productivity of these units compared with strict batch, the need for large vessels may decrease and smaller vessels, 20 – 40 m^3 in size may again become popular. In connection with the vessel size, consideration based on optimization theory and calculation is deemed necessary.

11.1.3. Bearing assemblies

Two types of bearing assemblies are used—outboard (outside vessel) and inboard (inside vessel). Figure 11.3 is an example of the external bearing assembly.[1] In the figure, a drawing of the seal is omitted, for convenience. Good practice requires at least two bearing assemblies. The seal should not be used as one of the bearings. When an inboard bearing is used, it is usually of the bottom type, i.e., a bearing support is located in the vessel bottom and the shaft extends through from the top. A bottom bearing is frequently required with a top-entering shaft because of the longer shaft extension and hence the tendency to whip. This can be avoided with a bottom-entering shaft.

Although bottom-entering shafts are not widely used yet in the fermentation industry because of potential contamination hazard encountered with bottom seals, recent developments with tangsten-carbon rotary seals, with steam-sealed units, etc. make their installation more feasible [see Fig. 11.1 (a) and Section 11.2.].

11.1.4. Motor drives

According to Solomons,[5] the power input to a fermentor is suggested as follows:

Size	Power input (W/l)
Laboratory	8 – 10
Pilot plant	3 – 5
Plant	1 – 3 (equivalent to 1 – 4 $\text{\scriptsize HP}/m^3$)

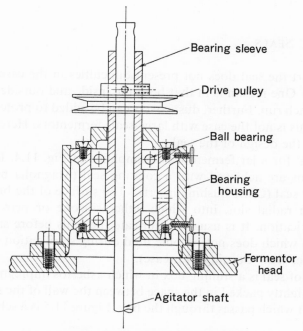

Bearing sleeve

Drive pulley

Ball bearing

Bearing housing

Fermentor head

Agitator shaft

Fig. 11.3. Bearing assembly for a 30-*l* jar fermentor.[1]

Obviously the agitator bearing resistance can make up 50 – 75% of the total power drain in small fermentors, whereas the relative magnitude of bearing to shaft loss is considered negligible in plant-scale fermentors. Accordingly, the use of watt-meters across the motor as a means of measuring the shaft power input in laboratory fermentors is not permitted; a torsion dynamometer (or the strain-gage dynamometer) is recommended particularly for small fermentors (*cf.* Section 12.2.3., Chapter 12). As discussed in Section 7.2.1., Chapter 7, the power input (shaft power input), 1– 4 ʜᴘ/m³ for large fermentors is in the category of strong agitation in general use in chemical processes.

For small-scale fermentors, small electric motors with silicon rectifiers for speed and power control have proved excellent. Motors controlled by rheostats should be avoided, because they are subject to gross variations. For intermediate drives (up to 10 ʜᴘ), variable speeds are provided by using pulleys; these consist of split V pulleys that can be separated to control the pulley diameter. For large drives (>100 ʜᴘ), variable-speed drives are costly; what is usually done is to install 2-speed drives. An alternate is to utilize a reversing mechanism that changes the impeller rotation direction. Another alternate is to use hinged impeller tips; an effective change in impeller diameter and hence in impeller power drain can be achieved.

11.2. ASEPTIC SEALS

For the most part the seal does not present difficulties in the case of laboratory-scale fermentors. One reason is in that both the inside and outside of the seal can be sterilized in each run. Further, dust covers can be added to protect the seal from dust and dirt. This is not the case with large-scale fermentors. Here close attention must be paid to the design of the seal.

An aseptic seal for a jar fermentor is exemplified in Fig. 11.4. In this example, aseptic conditions are achieved with a rotating seal (Magnolia bearing bronze)[1] and a stationary seal (Oilite bushing insert).[1] The surface of the bronze bearing is provided with a radial slot, into which either silicone or petroleum grease is charged for lubrication. It is usual to lubricate the seals before sterilization with a silicone grease which does not become fluid during sterilization at 120°C for 20 min (for instance, Dow Corning 44 grease).[1]

Another type of seal is a stuffing box in which asbestos impregnated with lubricating grease is tightly packed in the space between the wall of the stuffing box and the agitator shaft which passes through the box. Figure 11.5 is a schematic drawing

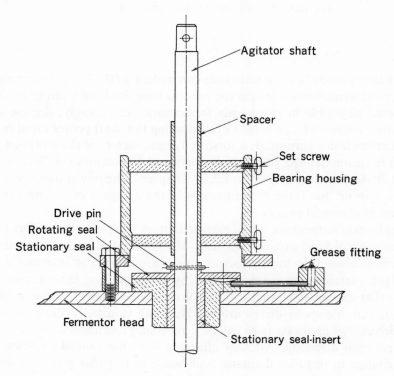

Fig. 11.4.   Shaft seal.[1]

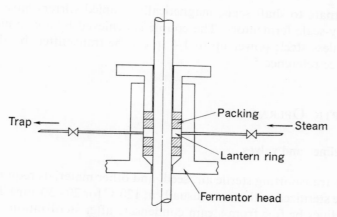

Fig. 11.5.   Steam-sealed gland.[4]

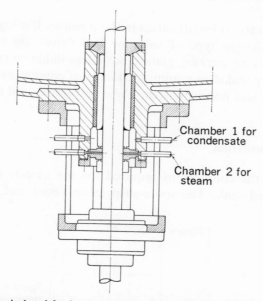

Fig. 11.6.   Mechanical seal for bottom-entering shaft[5] (Hedén, 1958). Three mechanical seals are used. Chambers 1 and 2 can be used for condensate and steam, respectively.

of a stuffing box-type seal, equipped with a lantern ring to allow steam to penetrate to the shaft. This type of seal is referred to as a mechanical seal and is used for large fermentors where the penetration of steam into the sealing materials of the gland may be inadequate. Hedén has described three mechanical seals for the bottom-entering shaft of a 1,000-*l* fermentor (see Fig. 11.6.).[5] During use the top chamber is exposed to sterile condensate, while the bottom chamber is exposed to steam.

As an alternate to shaft seals, magnetically coupled stirrers have been devised for laboratory-scale fermentors. The couple is achieved by a ceramic magnet encased in stainless steel; power up to 1/4 HP can be transmitted by these systems. For details, see reference 5.

11.3. ASEPTIC OPERATION

11.3.1. Pipelines and valves

The pipelines transporting sterile air, seed, and other materials required for aseptic use should be sterilized with steam (usually at 120°C for 20–30 min). It is important that the pipelines be free from steam condensate after sterilization; therefore, the line should be constructed as simply as possible. Except in particular fermentations which are vulnerable to metals, ordinary steel pipes may be employed for the transport of fluids.

It is usually necessary to install valves in the pipeline. If a high order of sterility is needed, the diaphragm type of valve or ball valves are recommended. The diaphragm valve has no packing gland and is less liable to contamination. Ball valves can be doubly sealed to minimize chances of contamination. Where cost is a factor, ball valves and nonrising-stem globe valves are used in the fermentation industry.

11.3.2. Aseptic inoculation

Figure 11.7 shows the connections required for the aseptic transfer of a spore suspension to a seed tank. The spore-suspension vessel and its piping are first

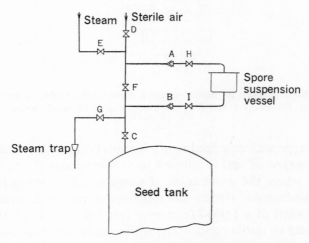

Fig. 11.7. The aseptic transfer of a spore suspension into a seed tank.[4]

sterilized, cooled, and then the spores introduced into the vessel; the system is connected at both A and B with a pipeline leading to a seed tank. The connections at A and B are first slackened so that steam bleeds from A and B when steam flows through open valves E, F, and G, with valves D, H, I, and C closed (see figure). After sterilization with steam at 120°C for about 20 min, valves E and G are closed and the connections at A and B are tightened. Valve D is then opened and the line is cooled to the desired level under positive pressure with sterile air.

At the correct temperature, valve F is closed, valves H, I, and C are opened, and sterile air is used to blow the spore suspension from the vessel to the seed tank. Valves D, C, H, and I are then closed and the spore suspension vessel is disconnected at A and B.

Figure 11.8 illustrates another technique for aseptic inoculation of a fermentor from a seed tank. Two vessels are connected with a flexible pipe at joints A and B. To sterilize the medium in the fermentor, steam from valves J and G passes through D, E, and F into the fermentor to heat the medium to 120°C for 20 min. During this period, valve C is closed, while valves H and I are slightly opened to bleed steam and to remove condensate. With the completion of steam sterilization, valves G, J, H, and I are closed, while valves F, E, and D are left open. The medium is cooled to the incubation temperature under positive pressure using sterile air (connections to the sterile air are not shown in Fig. 11.8). Valve C is then opened and the seed culture is transported to the fermentor either by gravity or by the pressure difference between the seed tank and the fermentor. Finally, valves C and F are closed and the inoculation line is re-sterilized with steam before the flexible pipe is detached from A and B.

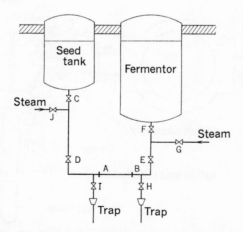

Fig. 11.8. Services to allow aseptic inoculation of a fermentor.[4]

11.3.3. Aseptic sampling

It is frequently necessary to sample the broth during fermentation; Fig. 11.9

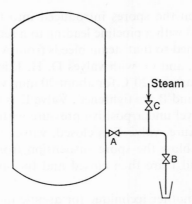

Fig. 11.9.   Sampling point.[4]

shows a sampling point which can be steam sterilized. Normally valves A, B, and C are closed and the end of the sampling pipe is immersed, for example, in 40% formaldehyde solution.

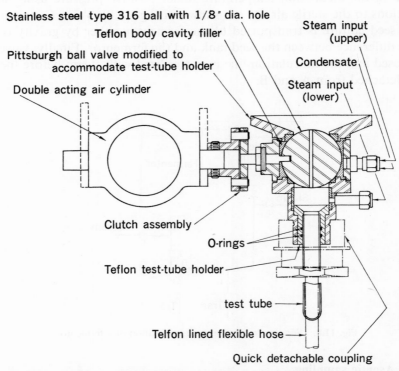

Fig. 11.10.   Quick sampling assembly. Operation example: Vessel back pressure = 0.35 – 1.05 Kg/cm² gage, Minimum sample-time setting = 0.1 to 0.2 sec. (*Courtesy* Fermentation Design, Inc., Allentown, Pa., U.S.A.)

When a sample is required, the vessel containing the germicide is removed and steam is blown through C and B long enough to sterilize the section. After that, C and B are partly closed to bleed some steam and condensate through B. Valve A is then opened slightly to let some broth pass to waste and to cool the line. Valve C is closed and broth from the fermentor is collected in a sterile bottle. After sampling, the outlet is re-sterilized with steam by closing A and then put in the out-of-use arrangement.

Recently a time-controlled and air-activated ball valve which can be independently sterilized with steam and cooled has been used for sampling. Figure 11.10 shows a picture of this valve. Since the valve can be completely flushed prior to operation and since it projects into the fermentation broth, sampling errors due to improperly flushed lines can be minimized.

11.4. SUMMARY

1. The fermentor is the heart of any fermentation process. The design may be very simple (aeration tanks in the activated sludge process of waste treatment) or very sophisticated (highly instrumented fermentor for antibiotic production).

2. Good practice in fermentor design embodies such cardinal rules as:
   a. No direct connections should be made between sterile and nonsterile parts,
   b. Flange connections should be minimized,
   c. Welded construction only should be used,
   d. The vessel should be capable of operating under positive pressure,
   e. Various parts should be independently sterilizable,
   f. Dead spaces should be avoided,
   g. Valves should be easy to clean and maintain, and
   h. All vessel connections should be steam-sealed.

3. The agitator assembly, including the seal, is often a source of contamination. Care must be used in its design and maintenance.

4. Sizes of typical agitator motor drives are:

Size	Power (W/l)
Laboratory	8 – 10
Pilot plant	3 – 5
Plant	1 – 3 (1 – 4 HP/m^3)

5. Aseptic operation demands that pipelines transporting sterile air, seed, and other materials required for the fermentation be sterilized by exposure to steam (usually at 120°C) for at least 20 to 30 min.

6. Ball valves have proved easy to maintain and operate aseptically in the fermentation industry; they are most useful for seed and sample lines.

REFERENCES

1. Friedland, W. C., Peterson, M. H., and Sylvester, J. C. (1956). "Fermentor design for small-scale submerged fermentation." *Ind. Eng. Chem.* **48**, 2180.
2. Greenshields, R. N., Alagaratnam, R., Coote, S. D. J., Daunter, B., Morris, G. G., Imrie, F. K. E., and Smith, E. L. "The growth and morphology of microorganisms in tower fermentors." (1972). *Abstracts, IV Inter. Ferm. Symposium, Kyoto, Japan* p. 270.
3. Kitai, A., Tone, H., and Ozaki, A. (1969). "The performance of perforated plate column as a multi-stage continuous fermentor. (I) Wash-out and growth phase differentiation in the column." *J. Ferm. Tech. (Japan)* **47**, 333.
4. Parker, A. (1958). "Sterilization of equipment, air and media." *Biochemical Engineering* p. 97. (Ed.) Steel, R. Heywood, London.
5. Solomons, G. L. (1969). *Materials and methods in fermentation* Academic Press, London and New York.

Chapter 12

Instrumentation for Environmental Control

In scaling up any successful fermentation, the molecular biologist would suggest the necessity for controlling the environment and hence regulating the fermentation. However, two problems arise. Firstly, knowledge may not exist of the regulation mechanism of metabolic pathways which produce the desired product. In fact, the metabolic pathways may not be fully known or understood. Secondly, even if the pathways and the regulatory mechanism are known, the necessary instrumentation to detect regulatory metabolites may not exist. The engineering approach is, therefore, presented with a dilemma in developing a new fermentation.

Should the biochemical engineer focus his efforts on researching the mechanisms of product regulation and control, developing needed analysis and sensing instrumentation to provide this control or should he use trial-and-error development procedures for strain and medium selection to evolve an apparent optimal environment for plant-scale production? The answer at this time is that he must do both.

In the past, metabolic controls were simply bred into or out of fermentation organisms through mutation and strain selection. However, recently tremendous strides have been and are being made in sensor development. The biochemical engineer will soon be able to rely more on environmental control to gain economic fermentation results. Figure 12.1 demonstrates a highly instrumented fermentor which is designed to secure basic information on environmental control and on coupling with computers.

To achieve meaningful fermentation control, the following three steps are necessary.

a. Carry out fermentation research on *fully monitored* environmental systems.
b. *Correlate* the environmental observations with existing knowledge of cellular control mechanisms.
c. Reproduce the desired environmental control conditions through continuous *computer monitoring, analysis,* and *feedback control* of the fermentation environment.

Step *c*, without proceeding through the previous, *a* and *b*, has frequently been taken by the biochemical engineer. This is a kind of "black box" approach (*cf.* Section 5.2.1., Chapter 5). Control has been established by perturbing the system,

317

Fig. 12.1. A highly instrumented fermentor (working volume=40 *l*) (*Courtesy* Fermentation Design Inc., Allentown, Pa., U.S.A.)

i.e., fermentation, and then looking at the response. From the observed responses a model is constructed and a computer-based feedback control loop is designed to operate the fermentation based on the model. This approach has been a vital evolutionary step simply because adequate environmental monitoring systems have not been available. A more scientific approach to computer control is now possible because of advances that are taking place in sensor development.

12.1. STATUS OF ENVIRONMENT SENSORS

In general, fermentor sensors can be divided into those involved with the physical and the chemical environment. Table 12.1 lists the physical environment factors, while Table 12.2 is concerned with those related to the chemical environment.

For the most part, adequate sensors for monitoring and controlling the physical environment do exist. Perhaps the only exceptions are good sensors for monitoring broth viscosity and cell density. Though ways do exist for measuring these factors, they are far from adequate. With regard to chemical environment sensors the sta-

TABLE 12.1

SYSTEMS INVOLVED IN CONTROLLING THE PHYSICAL ENVIRONMENT.

Temperature
Pressure
Agitator shaft power
Foam
Flow rate (gas and liquid)
Turbidity
Viscosity

TABLE 12.2

SYSTEMS INVOLVED IN CONTROLLING THE CHEMICAL ENVIRONMENT.

Existing systems

pH
Redox
Dissolved oxygen
Dissolved carbon dioxide
Oxygen in exit gas
Carbon dioxide in exit gas
Precursor level and feed rate
Sugar (carbon) level and feed rate
Protein (nitrogen) level and feed rate

Potential systems

Mineral ion level
Mg^{2+}, K^+, Ca^{2+}, Na^+, Fe^{3+}, SO_4^{2-}, PO_4^{3-}
RNA
DNA
NAD, NADH
ATP, ADP, AMP

tus is quite different. Adequate sensors do not exist except for pH, redox, dissolved oxygen, and exit gas (oxygen and carbon dioxide).

One major difficulty with developing fermentor sensors is that they are to be used "in tank," i.e., they must be capable of withstanding sterilization conditions.

12.2. PHYSICAL ENVIRONMENT SENSORS[11]

12.2.1. Temperature

Temperature can be monitored by: *a.* Thermometer bulbs (Hg in steel), *b.* Thermo-couples, *c.* Thermistors, and *d.* Metal resistance thermometers.

In recent years a thermistor coupled to a cascade control system appears to be the preferred device for temperature regulation.

12.2.2. Pressure

Pressure can be detected by relatively simple diaphragm gages. Their utility stems from the ease in operating them in a sterile manner. Pressure in the fermentor vessels can be regulated by a simple back-pressure regulator valve in the gas exit line.

12.2.3. Agitator shaft power

It is usual to measure the power consumption with a wattmeter attached to the motor of the agitator in a large-scale formentor. Although this approximation in measuring the power is permissible in commercial equipment, it cannot be adopted in a bench-scale fermentor, primarily because friction in the stuffing box contributes significantly to the loading of the motor (*cf.* Section 11.2.).

Two main systems have been used for measurement of agitator shaft power in bench-scale fermentors, i.e., *a.* Torsion dynamometer, *b.* Strain gage. Because the dynamometer system is necessarily external to the vessel, its signal includes bearing

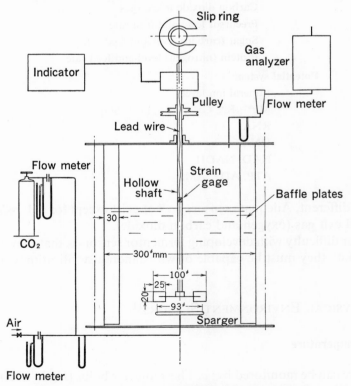

Fig. 12.2.   Fermentor equipped with strain gage system.[1]

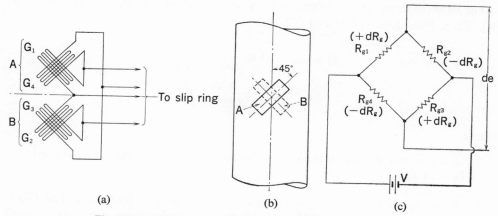

Fig. 12.3. Strain gage mounting and monitoring arrangements.[1]

and seal frictional losses. Hence, the strain gage is preferred because of its greater accuracy. In this system, balancing strain gages are mounted on the agitator shaft inside the fermentor vessel. Lead wires from the gages pass out of the vessel via an axial hole in the shaft. The electric signal is picked up from the rotating shaft by an electrical slip ring arrangement. For details of the strain gage system for monitoring shaft power input, see Figs. 12.2 and 12.3.

If the four identical strain gages are connected as shown in Fig. 12.3 (c) and each gage is exposed to a longitudinal stress, σ, resulting in a strain, $\pm\varepsilon$, in each gage, it is readily shown that the value of the potential, de, generated between the terminals of the bridge is:

$$de = VK_s\varepsilon$$

(12.1)

where

V = electric potential applied
K_s = gage factor

To measure the torque of a rotating shaft, the gages are mounted on the shaft so that they are tilted to the axis by 45°, as shown in (a) and (b) of Fig. 12.3. The angle of 45° is selected because the shearing strain, ε, becomes maximum, ε_{max}, in this arrangement. Further, for the specific arrangement of strain gages, the value of ε_{max} can be equated to:

$$\varepsilon_{max} = \frac{8Md_o}{\pi(d_o^4 - d_i^4)G}$$

(12.2)[13]

where

G = modulus of rigidity
M = torque
d_o = outer diameter of hollow shaft
d_i = inner diameter of hollow shaft

The value of de in Eq. (12.1) can be obtained by amplifying the indicator reading, $[R]$. Thus, from Eqs. (12.1) and (12.2),

$$[R] = \kappa de = \kappa V K_s \frac{8 M d_o}{\pi(d_o{}^4 - d_i{}^4)G}$$

$$= \frac{1}{\alpha} \frac{8 M d_o}{\pi(d_o{}^4 - d_i{}^4)G} \qquad (12.3)$$

where

κ = conversion factor which varies depending on the specific indicator used

$1/\alpha = \kappa V K_s$

If the gages are coated with polyester material, the shaft can be sterilized at 120°C without disturbing the gages. This type of dynamometer allows the agitator shaft power, which is obtained by multiplying the observed value of M by the angular velocity of the impeller, to be recorded continuously in a fermentation.

12.2.4. Foam

Foam detection and control can be achieved either by: *a*. Mechanical foam destruction, *b*. Chemical anti-foaming agents, *c*. A combination of both.

Most mechanical systems are self-regulating and need no sensors. This is true of such systems as the centrifugal defoamer, etc. When there are severe foaming problems, mechanical systems can be overwhelmed, when they act more like foam densifiers. For this reason fermentors with mechanical defoamers incorporate an anti-foam addition system as a stand-by safety device.

Figure 12.4 shows an anti-foam system; if the foam contacts a rubber-sheathed electrode, an electric unit actuates a solenoid valve to allow the passage of a sterile anti-foaming agent into the fermentor. To secure uniformity of dispersion of the agent on the surface of the broth, a deflection trough is provided as shown in Fig. 12.4. The anti-foaming agent is thus distributed uniformly over the foaming surface; its action is supplemented by the centrifugal effect of the foam breaker. The amount of anti-foaming agent entering the fermentor is usually controlled by a timer in the circuit to the solenoid valve.

Silicone emulsion anti-foams are considered to be high performance, easy-to-sterilize defoamers. However, they are usually prohibitive in cost for commercial fermentations. Oil-based anti-foams are utilized for most industrial fermentations, but have a severe sterility requirement. Often such defoamers need exposure to dry temperature conditions at 160°C for 2 – 3 hr to ensure sterility.

12.2.5. Flow rate (gas and liquid)

A number of devices can be used to measure and control gas flow rate. Variable-

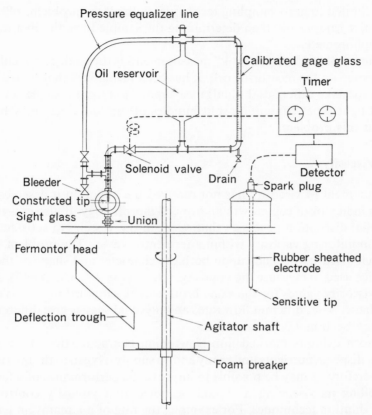

Fig. 12.4. Ancillary equipment to control foaming in fermentation broths.[7]

area flow meters such as rotameters appear to be preferred from an economical point of view. The position of the float in the meter is converted to an electric signal through either a capacitance or resistance principle. This signal is then used, with the aid of a controller, to regulate a flow valve in the gas line.

Numerous systems have been proposed to monitor and control liquid feed rate. An electromagnetic flow meter can be used to monitor liquid addition. In general, this system is expensive and unsatisfactory for small flow rates. Mounting the addition vessels on load-cell systems permits continuous weighing of the contents, and hence, measures the additions. Although this system tends to be insensitive to short-term variations, it has the advantage of long-term accuracy.

12.2.6.   Turbidity

Measurement of fermentation broth turbidity inside the vessel, although technically possible, gives many operational difficulties. This measurement is most important, because it is a rapid means of following cell growth. At present turbidity measurements are limited to grab-sampling techniques with all the problems of obtaining a representative sample and then determining the sample turbidity in a turbidimeter or spectrophotometer.

The primary obstacle to "in tank" measurements is the tendency of culture debris to be deposited on the measuring cell. It has been recognized that to overcome this, one needs to make differential (dual) measurement(s) from a single source. Newly developed light pipes containing split bundles of random fibers may be useful to circumvent this problem.

12.2.7. Viscosity

The measurement of viscosity has not received the attention it deserves. It can be used as an indicator of cell growth and/or cell morphology. It also can yield insight into mycelial disruption processes that occur in the vessel. No satisfactory system exists for monitoring viscosity within a fermentor vessel. The problem stems from the fact that viscosity depends upon both the character and shape of the flow field in the device used to measure the viscosity. (*cf*. Section 7.3., Chapter 7). As a result, different viscosity values for the same broth can be obtained in vessels differing in size and shape. Also, different flow rates and mycelial shapes give different viscosity readings (*cf*. Section 7.3.2.).

As a broth exhibits marked non-Newtonian characteristics, it becomes more difficult to disperse nutrients uniformly and to supply oxygen to the microbial population. Therefore, it may be possible to improve the performance of a fermentation by controlling its viscosity; it should be noted that viscosity control has been limited to dilution techniques. For example, the rate of a kanamycin fermentation could be improved by diluting the broth (the mycelial age = 65-th to 80-th hr) with an addition of 15% of nutrient, thus reducing the value of apparent viscosity by about 50% without decreasing the final potency of the kanamycin.[10]

12.3. CHEMICAL ENVIRONMENT SENSORS

12.3.1. pH

pH measurement is readily accomplished by combined glass-reference electrodes. Figure 12.5 shows a sterilizable pH electrode with which Gualandi *et al*.[5] measured the pH of fermenting broth, using a device placed directly into the fermentor. The half-cell of the glass electrode was composed of Ag/AgCl saturated with solid KCl.

The half-cell of the reference electrode consisted of the same material as the glass electrode, an asbestos or porcelain cylinder being used as the junction material.

To ensure good insulation, both the glass and reference electrodes were mounted in Teflon gaskets and silicone rubber washers (see Fig. 12.5). The internal resistance

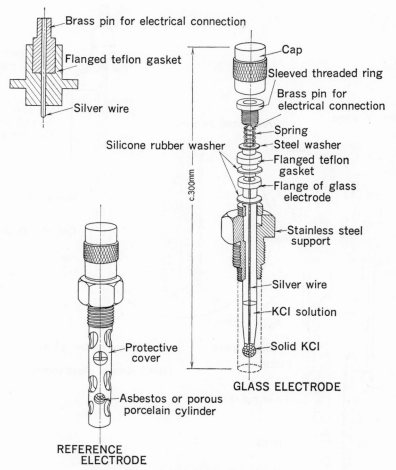

Brass pin for electrical connection

Flanged teflon gasket

Silver wire

Cap

Sleeved threaded ring

Brass pin for electrical connection

Spring

Steel washer

Flanged teflon gasket

Flange of glass electrode

Silicone rubber washer

c.300mm

Stainless steel support

Silver wire

KCl solution

Solid KCl

GLASS ELECTRODE

Protective cover

Asbestos or porous porcelain cylinder

REFERENCE ELECTRODE

Fig. 12.5.   Sterilizable pH electrode.[5]

of the electrode was 300 – 500 MΩ. Both electrodes were protected by steel sleeves provided with several holes to allow free passage of the broth, as shown in the figure.

Although the chemical composition of the glass membrane was not given in the publication of Gualandi *et al.*, it is noted that these electrodes were used successfully and continuously without replacing the KCl for 200 hr, and during this period, the electrodes were sterilized with steam three times.

Figure 12.6 shows the "3-in-1" type of pH electrode which is widely used in the fermentation industry. The glass electrode (specific silver chloride electrode), reference electrode and thermal compensator are assembled to form the single pH electrode. This holder is characterized by forced flow of a small amount of KCl solution in the reference electrode into the broth through the ceramic junction at a

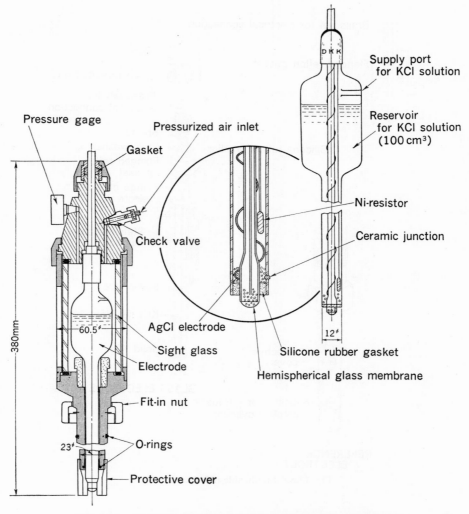

Fig. 12.6. "3-in-1" pH electrode[14] (DKK, Musashino-shi, Tokyo).

rate of about 0.2 ml/hr when the pressure difference between the inside and outside of the electrode is 1.0 Kg/cm². This forced flow of KCl solution keeps the junction surface to the broth always clean and guarantees a very stable potential between the electrode and the broth.[14] Even after 50 sterilizations with steam at 130°C for 30 min, this electrode is claimed to exhibit no deterioration of its performance in terms of zero point shift (within ± 15mV), rapid response to pH (within 1 – 2 min) and mV/pH variations (98 – 99%).

A problem exists with some pH systems in that the reading may be referenced to the instrument ground, but in fact, may be interacting with a different ground through the temperature control system. Sometimes an isolating transformer on the power supply plus the grounding of the instrument panel and fermentor to a common source helps, but this does not completely eliminate the problem.

12.3.2. Redox

Redox control has been achieved by gaseous sparging with nitrogen or oxygen or by chemical control with cystine, ascorbic acid, or sodium thioglycolate. Redox measurements are almost universally made with a combined platinum-reference electrode system. The problem lies in interpretation of the reading. The cellular system can be redox-poised differently from the bulk solution. Further, the dissolved oxygen strongly interacts with the redox system. Since oxygen can be a growth-regulating substrate, a complicated response to redox can exist in aerobic fermentations.

12.3.3. Dissolved oxygen

Perhaps the most significant advance in the area of fermentation instrumentation has been the development of the "in place," (steam-sterilizable) dissolved oxygen sensor. Figure 12.7 illustrates the typical oxygen sensors [Beckman, Au(cathode)-Ag(anode)(polarographic); Mackereth, Ag(cathode)-Pb(anode)(voltametric)]. Each electrode is covered by a polymer membrane and an electrolyte is sandwiched between the cathode and the membrane.

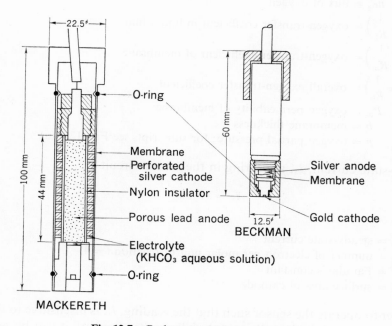

Fig. 12.7.　Beckman and Mackereth probes.

Suppose that a membrane-covered electrode is immersed in a liquid medium, and that the oxygen reduced at the cathode surface is supplied from outside the electrode. Oxygen is reduced at the cathode as follows:

$$O_2 + 2H_2O + 4e \longrightarrow 4OH^- \tag{12.4}$$

Oxygen partial pressure profile in the oxygen sensor in the steady state is shown schematically in Fig. 12.8 (a). Resistances to oxygen flux are designated as R_L, R_b, R_m, R_0', and R_E in the figure. If the membrane is tightly attached to the cathode surface, values of R_0' and R_E are assumed negligible. The value of p_a at the cathode surface is equated to zero due to the rapid electrochemical reaction, and the resistance, R_b, at the medium-membrane interface is also assumed zero.

Oxygen flux, n_{O_2}, in the steady state can, then, be expressed by the following equations (cf. Section 6.1.1., Chapter 6).

$$n_{O_2} = k_L(p_L - p_m) \tag{12.5}$$

$$= k_m(p_m - p_0) \tag{12.6}$$

$$= (P_m/b)(p_m - p_0) \tag{12.7}$$

$$= K_0(p_L - p_0) \tag{12.8}$$

where

$$n_{O_2} = \text{flux of oxygen}$$
$$k_L\left(=\frac{1}{R_L}\right) = \text{oxygen-transfer coefficient in liquid film}$$
$$k_m\left(=\frac{1}{R_m}\right) = \text{oxygen-transfer coefficient of membrane}$$
$$k_0\left(=\frac{1}{R_0}\right) = \text{overall oxygen-transfer coefficient}$$
$$P_m = \text{oxygen permeability of membrane}$$
$$b = \text{membrane thickness}$$
$$p = \text{oxygen partial pressure; for subscript, see Fig. 12.8 (a)}$$

Steady-state current, i_∞, appearing in the sensor circuit is:[2]

$$i_\infty = nFAn_{O_2} \tag{12.9}$$

where

$$i_\infty = \text{steady-state current}$$
$$n = \text{number of electron per molar unit of reaction } (= 4)$$
$$F = \text{Faraday's constant}$$
$$A = \text{surface area of cathode}$$

In order to operate the sensor such that the reading, i_∞, is insensitive to the physical condition of a liquid medium, especially to its viscosity, it can be seen from

Eqs. (12.5) to (12.9) that the resistance, $R_L(=1/k_L)$, of the liquid film outside the membrane must be minimized as a fraction of the overall resistance, R_0, to the oxygen flux $(R_L/R_0 \leq 0.05)^4$ (*cf.* Section 6.1.1., Chapter 6).

Figure 12.8 (b) shows that the value of i_∞ measured with a specific sensor (not shown in Fig. 12.8) deteriorates with increase of liquid viscosity. On replacing the Teflon membrane with a thicker one or with one which had a lower oxygen permeability, the sensor became less sensitive to the liquid viscosity [see Fig. 12.8 (c)].[4]

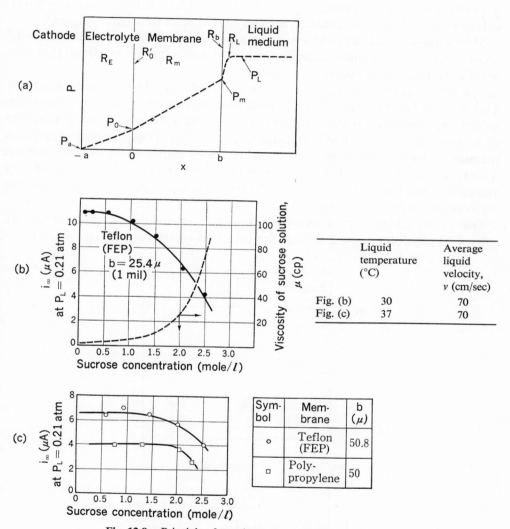

	Liquid temperature (°C)	Average liquid velocity, v (cm/sec)
Fig. (b)	30	70
Fig. (c)	37	70

Symbol	Membrane	b (μ)
o	Teflon (FEP)	50.8
□	Polypropylene	50

Fig. 12.8. Principle of membrane-covered oxygen sensor.

(a) Oxygen partial pressure profile in steady state.

(b) Example of steady-state current, i_∞, decreasing with an increase of liquid viscosity.

(c) Example to correct for the previous decrease of i_∞ by replacing the membrane.

However, it should be noted that a sensor corrected for liquid viscosity became less sensitive to changes in oxygen concentration in the medium; i.e., the response time became longer.

In general, the response of these membrane-covered sensors is slow, i.e., the 90% response is of the order of 10 to 100 sec. While this is adequate for monitoring average oxygen levels and long-term uptake, these sensors only give a qualitative picture of the short-term oxygen fluctuations. A response correction factor must be introduced when instantaneous oxygen uptake rates of microbes are to be measured with the sensors by dynamic methods (see Section 12.4.1.). For response time calculations and membrane selection, readers should refer to original papers.[2-4]

Membrane-covered sensors are quite sensitive to temperature fluctuations;[2] these sensors must be temperature compensated with a thermistor circuit that electrically matches the temperature behavior.

The tubing method presents another means for dissolved oxygen determination. This method has been described for measuring the concentration of various dissolved gases including oxygen. It consists of passing a stream of inert gas (helium or nitrogen) through a coil of permeable tubing (Teflon or polypropylene) and then measuring the amount of gas that is picked up by diffusion through the tubing. This system can be useful for long-term monitoring purposes. However, it is a very sluggish monitoring system, i.e., it has 2 - 10 min delays; hence it has limited value for control purposes (see Fig. 12.9).

Due to the attention paid recently to the effect of dissolved carbon dioxide on microbial physiology, some sensors for dissolved carbon dioxide are available.[6,8]

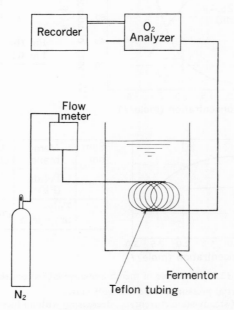

Fig. 12.9. A probe of Teflon tubing coupled to a gas analyzer for the measurement of dissolved oxygen.[9]

12.3.4. Exit gas composition

Exit gas composition can be monitored fairly easily. Long path infra-red analyzers or thermal conductivity devices have been successfully used for carbon dioxide measurement. Paramagnetic analyzers have been utilized for oxygen analysis. A much cheaper, although slower responding, alternative to the paramagnetic analyzer is a system in which the exit gas is bubbled through slowly replenished water; the dissolved oxygen in the solution is continuously monitored. When proper temperature control is exercised, this system is as reliable and certainly easier to maintain than the paramagnetic analyzers for exit oxygen analysis.

12.3.5. Intermediary metabolites

One way of following cellular activities is to monitor the level of key intermediates. Two such key intermediary systems are: a. NAD - NADH$_2$, b. ATP - ADP - AMP. The former is an indicator of catabolic activity and the latter of anabolic activity (cf. Chapter 3). Fluorimetric techniques have been recently developed for monitoring NADH$_2$ levels in whole fermentation broths. A combined enzyme and fluorimetric analysis has been developed for ATP assay. However, to date ATP assay has not been applied to a continuous monitoring set-up.

12.3.6. Other chemical factors

If environmental control of fermentations is to be achieved, then all chemical factors that regulate or inhibit growth and/or product formation must be monitored. This could include substrates containing C, N, P, S, Mg, K, Ca, Na, Fe, plus certain complex growth regulators and product precursors. A number of interesting developments are in the offing. For example, prototype enzyme sensors have been developed for glucose and urea detection (cf. Chapter 14). Ion-specific sensors are available for the measurement of NH_4^+, Mg^{2+}, Na^+, Ca^{2+}, PO_4^{3-}, etc. The auto-analyzer-type systems can be used for measurement of precursors and complex growth regulators. Unfortunately, none of these sensors are at a state where they can be used for "in tank" continuous monitoring. The primary reason is that these systems cannot withstand steam sterilization conditions necessary for the maintenance of sterile operation. An external chemical sterilization system coupled with an aseptic inserting mechanism remains to be developed to permit utilization of these sensors in fermentations.

12.4. GATEWAY SENSORS

Several of the monitoring systems described previously are "gateway" sensors,

TABLE 12.3

GATEWAY SENSORS.

Sensor	Information
pH	Acid product formation
Dissolved oxygen	Oxygen transfer rate
Oxygen in exit gas } Gas flow rate }	Oxygen uptake rate
Carbon dioxide in exit gas } Gas flow rate }	Carbon dioxide evolution rate
Oxygen uptake rate } Carbon dioxide evolution rate }	Respiratory quotient
Sugar level and feed rate } Carbon dioxide evolution rate }	Yield and cell density

i.e., they open the way, through combination with other sensor systems and data, to obtaining further information about the fermentation (see Table 12.3.). With these sensor systems, computer coupling, data reduction and analysis have the greatest attraction. For example, pH control coupled with monitoring of base or acid addition can give the instantaneous rate of product formation plus cumulative product data. Differentiation of the dissolved oxygen trace during brief interruption of aeration can lead to determination of the volumetric respiration rate, Q_{o_2}/X, and the volumetric oxygen-transfer coefficient, k_La.

12.4.1.   Oxygen uptake rate; volumetric oxygen-transfer coefficient; carbon dioxide evolution rate

The rate of change, $d\bar{C}/dt$, in dissolved oxygen concentration at a particular point in a fermentor vessel is given by the following equation (cf. Eq. (6.10), Chapter 6).

$$\frac{d\bar{C}}{dt} = k_La(C^* - \bar{C}) - Q_{o_2}X \qquad (12.10)$$

where

C^* = concentration of dissolved oxygen which is in equilibrium with partial pressure, $\bar{p}$ in bulk gas phase
$\bar{C}$ = concentration of dissolved oxygen in bulk liquid
k_La = volumetric oxygen-transfer coefficient
Q_{o_2} = specific rate of oxygen uptake (microbial respiration)
X = cell mass concentration

If the term, $C^* - \bar{C}$ is taken as an average rather than a point value in the fermentor vessel, the second term, $Q_{o_2}X$, in the right-hand side of Eq. (12.10) deals necessarily with an average for the cells throughout the vessel. For convenience, the logarithmic mean, $(C^* - \bar{C})_{\text{mean}}$ with respect to the vessel bottom and top may be taken in place of $(C^* - \bar{C})$ in Eq. (12.10). In small vessels where perfect mixing is approached, the above averaging treatment is not required.

The term, $Q_{O_2}X$, can then be evaluated from the monitoring of oxygen partial pressures in inlet and exit gas around the fermentor (see Sections 12.3.4. or 12.3.3.).

$$Q_{O_2}X = \frac{F}{V}\left(\frac{p_{\text{in}}}{P - p_{\text{in}}} - \frac{p_{\text{out}}}{P - p_{\text{out}}}\right) \tag{12.11}$$

$$= \text{OUR (oxygen uptake rate)} \tag{12.12}$$

where

p_{in} = oxygen partial pressure in inlet gas
p_{out} = oxygen partial pressure in exit gas
P = total pressure, disregarding the increase in total pressure in inlet gas due to liquid head in fermentor vessel
V = broth volume in vessel
F = inert gas flow rate (molar basis)

In the steady state, the value of $k_L a$ can be determined from:

$$k_L a = \frac{Q_{O_2}X}{(C^* - \bar{C})_{\text{mean}}}$$

$$= \frac{\text{OUR}}{(C^* - \bar{C})_{\text{mean}}} \tag{12.13}$$

Another means to assess the value of $k_L a$ is by using Eq. (12.10) in the unsteady state. The term, $-Q_{O_2}X$ in the equation can be determined from the decrease of $\bar{C}$ with time, if the air supply is turned off, and if the rate of oxygen consumption by the microbes is unaffected by the suspension of this air supply.

When aeration begins again, the value of $\bar{C}$ increases and levels off $(d\bar{C}/dt = 0)$, once the oxygen supply and consumption rates are balanced. The dynamic method of measuring the value of $k_L a$ from Eq. (12.10) is first to observe $-Q_{O_2}X$ (the rate of decrease in $\bar{C}$ values following the temporary suspension of aeration), and secondly, to observe the rate of increase in $\bar{C}$ values when aeration is resumed [see Fig. 12.10 (a)].

Rearranging Eq. (12.10),

$$\bar{C} = \left(-\frac{1}{k_L a}\right)\left(\frac{d\bar{C}}{dt} + Q_{O_2}X\right) + C^* \tag{12.14}$$

If the value of $\bar{C}$ observed during the latter period, i.e., the transient period associated with a resumption of aeration is plotted against the value of $(d\bar{C}/dt + Q_{O_2}X)$, a straight line should be obtained. The slope of this line represents the value of $-1/k_L a$.

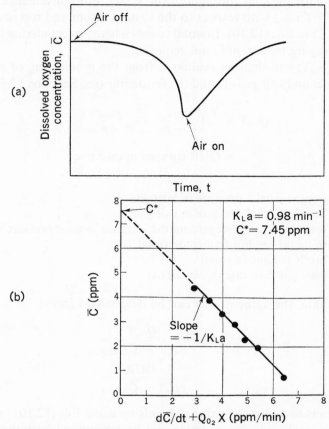

Fig. 12.10. Dynamic measurement of volumetric oxygen-transfer coefficient, $k_L a$.
(a) Typical pattern of dissolved oxygen concentration as varied with time during ungassed and regassed periods.
(b) Example of $k_L a$ determination with *Asp. niger* culture (aeration=1.2 vvm, apparent viscosity=300 cp, agitation speed of impeller=500 rpm).[12]

Figure 12.10 (b) shows an example of this dynamic method for the assessment of $k_L a$ in a mycelial broth. The Beckman oxygen sensor was used to measure the rates of decrease or increase of $\bar{C}$ values. It is seen from the figure that Eq. (12.14) can be used for the determination of $k_L a$. It is also noted that this example gives a point value of $k_L a$ rather than an average throughout the vessel, as is remarked from Eq. (12.13).

Generally, the sensor response characteristic must be considered in the dynamic technique; implicitly, the reponse is assumed rapid enough in the previous example. In order to illustrate the response characteristic of the oxygen sensor in connection with the assessment of $k_L a$, a simple aeration system where no cells are present will be considered.

The rate of change, $d\bar{C}/dt$, in dissolved oxygen concentration in this system is:

$$\frac{d\bar{C}}{dt} = k_La(C^* - \bar{C})$$

(12.15)

$$\bar{C} = 0 \quad \text{at} \quad t = 0$$

(12.16)

From Eqs. (12.15) and (12.16),

$$\bar{C} = C^*\{1 - \exp(-k_Lat)\}$$

(12.17)

The sensor is assumed to exhibit first-order response to the change in $\bar{C}$ values in bulk liquid.[3]

$$\frac{d\bar{C}_p}{dt} = k_p(\bar{C} - \bar{C}_p)$$

(12.18)

where

$\bar{C}_p$ = concentration of dissolved oxygen which corresponds to the sensor reading
k_p = sensor constant, including the conductance of membrane and liquid film outside the sensor (cf. Section 12.3.3.)

$$\bar{C}_p = 0 \quad \text{at} \quad t = 0$$

(12.19)

Solving Eqs. (12.15) and (12.18) under the conditions of Eqs. (12.16) and (12.19) to yield,

$$\bar{C}_p = C^* \left\{ 1 + \frac{k_La}{k_p - k_La} \exp(-k_pt) - \frac{k_p}{k_p - k_La} \exp(-k_Lat) \right\}$$

(12.20)

Now, by observing the values of $\bar{C}_p$ as a function of time and knowing k_p in advance, the value of k_La can be estimated through a trial-and-error procedure. This illustrates an advantage of on-line computer analysis which can solve Eq. (12.20) instantaneously.

If $k_p \gg k_La$ in Eq. (12.20), the equation is reduced to Eq. (12.17). This implies that the sensor response is quite rapid and the reading (as converted to dissolved oxygen concentration in bulk liquid) fairly accurately follows the change in $\bar{C}$ values.

The rate of carbon dioxide evolution can be obtained in a manner similar to the assessment of OUR [see Eq. (12.11)].

$$\text{carbon dioxide evolution rate} = \frac{F}{V} \left(\frac{p_{\text{out}}}{P - p_{\text{out}}} - \frac{p_{\text{in}}}{P - p_{\text{in}}} \right)$$

(12.21)

where

p_{out} = carbon dioxide partial pressure in exit gas
p_{in} = carbon dioxide partial pressure in inlet gas
P = total pressure of gas
V = broth volume in vessel
F = inert gas flow rate (molar basis)

12.4.2. Other control information

Knowing the carbon dioxide evolution rate and the oxygen uptake rate, the respiratory quotient can be determined from their ratio. This quotient is an indicator of unit cellular activity in the fermentor. From the volumetric respiration rate and the cell density, the specific rate of respiration and cell yield can be obtained [cf. Fig. 6.1 and Eq. (5.37)]. Under situations in which sugar is being continuously fed as the growth-limiting substrate, it is possible to calculate the cell mass from a carbon material balance around the fermentor.

To do this, the monitoring of sugar addition rate and broth sugar level must be combined with data concerning carbon dioxide evolution rate. The carbon which is unaccounted for is then assumed to be incorporated into the cell mass. Needless to say, some carbon is lost as volatile acids and alcohols. However, this loss can generally be related to the respiratory quotient. It is here where coupling of the computer to the fermentor sensors can be particularly useful in making this analysis essentially instantaneously.

12.5. DIRECT AND INDIRECT CONTROL

Control of fermentations can be exercised in several ways. The most obvious is direct control. Here the controlling environmental factors are monitored and directly manipulated. Examples of such systems can be found in the continuously operating turbidostat or oxystat. In the latter system, environmental conditions are selected so that dissolved oxygen controls the growth rate. Then, the dissolved oxygen is continuously monitored in the fermentor and the feed rate, along with those of other substrates, is manipulated to maintain the desired steady state.

An example of indirect control is the operation of a chemostat. No environmental factors are controlled directly; rather, fermentor feed rate and broth volume are controlled (see Section 5.1., Chapter 5). This physical control indirectly regulates the chemical environment. Regulation of this type usually provides stable operation, provided other environmental controls are not being manipulated independently of the primary control.

This latter point gives rise to a problem not fully appreciated in many fermentations. It is the problem of interaction between various physical and chemical environmental factors that may be controlled independently of one another, but which are highly interactive.

The problem of interaction among various control systems is important. Indeed, it is possible that improperly designed and highly controlled fermentors can become much less stable and productive than very simple systems. When this occurs, the advantage of control as a labor-saving device may be lost. In the operation of a highly controlled chemostat, the specific growth rate, μ, can be affected not only by the growth-limiting substrate but also by temperature and pH; the system is highly interactive. If substrate conditions are poorly selected so that there is multiple substrate limitation, then an even higher level of interaction is imposed on the system. The computer, with its rapid analysis capabilities, offers a way around these difficulties.

12.6. Computer-Coupled Fermentors

The complexity of fermentation processes requires better and more efficient methods for data collection and analysis as well as feedback control. Recent advance of computer techniques coupled with considerable reduction of computer costs has placed reliable data acquisition, analysis and control methods via computers within reach for fermentation processes.

It should be emphasized that the present limitations to these systems are not due to the computer; rather, adequate software (computer programs) for computerized fermentations and sensors have not been developed at this time to provide the needed information for evolving a control strategy.[15]

Ultimately, the time will come when a fermentation will be taken to the pilot plant in a fairly rudimentary state. The computer will evolve a dynamic strategy through dynamic optimization techniques to develop a better fermentation process. Hence, most of the process development work will be done by computers and not by a cadre of development engineers.

12.7. Basic Concepts of Computer Control

12.7.1. Layout of facilities

Figure 12.11 presents the major elements of the fermentor instrumentation-computation-logic complex with specific regard to the relationship between the fermentor and the computer.

Sensor signals from the fermentor are amplified (AMP) and displayed on digital panel meters (DPM). The latter converts the analog values to digital forms (BCD) and transmits them to the computer.

The computer has access to the fermentor via limit switches or set points on the individual controllers. The controllers actuate pumps or valves on a real-time basis to alter the controlled variables. In order to assure flexibility the operator can determine the control set points as desired.

This concept obviously requires a highly instrumented fermentor with perfected operation performance both from microbiological and instrumentational points of view (cf. Fig. 12.1).

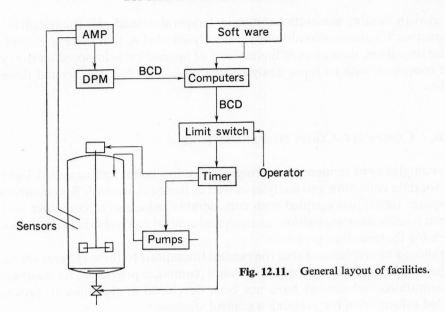

Fig. 12.11. General layout of facilities.

12.7.2. Computer

Figure 12.12 indicates the configuration of the computer with respect to its major functions. The computer performs on-line operation in three basic fields of function, i.e.,

a. Logging of data obtained through the data acquisition system,

b. Analysis of data on the basis of algorithms, and

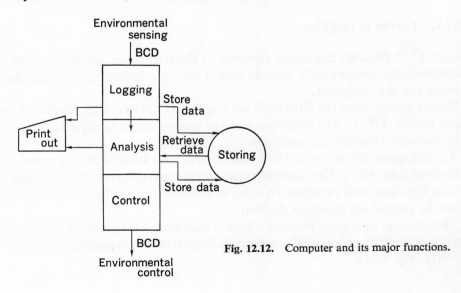

Fig. 12.12. Computer and its major functions.

c. Control of the function of pumps and valves which results in the control of certain process variables.

These functions include the operation of computer peripherals such as storage devices as well as the teletype for the printout of results. Considerations for the choice of a particular computer to perform the data logging, analysis and control functions include:

a. A suitable word capacity which assures the handling requirements of the whole system,

b. An expandable memory capacity which can accommodate future needs such as decision control loops,

c. A large selection of peripheral equipment, interface boards and software library, and

d. Adequate field service ability.

A computer with rapid access (50,000 words/sec) and reasonable memory (20K core memory size) is roughly the size needed for single fermentor coupling.

12.7.3. Computer program

A program to operate the computer must consider the hardware and software capa-

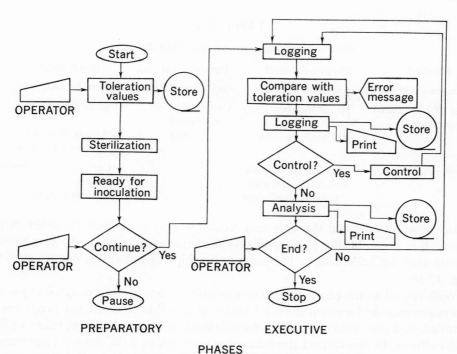

Fig. 12.13. Essentials of computer program for fermentation processes.

bilities, and the critical fact that the computer has to deal with a living system. A typical program structure is presented in Fig. 12.13. A program package consists of major portions, i.e.,

a. Preparatory phase for fermentation, and
b. Executive phase for the fermentation process.

In the preparatory phase, two basic functions are performed:

a. Calculation of toleration values of the process variables which are introduced through the operator's console, and
b. Sterilization.

In the preparatory phase the program sequence has no major program loops and its function can be suspended when the fermentor is ready for inoculation.

The executive phase contains the program routines for data logging, reduction and process control.

12.7.4. Data analysis

One of the objectives of the computer is to provide an insight into the physico-chemical, physiological and biochemical conditions of the process which would otherwise be either unrecognized or could be determined only after the actual condition had passed. A way of grouping algorithms for data analysis is listed in Table 12.4.

TABLE 12.4

GROUPING OF ALGORITHMS FOR DATA ANALYSIS.

Physical	Physico-chemical	Physiological	Biochemical
Agitator shaft power	Apparent viscosity	Oxygen uptake	Carbon balance
Rate of addition of ingredients	Power number	Carbon dioxide evolution	Organic energy yield
	Reynolds number	RQ	Cell mass based on Acid/base titration,
Broth volume	Flow characteristics Volumetric oxygen-transfer coefficient		Carbon dioxide evolution, Carbon balance

Information obtained through data reduction may be utilized for different purposes. For example, the physico-chemical characteristics might be combined to obtain scale-up information. The manner in which this might be done is shown in Fig. 12.14.

With regard to the physiological characteristics, application of reliable gas analysers (oxygen and carbon dioxide) makes it possible to construct programs for detailed, real-time study of carbon dioxide evolution and oxygen uptake of microbial cultures. In this respect the fermentor operates as a differential respirometer. Figure 12.15 shows the basic idea of the study of some physiological characteristics

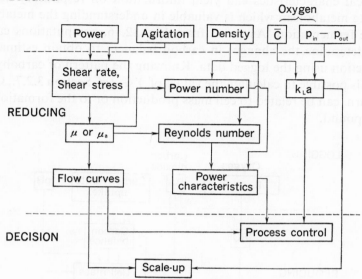

Fig. 12.14. Physico-chemical characteristics.

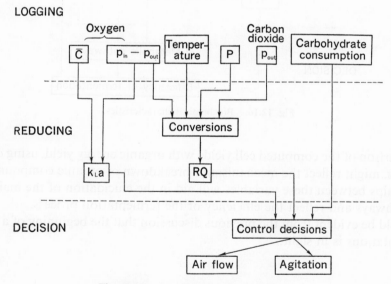

Fig. 12.15. Physiological characteristics.

of cells with the help of the computer. Oxygen uptake and carbon dioxide evolution by the culture can be used to obtain information on the actual respiratory activity of the culture. Respiratory quotient and data on the carbohydrate metabolism, as well as mass-transfer coefficient, can then be interrelated to give decisions that are utilized for process control.

Biochemical characteristics can yield information on respiratory activity and carbohydrate metabolism which is valuable in understanding the metabolic pathways of the fermentation. As noted from Fig. 12.16, computations can be performed to determine the changes in carbon balance; to obtain estimates of cell mass production using the logged data. Knowing the pattern of carbohydrate metabolism, it is possible to calculate the values of Y_{ATP} (cf. Section 3.2.7., Chapter 3), which, in turn, can be related to cell mass production or to the formation of a particular compound.

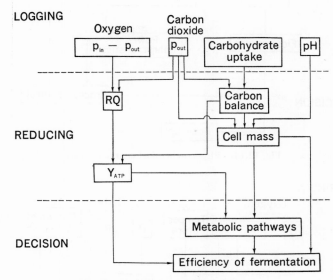

Fig. 12.16. Biochemical characteristics.

Comparison of the computed cell yields with organic energy yield, using different substrates, might reflect the mechanism of breakdown of organic compounds. The relationships between these variables will aid in the elucidation of the main metabolic pathways and reflect the efficiency of the fermentation process.

It should be evident from the previous discussion that the beginning of a new era in fermentations is in sight.

12.8. SUMMARY

1. To optimize the formation of product, fermentations must be carried out in fully monitored, highly instrumented fermentors.

2. Adequate sensors exist for monitoring such physical factors as:

       temperature,

       pressure,

agitator shaft power,
foam, and
gas flow rate.

3. Further development is needed to achieve satisfactory monitors for:
broth turbidity, and
broth viscosity.

4. In general, adequate sensors do not exist for monitoring the chemical environment. Exceptions to this are:
pH,
redox,
dissolved oxygen,
dissolved carbon dioxide, and
exit gas composition (oxygen and carbon dioxide).

5. There is a need to develop sensors for monitoring:
mineral ion levels, i.e., Mg^{2+}, K^+, Ca^{2+}, PO_4^{3-}, etc.,
RNA,
DNA,
$NADH_2$, $NADPH_2$, and
ATP, ADP, AMP.

6. Various sensor signals can be combined via computer analysis to provide
essentially instantaneous information on:
$k_L a$,
Q_{O_2},
Q_{CO_2},
RQ,
X,
ν_{sugar}, and
$\nu_{nitrogen}$.

7. It is now feasible to couple fermentors to computers for:
data logging,
data analysis, and
control.

NOMENCLATURE

A = surface area of cathode, cm^2
b = membrane thickness, cm
$\bar{C}$ = concentration of dissolved oxygen in bulk liquid, ppm; g mole/l
$\bar{C}_i$ = concentration of dissolved oxygen in bulk liquid at $t = 0$, degassing system,
ppm; g mole/l
$\bar{C}_p$ = concentration of dissolved oxygen which corresponds to the sensor reading,
ppm; g mole/l

C^* = concentration of dissolved oxygen which is in equilibrium with partial pressure, $\bar{p}$, in gas phase in bulk, ppm; g mole/l

d_o = outer diameter of hollow shaft, cm

d_i = inner diameter of hollow shaft, cm

e = electric potential

F = Faraday's constant, 96,500 coulombs/gram-equivalent; flow rate of inert gas, mole/hr

G = modulus of rigidity, Kg/cm^2

i_∞ = steady-state current, μA

K_0 = overall oxygen-transfer coefficient, g mole/cm^2 sec cmHg

K_s = gage factor

k_L = mass (oxygen)-transfer coefficient of liquid film, g mole/cm^2 sec cmHg

k_m = mass (oxygen)-transfer coefficient of membrane, g mole/cm^2 sec cmHg

$k_\text{L}a$ = volumetric oxygen-transfer coefficient, min^{-1}, hr^{-1}

k_p = sensor constant

M = torque, Kg·m

n = number of electrons per molar unit of reaction ($=4$)

n_{O_2} = oxygen flux, g mole/cm^2 sec

P = total pressure, atm

P_m = oxygen permeability in membrane, cm^3 at S.T.P./cm^2 sec cmHg

p_L = partial pressure of oxygen in bulk liquid phase, atm, cmHg

p_m = partial pressure of oxygen at interface between liquid and polymer membrane, atm, cmHg

p_0 = partial pressure of oxygen at interface between electrolyte and polymer membrane, atm, cmHg

p = partial pressure of oxygen, atm, cmHg

$\bar{p}$ = partial pressure of oxygen in bulk gas phase, atm, cmHg

p_out = partial pressure of oxygen or carbon dioxide in exit gas, atm, cmHg

p_in = partial pressure of oxygen or carbon dioxide in inlet gas, atm, cmHg

Q_{O_2} = specific rate of oxygen uptake (respiration) by microbes, mg mole/mg cell hr

Q_{CO_2} = specific rate of carbon dioxide evolution, mg mole/mg cell hr

[R] = indicator reading

R_b = oxygen-transfer resistance at interface between liquid and polymer membrane, cm^2 sec cmHg/g mole

R_E = oxygen-transfer resistance of electrolyte, cm^2 sec cmHg/g mole

R_g = strain gage resistance, Fig. 12.3

R_L = oxygen-transfer resistance of liquid film, cm^2 sec cmHg/g mole

R_m = oxygen-transfer resistance of polymer membrane, cm^2 sec cm Hg/g mole

R_0 = overall oxygen transfer resistance, cm^2 sec cmHg/g mole

R'_0 = oxygen-transfer resistance at interface between electrolyte and polymer membrane, cm^2 sec cmHg/g mole

RQ = respiratory quotient ($= Q_{\text{CO}_2}/Q_{\text{O}_2}$)

t = time, min, sec

V = broth volume, m^3, or electrical potential, Eq. (12.1)

$\bar{v}$ = average liquid velocity, cm/sec, Fig. 12.8 (b)

X = cell mass concentration, mg/ml, g/l

x = coordinate, Fig. 12.8 (a)

Greek letters

$\alpha = 1/(\kappa V K_s)$

$\varepsilon = $ strain

$\varepsilon_{max} = $ maximum strain

$\kappa = $ factor, Eq. (12.3)

$\mu = $ viscosity of liquid (sucrose solution), cp, Fig. 12.8 (b); specific growth rate, hr^{-1}

$\mu_a = $ apparent viscosity, cp, Fig. 12.14

$\nu_{sugar} = $ specific rate of consumption of carbonaceous substance, mg mole/mg cell hr

$\nu_{nitrogen} = $ specific rate of consumption of nitrogenous material, mg mole/mg cell hr

$\sigma = $ stress, Kg/cm^2

REFERENCES

1. Aiba, S., Okamoto, R., and Satoh, K. (1965). "Two sorts of measurements with a jar type of fermentor—power requirements of agitations and capacity coefficient of mass transfer in bubble aeration." *J. Ferm. Tech. (Japan)* **43**, 137.
2. Aiba, S., Ohashi, M., and Huang, S. Y. (1968). "Rapid determination of oxygen permeability of polymer membranes." *I & EC Fundam.* **7**, 497.
3. Aiba, S., and Huang, S. Y. (1969). "Oxygen permeability and diffusivity in polymer membranes immersed in liquids." *Chem. Eng. Sci.* **24**, 1149.
4. Aiba, S., and Huang, S. Y. (1969). "Some consideration on the membrane-covered electrode." *J. Ferm. Tech. (Japan)* **47**, 372.
5. Gualandi, G., Caldarola, E., and Chain, E. B. (1959). "Automatic continuous measure and control of pH by means of steam-sterilizable glass electrode." *Sci. Repts. Istituto Superiore di Sanita* **2**, 50.
6. Ishizaki, A., Shibai, H., Hirose, Y., and Shiro, T. (1971). "Studies on ventilation in submerged fermentations. Part I. Dissolution and dissociation of carbon dioxide in a model system." *Agr. Biol. Chem. (Japan)* **35**, 1733.
7. Nelson, H. A., Maxon, W. D., and Elferdink, T. H. (1956). "Equipment for detailed fermentation studies." *Ind. Eng. Chem.* **48**, 2183.
8. Nyiri, L., and Lengyel, Z. L. (1968). "Studies on ventilation of culture broths. I. Behavior of CO_2 in model systems." *Biotech. & Bioeng.* **10**, 133.
9. Phillips, D. H., and Johnson, M. J. (1961). "Measurement of dissolved oxygen in fermentations." *J. Biochem. Microbiol. Tech. & Eng.* **3**, 261.
10. Sato, K. (1961). "Rheological studies on some fermentation broths (I and IV). Rheological analysis of kanamycin and streptomycin fermentation broths. Effect of dilution rate on rheological properties of fermentation broths." *J. Ferm. Tech. (Japan)* **39**, 347, 517.
11. Solomons, G. L. (1969). *Materials and methods in fermentation* Academic Press, New York.
12. Taguchi, H., and Humphrey, A. E. (1966). "Dynamic measurement of the volumetric oxygen transfer coefficient in fermentation systems." *J. Ferm. Tech. (Japan)* **44**, 881.
13. Timoshenko, S., and MacCullough, G. H. (1949). *Elements of strength of materials* 3rd Ed. D. van Nostrand, Princeton, N. J.
14. Yamashita, S. (1971). "pH electrode holder with special ease of maintenance." *Automation (Japan)* **16**, 62.
15. Yamashita, S., Hoshi, H., and Inagaki, T. (1969). "Automatic control and optimization of fermentation processes: glutamic acid." *Fermentation Advances* p. 441. (Ed.) Perlman, D. Academic Press, New York.

Chapter 13

Recovery of Fermentation Products

From the amount of space devoted to fermentor design and scale-up, one might gather that the recovery processes of fermentation products are rather straight-forward and relatively simple. Nothing could be further from the truth. In one case of an antibiotic production plant, the investment for the recovery facilities is claimed to be about 4 times greater than that for the fermentor vessels and their auxiliary equipment. Often as much as 60% of the fixed costs of fermentation are attributable to the recovery stage in organic acid and amino acid productions.

Figure 13.1 shows a typical recovery process for antibiotics, while Fig. 13.2 presents another flow sheet for an enzyme plant. It is apparent from these diagrams that most recovery processes involve combinations of the following procedures:

    a. Mechanical separation of cells from fermentation broth,
    b. Disruption of cells,
    c. Extraction,
    d. Preliminary fractionation procedures, and
    e. High resolution steps.

In addition, crystallization, evaporation, drying, and solvent recovery are neces-sarily involved in the recovery process. However, in view of the fact that the equipment needed for these adjunct steps is nearly the same as that in other chemical industries, no mention will be made except for a brief reference to extraction.

Emphasis will be placed on a broad review of the mechanical separation of cells, disruption of cells, followed by preliminary fractionation and high resolution of enzymes. The principal reason to stress preliminary fractionation and high resolu-tion of enzymes later on in this chapter, without referring to the recovery of an actual product from case to case lies in that these processes are partly shared by general fermentation products such as antibiotics. Further, these fractionation and high resolution techniques are indispensable for recovering single cell protein and biologically active substances, both of which will become more significant in the near future.

In this comprehensive review, the discussion of working principles involved in each recovery procedure will be stressed, minimizing the details in connection with

346

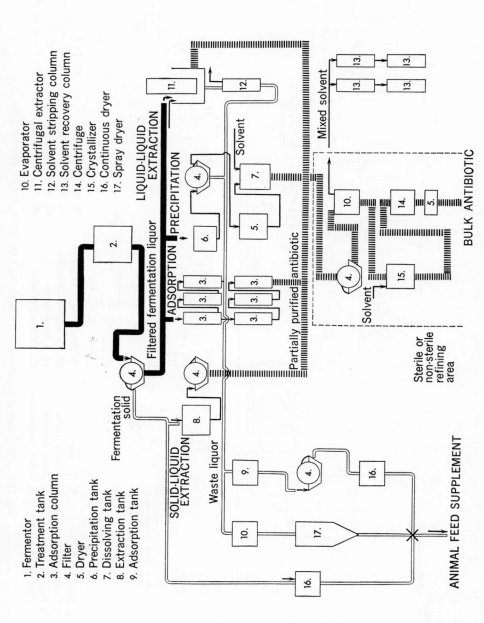

1. Fermentor
2. Treatment tank
3. Adsorption column
4. Filter
5. Dryer
6. Precipitation tank
7. Dissolving tank
8. Extraction tank
9. Adsorption tank

10. Evaporator
11. Centrifugal extractor
12. Solvent stripping column
13. Solvent recovery column
14. Centrifuge
15. Crystallizer
16. Continuous dryer
17. Spray dryer

Fig. 13.1. Basic flow-sheet for the recovery of antibiotics. (from Beesch, S.C., and Shull, G.M. (1957). *Ind. Eng. Chem.* **49**, 1491.)

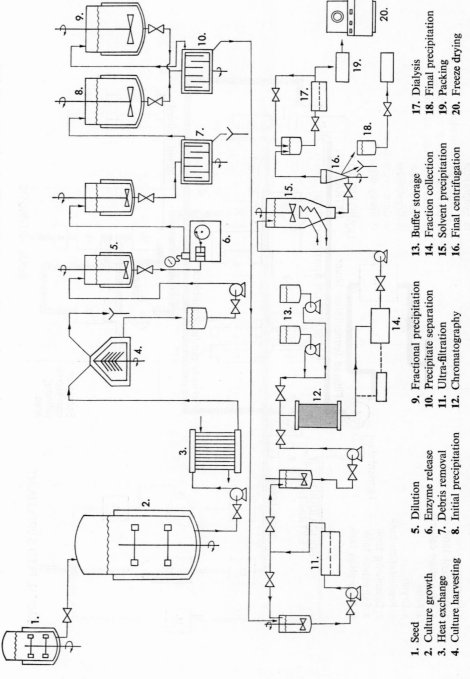

Fig. 13.2. Flow-sheet of enzyme plant. (from Malby, P.G. (1970). *Process Biochem.* **5** (8), 22.)

1. Seed
2. Culture growth
3. Heat exchange
4. Culture harvesting

5. Dilution
6. Enzyme release
7. Debris removal
8. Initial precipitation

9. Fractional precipitation
10. Precipitate separation
11. Ultra-filtration
12. Chromatography

13. Buffer storage
14. Fraction collection
15. Solvent precipitation
16. Final centrifugation

17. Dialysis
18. Final precipitation
19. Packing
20. Freeze drying

design and operation, because any single item of the above recovery procedures, e.g., filtration, extraction, preliminary fractionation of enzymes, chromatography, etc., has been discussed extensively elsewhere by many workers.

13.1. Principles of Mechanical Separation

13.1.1. Hindered settling in gravitational and centrifugal fields

According to Newton's law, the resistance, R, of single particles (diameter $= d_p$) in a flowing fluid (viscosity $= \mu$) of infinite extent is expressed by:

$$R = \frac{1}{2g_c} C_D \rho_m U_0{}^2 A \tag{13.1}$$

where

  C_D = drag coefficient
  ρ_m = fluid density
  U_0 = relative velocity between fluid and single particles
  A = sectional area of particles perpendicular to the direction of fluid motion
  g_c = conversion factor

In a range of $N_{Re} (= d_p U_0 \rho_m / \mu) < 0.3$ for spherical particles,

$$C_D = \frac{24}{N_{Re}} \tag{13.2}$$

This equation has been derived theoretically from

$$R g_c = 3\pi \mu d_p U_0 \tag{13.2'}$$

Equation (13.2) or (13.2)′ is Stokes' law of resistance.

For microbial suspensions which are diluted so that the effect of neighboring particles on the motion of single particles can be neglected, Eq. (13.2) may generally be applied; the effect of deviation from the spherical shape must be considered separately.

Since particles like microbes almost instantaneously attain an equilibrium state in which the resistance exerted on the microbes becomes equal to the driving force of motion, the settling velocity, U_0, of single particles also represents the terminal velocity in the following discussions.

In a gravitational field, therefore, from Eq. (13.2)′,

$$3\pi \mu d_p U_0 = \frac{\pi}{6} d_p{}^3 (\rho_y - \rho_m) g$$

$$U_0 = \frac{g d_p{}^2 (\rho_y - \rho_m)}{18\mu} \tag{13.3}$$

where

  ρ_y = density of microbial particles
  g = acceleration due to gravity

In a centrifugal field, likewise,

$$U_{c0} = \frac{gZd_p^2(\rho_y - \rho_m)}{18\mu} = ZU_0 \tag{13.4}$$

where

U_{c0} = velocity of single particles in the centrifugal field
Z = centrifugal effect ($= r\omega^2/g$)
r = radial distance from center of rotation
ω = angular velocity of rotation

As the cell concentration is increased, the interfering effect of adjacent particles on the motion of single particles cannot be ignored. The value of U_0 is decreased to U in these cases, and the settling of a swarm of such particles is called "hindered settling." The separation of a suspension of microbial cells in either a gravitational or a centrifugal field belongs to this category of hindered settling.

The resistance, R', exerted on spherical particles which move with velocity, U, in hindered settling will be shown to be as follows:

$$R'g_c = 3\pi\mu d_p U\{1 + \beta_0(d_p/L)\} \tag{13.5}$$

provided:

L = distance between neighboring particles
β_0 = coefficient

According to Smoluchowski,[50] the value of β_0 was calculated to be 1.6 for a rectangular arrangement of particles. It is evident from Eq. (13.5) that in a dilute suspension in which the value of d_p/L becomes approximately zero, the value of R' approaches that of R in Eq. (13.2)'.

The following equation will hold for various shapes of particles in cubic arrangements if appropriate substitutions for the value of d_p are made in cases where particles are not spherical:[1]

$$d_p/L = \beta\alpha^{1/3} \tag{13.6}$$

where

α = volume fraction of particles
β = geometrical factor

Although Eq. (13.5) should strictly apply only to a dilute suspension of particles, the assumption that β_0 is a function of α will be made, as in the following equation:

$$\beta_0 = \beta_0' + f(\alpha) \tag{13.7}$$

where

$\beta_0 = \beta_0'$ when α approaches zero

From Eqs. (13.5), (13.6), and (13.7),

$$R'g_c = 3\pi\mu d_p U[1 + \{\beta_0' + f(\alpha)\}\beta\alpha^{1/3}]$$
$$= 3\pi\mu d_p U(1 + \alpha'\alpha^{1/3})$$

provided:

$$\alpha' = \{\beta_0' + f(\alpha)\}\beta \tag{13.8}$$

Since $R=R'$ in the case of settling due to the fact that the driving force, $(\pi/6)d_p^3(\rho_y - \rho_m)g$ remains unchanged,

$$\frac{U}{U_0} = \frac{1}{1 + \alpha'\alpha^{1/3}} \tag{13.9}$$

Referring to the extensive data published by many investigators on the relation between U/U_0 and α, the value of α' is empirically correlated with α as follows:[1] For irregular particles,

$$\alpha' = 1 + 305\alpha^{2.84} \tag{13.10}$$

provided:

$$0.5 > \alpha > 0.15$$

For spherical particles,

$$\alpha' = 1 + 229\alpha^{3.43} \tag{13.11}$$

provided:

$$0.5 > \alpha > 0.2 \tag{13.12}$$

For dilute suspensions,

$$\alpha' = 1 \sim 2$$

provided:

$$0.15 > \alpha$$

Here some reference will be made to centrifuges. The right-hand diagram of Fig. 13.3 is a schematic drawing of a tubular-type centrifuge (Sharples), while the left-hand one represents the separator bowl of another type of centrifuge (De Laval).

The settling of single particles in centrifuges will be considered. The distance of travel, x, of single particles in a centrifugal field is given by the following equation [see Eq. (13.4)].

$$x = U_{c0}t$$
$$= \frac{r\omega^2 d_p^2(\rho_y - \rho_m)}{18\mu} \cdot \frac{V}{F} \tag{13.13}$$

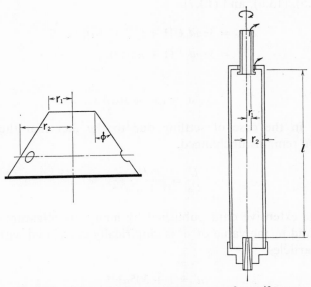

Fig. 13.3. Schematic diagram of two types of centrifuges.

where

V = liquid volume in centrifuge
F = rate of liquid flow through centrifuge

Substituting

$x = L_e/2$ and $r = r_e$ into Eq. (13.13),

$$F = 2\left\{\frac{d_p{}^2(\rho_y - \rho_m)}{18\mu}\right\}(r_e\omega^2 V/L_e)$$

$$= 2\left\{\frac{g d_p{}^2(\rho_y - \rho_m)}{18\mu}\right\}(r_e\omega^2 V/gL_e)$$

$$= 2U_0\Sigma \tag{13.14}$$

provided:

L_e = effective distance of settling
r_e = effective radius of rotation in centrifuge
$\Sigma = r_e\omega^2 V/gL_e$ [3]

The value of Σ corresponds to the surface area of the gravity settling basin whose capacity to handle the suspension is equal to that of the particular centrifuge.

It is clear that the value of Σ depends on the type and operating conditions of the centrifuge. It has been shown that

$$\Sigma = \frac{2\pi l\omega^2}{g}\left(\frac{3}{4}r_2{}^2 + \frac{1}{4}r_1{}^2\right) \tag{13.15}$$

for a tubular type of centrifuge,[3,32] where

l = length of cylindrical separator of the Sharples centrifuge

and

$$\Sigma = \frac{2\pi n' \omega^2}{3g \tan \phi} (r_2{}^3 - r_1{}^3) \qquad (13.16)$$

for the bowl type of centrifuge, where

n' = number of separator bowls (see Fig. 13.3)[3,32]

It has also been claimed by Ambler that the following equation holds when treating the same kind of suspension in centrifuges that differ from each other only in size:[3]

$$F_2 = F_1 \frac{\Sigma_2}{\Sigma_1} \qquad (13.17)$$

provided that subscripts 1 and 2 refer to small and large centrifuges, respectively.

Although Eq. (13.17) is an approximation, it is useful for estimating the value of F_2 in scale-up.

Obviously, densities and sizes of microbial cells suspended in liquids must be given to permit the calculation of their settling velocity either in gravitational or in centrifugal fields.[1] If cells are to be recovered from very large suspensions, as in the activated sludge process of waste treatment, flocculation of the cells is advisable as a pretreatment to increase the rate of separation. Those who are interested in a basic analysis of flocculation (and deflocculation) of cell suspensions should consult reference 2.

13.1.2. Filtration

The flow of liquid through the network of a filter cake and through a filter medium is considered to be "viscosity controlled," so the rate equation may be formulated as follows:

$$\frac{dv}{dt} = \frac{\Delta P g_c}{(r_m + r_c)\mu} \qquad (13.18)$$

where

v = volume of filtrate per unit area of filter
t = time of filtration
ΔP = driving force of filtration, pressure drop through the filter medium and the filter cake
g_c = conversion factor
r_m = resistance coefficient of the medium
r_c = resistance coefficient of the cake
μ = viscosity of the filtrate

Although the value of r_m is characteristic of the filter medium and is independent of filtration period, the value of r_c will increase during filtration. Generally, after a certain period of operation, the value of r_c far exceeds that of r_m.

Denoting A as the cross-sectional area of the filter and W as the total amount of solids contained in the original suspension to be filtered, then,

$$r_c = \alpha_R W/A \tag{13.19}$$

provided:

α_R = proportionality constant (=specific resistance of cake)

The value of α_R may depend on the pressure imposed ("compressible") or may be independent of pressure exerted on the cake ("incompressible"). The filter cakes encountered in the fermentation industry are primarily composed of cells and other organic material, and generally show compressible chracteristics.

V is designated as the total volume of filtrate corresponding to the total mass of solid, W, obtained from the original suspension with a filter whose cross-sectional area is A. Then,

$$W = \alpha''V \tag{13.20}$$

provided:

α'' = proportionality constant

Substituting Eqs. (13.19) and (13.20) into Eq. (13.18) and integrating from $t=0$ to $t=t$ with respect to the total area, A, of the filter ($dv=(1/A)dV$ and $V=0$ to $V=V$), assuming ΔP=constant due to the usual practice of "constant-pressure filtration" in the fermentation industry,

$$V^2 + 2VV_0 = Kt \tag{13.21}$$

provided:

$$V_0 = (r_m/\alpha_R\alpha'')A \tag{13.22}$$

$$K = (2A^2/\alpha_R\alpha''\mu)\Delta Pg_c \tag{13.23}$$

Equation (13.21) indicates that the relation between V and t is parabolic; this relationship was derived by Ruth et al. and so is known as Ruth's equation for constant pressure filtration.[46]

In a rotating drum filter, the drum is covered with a filter cloth and rotated at a constant speed (n rps). Since the period of time during which filtration is carried out is ψ_0/n sec per revolution of the drum, where ψ_0 =radians of drum surface immersed in suspension reservoir, Eq. (13.21) will be modified as follows:

$$(V_u/n)^2 + 2(V_u/n)V_0 = K(\psi_0/n) \tag{13.24}$$

provided:

V_u = filtrate volume per unit time

In Eq. (13.24), (V_u/n) represents the value for one revolution of the filter drum. Equations (13.22) and (13.23) are applicable to the drum type of filter, provided that $A = 2\pi R_0 L$, R_0 = radius of drum, and L = axial length of drum.

The driving force for filtration in the continuous filter is obtained by reducing the pressure inside the drum. After the filter cake formed on the surface of the drum is washed, the cake is peeled off with a "doctor's blade." The Oliver type of drum filter and batch filter presses are commonly used in the fermentation industry.

13.1.3. Pretreatment of cells to alleviate filtration resistance

It is well known that broth cultures of actinomycetes, e.g., *Streptomyces griseus*, exhibit tremendous resistance to filtration. Many attempts have been made to overcome the difficulty of separating the mycelium from broth on an industrial scale. In the recovery of streptomycin, the resistance of the mycelium to filtration has been reduced by heating the fermentation broth.[48] The coagulation of mycelial protein with heat to accelerate filtration may have wide application in the fermentation industry.

The filter used to obtain the data in Fig. 13.4 had an area of 110 cm² and was equipped with a cotton cloth; the filter aid was a diatomaceous earth (Radiolite), and pressure was applied constantly at 2 Kg/cm² (gage). The sample of broth used in these experiments was taken from a large batch fermentor (60 m³) which had been operated for 3–4 days. The original culture medium consisted of glucose and soy bean powder, supplemented by inorganic salts and dried yeast.

In Fig. 13.4(a) t/V is plotted against V, parameters being pH values. It can be seen that the data points for each pH did not lie on a straight line starting from the origin, indicating that the mycelial cake was compressible. However, assuming that the relation between t/V and V is nearly linear, it is apparent that the specific resistance to filtration of the mycelium is markedly affected by the pH. In addition, the specific resistance is affected by the amount of filter aid added (Fig. 13.5), since the slope of each line is proportional to α_R [*cf.* Eqs. (13.21) to (13.23)].

Although the data points of Figs. 13.4(a) and 13.5 were obtained at 90°C and the time required to raise the broth temperature was 30 min, it is evident from Fig. 13.4(b) that the period of time during which the broth is exposed to elevated temperatures is also of prime importance. The coagulation of mycelial protein has apparently been achieved after 30–40 min at 100°C, but a longer exposure of the broth to heat has adversely affected the filtration rate, presumably due to disintegration of coagulated protein.

It was found that the rate of decomposition of streptomycin at 80°C followed a monomolecular reaction pattern. By varying the pH values, the half-life (= 0.693/ reaction rate constant) was determined as shown in the right-hand side of Fig.

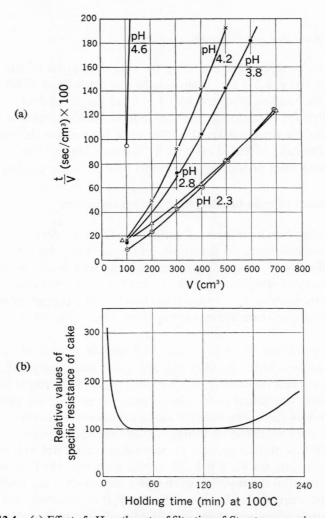

Fig. 13.4. (a) Effect of pH on the rate of filtration of *Streptomyces griseus* (filter aid 2%, 30 min required to raise the temperature of the broth to 90°C).
(b) Effect of holding time at elevated temperature (100°C) on the specific resistance of *Streptomyces griseus* broth to filtration.[48]

13.6. It is apparent that the fraction of streptomycin decomposed at 80°C at pH = 3.7 to 4.3 in 30 min will be negligible. On the other hand, the elution of streptomycin from the mycelium at pH = 4.0 shows that the antibiotic will be almost completely released into the broth after 30 min (left-hand side of Fig. 13.6); the streptomycin would be stable for this period. It is apparent from the figure that about 75% of the antibiotic had been liberated into the broth prior to mycelium filtration.

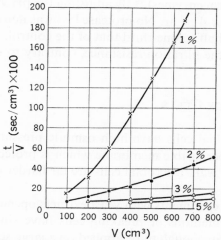

Fig. 13.5. Effect of filter aid on the specific resistance to filtration on *Streptomyces griseus* broth[48] (pH = 3.7 to 3.8; 30 min required to raise the temperature of broth to 90°C).

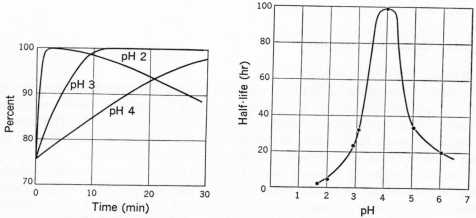

Fig. 13.6. Left: the elution rate of streptomycin from the mycelium at 80°C; Right: the effect of pH on the decomposition of streptomycin at 80°C.[48]

To summarize, the industrial filtration of *Streptomyces griseus* for the recovery of streptomycin via that of the cells was successful with the pH at 3.7 – 4.3 and the temperature at 80 – 90°C, using 30 – 60 min to raise the broth to the desired temperature and 2 – 3 % filter aid. These data were obtained from Figs. 13.4 to 13.6.

If the material to be released into a fermentation broth is highly heat sensitive, such as protease in the broth of *Streptomyces griseus*, clearly heating cannot be used to facilitate filtration of mycelia. Shirato *et al.* succeeded in decreasing the mycelial filtration resistance by adding 0.02 % of toluene at 40°C to cause autolysis

(lysis of cells by their own enzymes).[49] By adjusting the pH value at 6.0, the autolysis process continued for 4–6 hr. No protease loss was noted. The filtration resistance decreased by one-half to one-third that of the control. It is interesting to find that autolysis reduced the filtration resistance, contrary to expectation.

13.2. DISRUPTION OF CELLS

The contents of microbial cells have high osmotic pressure and are constrained within a fragile, semi-permeable membrane which is protected from rupture by a strong, rigid, outer cell wall. Microbial cells are far harder to break than are most animal or plant cells.

Buchner's classical experiments with yeast extracts opened the study of intracellular products. Many methods of breaking cells have since been published,[13,22] [29,55,56] but methods cheap enough to be applied on a large scale must be developed before the full potential of intracellular products can be realized. The cells must be handled in such a way that labile materials are not denatured by the process nor hydrolyzed by enzymes present in the cell; preferably, the operation should be continuous.

13.2.1. Mechanical methods

13.2.1.1. *Ultrasonic vibrations*
Ultrasonic waves of about 20 kc/sec applied to bacterial suspensions disrupt both cell wall and cell membrane; rods are more readily broken than cocci, and gram-negative cells disintegrate more readily than gram-positive ones.[27,28,38] The method is not as successful for breaking fungal cells; Zetelaki[57] found ultrasound to be the least effective of six methods tested on *Aspergillus niger* mycelium.

Ultrasonic waves in a liquid cause fluctuations in pressure forming bubbles which grow to about 10μ, begin to oscillate, and then implode violently generating shock waves of several thousand atmospheres and localized high temperatures. Free radicals are also generated in the solution. Neppiras and Hughes[38] believe rapid oscillations of the bubbles are responsible for cell rupture rather than damage from shock waves or free radicals. Nonetheless free radicals are important since they damage enzymes released into solution.

An electronic generator is used to produce ultrasonic waves of 20 kc/sec. A transducer converts these to mechanical oscillations in a titanium probe immersed in the cell suspension, cooled in ice. Figure 13.7 shows the rapid release of alcohol dehydrogenase (a typical cytoplasmic enzyme) and the delayed, stepwise release of aconitase, fumarase, and succinic dehydrogenase (mitochondrial enzymes) from yeast cells treated by ultrasound.[27] The apparatus can be modified to allow continuous operation.[38] In principle, ultrasonic disruption would seem a possible industrial method, but in practice, there are difficulties in dissipating the heat generated, the free radicals tend to denature sensitive enzymes and the cell debris is extensively fragmented.

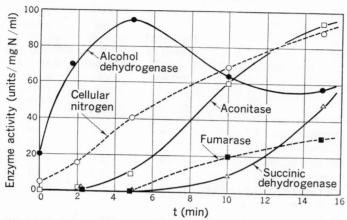

Fig. 13.7. The release of cellular nitrogen, aconitase, alcohol dehydrogenase, succinic dehydrogenase, and fumarase from yeast (30 ml of a 1 : 6 suspension) after exposure for different lengths of time to ultrasound using a 50-W M.S.E.-Mullard disintegrator.[27]

13.2.1.2. *Grinding and mechanical shaking*

Various methods of grinding cell pastes or frozen cells have been reviewed by Hugo;[29] they range from hand grinding in a pestle and mortar with powdered glass, sand, or alumina to mechanically driven mills such as those of Booth and Green or Gifford and Wood. On the laboratory scale, these methods are slow, laborious and relatively inefficient, but grinding techniques are attracting attention because they can be scaled up readily. Garver and Epstein disintegrated yeast in 10 min in a colloid mill which handled up to 600 ml of a slurry of cells and beads.[18] Réháček *et al.* have designed a novel cell disintegrator that used a disc rotating in a cooled chamber containing cells and ballotini beads.[41] The results of disintegrating suspensions of yeast that have different dry weights, at different stirrer speeds and with different designs of stirrer are shown in Table 13.1. With a cellular volume fraction of 3.5% and a tip speed of 18 m/sec, complete disintegration is achieved in half the time required at 9 m/sec, but at 17.5% volume fraction, the stirrer slips at the higher speed. The prototype had a capacity of 5 *l*, but models that operate continuously have been developed with capacities of 50 and 200 *l*.

Lilly and Dunnill reported that a Manton-Gaulin homogenizer, Model 15M-8 BA, operating at 350 Kg/cm² and recycling the suspension at 0.8 *l*/min, satisfactorily breaks yeast, bacteria and *Aspergillus niger*.[33] Ross developed a shaker-disintegrator where a slurry of plastic beads (20 – 50 mesh) and cells could be fed continuously at 0.8 – 4.5 *l*/hr through a cooled chamber.[45] This apparatus gave complete breakage of four species of gram-negative bacteria, a penicillium, a yeast, and a streptomyces, with hold-up times from 17 to 95 sec, depending on the organism.

TABLE 13.1

BREAKAGE OF YEAST CELLS IN A ROTARY DISINTEGRATOR OPERATED
UNDER DIFFERENT CONDITIONS.[41]

Concentration of suspension (%)	Disintegrator	Peripheral speed of stirrer (m/sec)	Time required for 100% disintegration (min)	Amount of 100% disintegrated material g/l of working volume	Average productivity (g/l min)
3.5	Vertical, open, 1-l, smooth stirrer	9	9	18	2
		18	5	18	3.6
	Horizontal, closed, 1.2-l, smooth stirrer	9	2	18	9
		18	1.2	18	15
	Horizontal, closed, 1.45-l, grooved stirrer	18	1.2	18	15
10.5	Vertical, open, 1-l, smooth stirrer	9	18	53	2.9
		18	12	53	4.4
	Horizontal, closed, 1.2-l, smooth stirrer	18	2.5	53	21.2 (17.5)
	Horizontal, closed, 1.45-l, grooved stirrer	18	1.6	53	33
		22	1.6	53	33
17.5	Vertical, open, 1-l, smooth stirrer	9	14	87	6.2
		18	>30	87	2.9
	Horizontal, closed, 1.2-l, smooth stirrer	9	4	87	22.0 (16.1)
		18	8	87	10.9
	Horizontal, closed, 1.45-l, grooved stirrer	18	5	87	17.5

The Sonomec wave-pulse generator is another device for imparting high shear to a slurry of beads and cells; sinusoidal motion at frequencies of 100–200 c/sec is imparted to a piston connected to a vessel containing the slurry. The shape of the container, the proportion of cells to beads, the height of the air space in the container, and the frequency and amplitude of vibration all affect the efficiency of cell rupture and can be controlled accurately.[43] The efficiency of breakage is similar to that of the Hughes press but the enzymatic activity of the extracts is a little less.

13.2.1.3. *Shearing by pressure*

There are several variants of laboratory presses to disrupt cells, the Hughes press and the French press being especially successful. The original Hughes press[26] consisted of a split block with a half cylinder hollowed in each face; the frozen cell paste (with or without abrasive) was placed in the hollow. The block was clamped together and a plunger driven by a fly-press forced the frozen paste from the cylinder into channels cut in the block. The frozen, disrupted cells were then scraped from the block. A cylindrical form of the Hughes press is now available which could potentially be scaled up.

With the French press, a hollow cylinder in a stainless steel block is filled with cell paste and subjected to a high pressure. The cylinder has a needle valve at the base and the cells burst as they are extruded through the valve to atmospheric pressure.[36] These presses are not readily scaled up, but they are widely used in laboratory studies and have special advantages when preparing cell membranes, or very labile products.

Two continuous presses have been scaled up, the Ribi fractionator and the X-press. The Servall-Ribi fractionator is similar in principle to the French press but circumvents the heating problems encountered with the French press;[40,42] either the needle valve is chilled with nitrogen gas or the broken cells are ejected directly into chilled liquid. Duerre and Ribi[10] found that with *Escherichia coli* most cells were ruptured at 1,000 Kg/cm² and all the cells were broken at 1,700 Kg/cm² with little loss of enzyme activity but, at 2,400 to 3,800 Kg/cm², the stability of some enzymes was affected; arginine decarboxylase and formic hydrogen lyase were destroyed, formic oxidase was partially inactivated, while lysine decarboxylase and glucose oxidase remained active.

In the X-press developed by Edebo, frozen cells are forced to and fro through a small hole in a disc between two cylinders at low temperature and pressure.[11,12] Disruption of the cells is probably caused by deformation of organisms embedded in the ice. The ice undergoes transformations in crystal structure when it passes through the hole since the temperature and pressure then change. Figure 13.8 shows the different forms of ice at different temperatures and pressures. When the phase boundary changes from I to III, the volume variation of 0.185 cm³/g is greatest.[14] It was found that cell disruption was greatest when phase boundaries were crossed repeatedly.

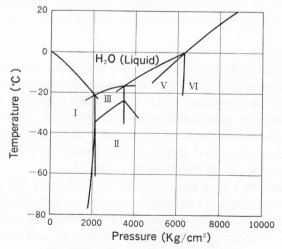

Fig. 13.8. Change of ice structure.[14]

The original X-press was designed to deliver up to 2,000 Kg/cm², while the high pressure X-press has a spring-loaded piston which can apply pressures above 2,100 Kg/cm².[13] These pressures are needed to disrupt heavy cell pastes, e.g., at −22°C, 4,000–6,000 Kg/cm² is needed for 75% breakage of a yeast paste with 27% volume fraction, or 90% breakage of a paste with 13.5% volume fraction.

13.2.2. Induction of lysis

Cells can also be disrupted by nonmechanical methods, either by disrupting the cell membrane itself or by hydrolyzing the cell wall, so that the fragile cell membrane bursts.

13.2.2.1. *Physical methods*
The membrane can be broken by physical stress such as violent depression, osmotic shock or rupture with ice crystals. In early studies, enzyme preparations were made from bacteria by slowly freezing and then thawing a cell paste; ice crystals are responsible for breaking the membrane. This method is slow and inefficient and has no commercial application.

There is a report that *Escherichia coli* is lyzed when cells are decompressed rapidly after prior compression at 35–63 Kg/cm² with a soluble gas,[17] but the method does not seem to have been exploited extensively. Lysis of whole cells by osmotic shock is not a practical proposition on a large scale, except perhaps for obligate osmophilic species.[47] There is an interesting finding that exponential cells treated first with 20% sucrose and $1 \times 10^{-4} M$ ethylene diamine tetra acetic acid (EDTA) and then dispersed in $5 \times 10^{-4} M$ $MgCl_2$ leak certain enzymes while remaining viable and retaining other enzymes.[39] The selective release of enzymes would have considerable commercial appeal, if it could be scaled up. Cell wall preparations can be made by heat-shocking suspensions of certain bacteria,[47] but this method is clearly not applicable to the release of heat-labile products.

13.2.2.2. *Lytic agents*
Cationic and anionic detergents, alkalis, bile salts and solvents are effective in damaging the lipoprotein of the cell membrane, but these chemicals frequently destroy intracellular materials as well.[23] Hence, this approach is only possible if molecules or enzymes of high stability are required; e.g., invertase can be transferred from papain-treated yeast into toluene.[25] Some antibiotics (polymyxins, tyrocidins, amphotericin B and nystatin) also increase the permeability of bacteria,[23] but they are too costly for most processes.

Another approach to lyzing bacterial cells is to destroy the cell wall either by interfering with its synthesis or by adding enzymes which attack specific linkages in the wall. Without the strong cell wall, the resulting protoplast (lacking a cell wall completely) or spheroplast (having remnants of cell wall but osmotically fragile) lyzes because of the high osmotic pressure of the cell contents (5–25 Kg/cm²).

There are a large number of enzymes produced by microorganisms and eucaryotic cells that hydrolyze different bonds in bacterial cell walls.[51] The enzymic hydrolysis of bacterial cell walls is best explained by reference to Figs. 13.9, 13.10, and Table 13.2. The structural part of bacterial walls consists of a mucopeptide made of alternating molecules of *N*-acetylglucosamine (G) linked β, 1–4 to *N*-acetylmuramic acid (M). The muramic acid has a tetrapeptide attached—most commonly this consists of D- and L-alanine, D-glutamic acid and either L-lysine or diaminopimelic acid. In addition, a variable number of pentapeptides cross-link the strands of –G–M– units via one of the amino acids of the tetrapeptides attached to the muramic acid. The composition of the pentapeptide varies with species; e.g., it may consist of glycine only or of alanine and threonine. Table 13.2 indicates

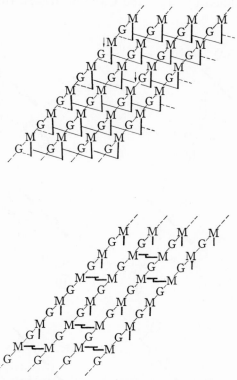

Fig. 13.9. Above: a schematic representation of the cell walls of *Staphylococcus aureus* and *Micrococcus roseus* showing extensive cross-linking between the polysaccharide chains by polypeptides. G = *N*-acetylglucosamine; M = *N*-acetylmuramic acid; —— = chemical bonds; —— = tetrapeptide; —— = pentaglycine (*S. aureus*) or tri-L-alanyl-L-threonine (*M. roseus*). Below: cell wall of *Escherichia coli* showing a loose network in which all of the acetylmuramic residues are substituted either by tetrapeptide monomers or peptide-linked dimers.[51]

some of the enzymes capable of hydrolyzing cell walls and their point of attack on the polymer is shown in Fig. 13.10.

The gram-positive bacteria have cell walls composed largely of muramic peptide and these are far more susceptible to lysis by lysozyme and similar enzymes. However, it is possible to lyze gram-negative bacteria with lysozyme if they are first treated with EDTA or by freezing and thawing. Enzymic hydrolysis is, at present, too expensive for commercial application, but a review by Kruf and Smekal indicates that research into the production of lytic enzymes is increasing and that, in future, this mild method of disruption may become economic.[31]

Cultures growing actively can be treated with one of the antibiotics that interfere with wall synthesis, such as penicillin or cycloserine; cells with such aberrant walls lyze readily, but this technique is rather costly.

Autolysis has long been used in the preparation of yeast extracts for human use.[25] The autolysis of yeast primarily depends on proteases normally contained in a vacuole, and is stimulated by Mg^{2+} and Ca^{2+}; proteases, ribonucleases and ester-

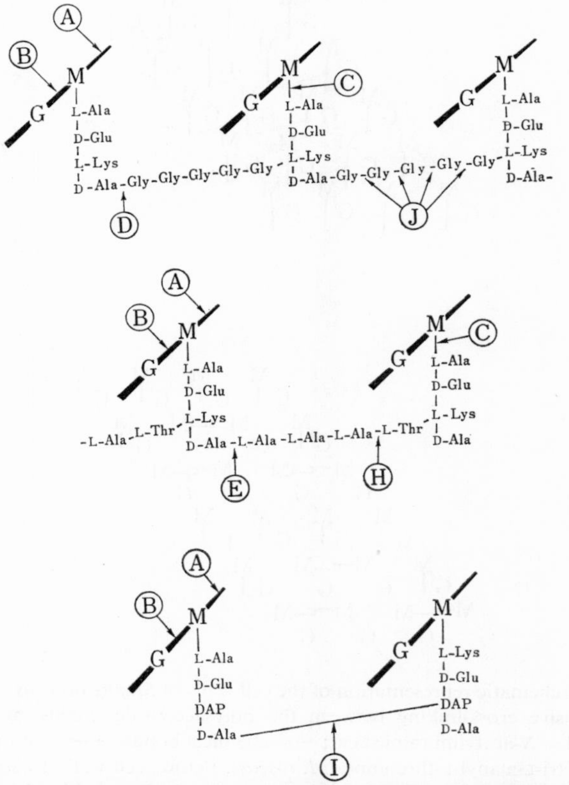

Fig. 13.10. Portions of bacterial cell walls showing the linkages attacked by different enzymes (A,B, ... as in Table 13.2). G=N-acetylglucosamine; M=N-acetylmuramic acid; L-Ala=L-alanine; D-Glu= D-glutamic acid; L-Lys=L-lysine; L-Thr=L-threonine; Gly=glycine; DAP=diaminopimelic acid.[51]

TABLE 13.2

ENZYMES DEGRADING THE MUCOPEPTIDES OF BACTERIAL CELL WALLS.*

Enzyme and Source	Type of linkage split
Carbohydrases	
Endoacetylmuramidases	
Plant, animal, T_2 phage lysozymes	N-acetylmuranmyl-N-acetylglucosamine (A)†
Streptomyces albus G	
Chalaropsis	
Streptococcus faecalis	
Endoacetylglucoasminidases	
Streptococcus	N-acetylglucosaminyl-N-acetylmuramic acid
Staphylococcus	(B)
Exoacetylglucosaminidases	
Pig epididymis	
Escherichia coli	
Acetylmuramyl-L-alanine amidases	
Streptomyces albus G	N-acetylmuramyl-L-alanine (C)
Bacillus subtilis	
Escherichia coli	
Listeria monocytogenes	
Staphylococcus aureus	
Myxobacterium	
Sorangium	
Peptidases	
Endopeptidases (bridge-splitting enzymes)	
Streptomyces albus G	D-alanylglycine (D); D-alanyl-L-alanine (E); D-alanyl-N^ε-L-lysine (F); L-alanyl-L-threonine (H); L-alanyl-L-alanine
Escherichia coli	D-alanyl-meso-diaminopimelic (I)
Streptomyces	D-alanyl-meso-diaminopimelic (I)
Staphylococcus	Glycylglycine (J)
Streptomyces albus G	Glycylglycine (J)
Flavobacterium	Glycylglycine (J); D-alanylglycine (D)
Myxobacterium	Glycylglycine (J); D-alanylglycine (D)
Sorangium	D-alanyl-N^ε-L-lysine (F)
Exopeptidases	
Streptomyces albus G (L-alanine and glycine aminopeptidase, D-alanine and glycine carboxypeptidases)	
Escherichia coli (D-alanine carboxypeptidase)	

* Modified from Strominger, J.L., and Ghuysen, J-M. (1967). *Science* **156**, 213.
† Letter in ( ) refers to the bond attacked as shown in Fig. 13.10.

ases are released from the vacuoles and cause autolysis.[34] Little carbohydrate is released during autolysis and the cell wall appears essentially intact by light microscopy. Yeast protoplasts which lyze readily can be formed by hydrolysis of the glucans and mannans of the wall by enzymes of the snail gut.

13.2.3. Desiccation

Enzymatically active materials have been liberated from air-dried or freeze-dried bacteria treated with acetone, butanol and buffers, but not all enzymes are stable to these methods.[22,29] Active preparations of labile enzymes can be obtained from molds by freeze-drying the mycelium, grinding it to a powder while still frozen, and then treating it with cold buffer.[7]

13.2.4. Increasing the fragility of cells

It has already been mentioned in Chapters 2 and 3 that the composition of the growth medium can greatly influence the composition of cell walls and the rate at which metabolites diffuse from cells. In some bacteria, the concentration of magnesium ions greatly influences the formation of septa; when deficient, long filaments form. Filament formation occurs also when mutant bacteria lack the capacity to form some component of the cell wall;[44] osmotically fragile molds have been detected which lack chitin in their cell walls.[30] Bacterial strains have also been isolated that form rods at one temperature but filaments at others. These findings could well be exploited in industrial processes. For instance, cells could be grown at a temperature or in a medium that allows rod formation and then switched to conditions where filaments form (by changing temperature, by adding antibiotics, or by withholding nutrients). Filamentous cells are more readily harvested by centrifugation as well as being more easily broken. Fermentations which require a change in environment and critical control of medium constituents lend themselves to production in continuous systems with two stages. So far, most research effort has been directed to improving the efficiency of breaking cells. The time is now ripe for exploring the possibilities of increasing the fragility of cells by biological methods.

13.3. EXTRACTION

Counter-current and multi-stage liquid-liquid extraction is a most efficient procedure to recover fermentation products. Most antibiotics production plants employ liquid-liquid extraction as the first and sometimes second stage in purifying and concentrating the product (cf. Fig. 13.1). The product is first extracted from the broth by continuous centrifugal extractors into an organic ester such as amyl acetate. Then by a pH adjustment the partition coefficient is shifted in favor of the aqueous phase and the product is transferred back to the water phase. Usually a 4-fold concentration is achieved by this means with a considerable product purification. Figure 13.11 shows a schematic diagram of the Podbielniak extractor, which has been employed ubiquitously in the fermentation industry.

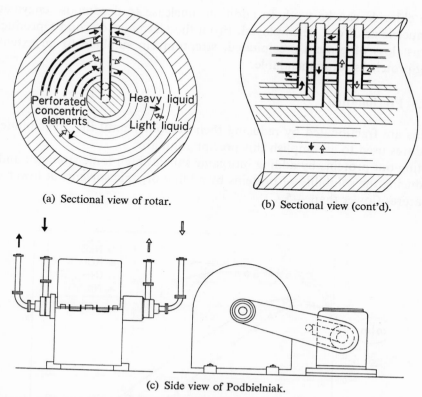

(a) Sectional view of rotar.

(b) Sectional view (cont'd).

(c) Side view of Podbielniak.

Fig. 13.11. Liquid-liquid extraction equipment (Podbielniak) (schematic diagram). Bold arrow indicates heavy liquid flow, while open arrow represents the flow of a mixture or light liquid.

Salt solution, buffer solution, acetone, perchloroacetic acid, butanol, esters, etc. are commonly used as extraction media.

13.4. PRELIMINARY FRACTIONATION PROCEDURES

13.4.1. Removal of nucleic acids

The preliminary fractionation procedures which follow are preceded by the removal of nucleic acids and other contaminating proteins, because these substances interfere with the subsequent fractionation and purification procedures that are taken after related fermentation products (or proteins) have been extracted into appropriate solutions. Particularly, nucleic acids content in extracts from bacteria runs as high as 20%, considerably higher than the content in extracts from animal or plant tissues.

Selective precipitants such as streptomycin sulfate, protamine sulfate, and manganese chloride can be used. In addition, nuclease to decompose enzymatically these nucleic acids can be employed. From the viewpoint of good reproducibility and least loss of protein precipitated, streptomycin sulphate to an extent of 1 mg/mg protein is recommendable.

13.4.2. Precipitation

Proteins are fractionated by reducing their solubility until they precipitate. Two procedures used to accomplish this precipitation are:
a. Salting-out proteins by adding inorganic salts at high ionic strength, and
b. Reducing the solubility of proteins by adding organic solvents at lower temperatures (lower than $-5°C$).

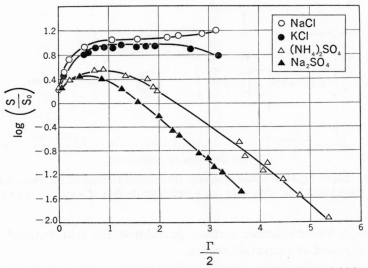

Fig. 13.12. Effect of inorganic salts on protein solubility (hemoglobin)[6]

Figure 13.12 is an example to show the effect of inorganic salts addition on the solubility of hemoglobin. The ordinate of this diagram is the logarithm of S/S_0, where S = solubility of protein in solution, g/l, and S_0 = hypothetical solubility when ionic strength, $\Gamma/2$, is zero. The abscissa, on the other hand, represents ionic strength, $\Gamma/2$, defined by

$$\frac{\Gamma}{2} = \frac{1}{2}\, \Sigma C_i Z_i{}^2 \qquad (13.25)$$

provided:

C_i = molar concentration of cationic and anionic species, mole/l

Z_i = cationic and anionic valences, respectively

For instance, $1M$ Na₂SO₄ solution gives the value of $\Gamma/2$ as follows:

$$\frac{\Gamma}{2} = \frac{1}{2}(2 \times 1^2 + 1 \times 2^2) = 3$$

It is clear from Fig. 13.12 that the solubility of hemoglobin decreased logarithmically in a higher range of ionic strength with respect to $(NH_4)_2SO_4$ and Na_2SO_4, though a slight increase of solubility was observed at lower ionic strength, especially on addition of NaCl and KCl solutions in this example. For enzymatic proteins other than this nonenzymatic hemoglobin the decrease in solubility at high ionic strength has been utilized for the fractionation of proteins by precipitation.[21] In practice, due to a large solubility in water, ammonium sulfate is extensively used for the salting-out procedure, i.e., the salt is added into a solution to a certain fraction of saturation, allowing the solution to come to equilibrium, and removing the precipitate that occurs. Then additional salt may be used and the procedure repeated.

As can be expected from Fig. 13.12, at high ionic strength the salting-out of proteins is represented by the following empirical equation.

$$\log S = \beta' - K_s'(\Gamma/2) \tag{13.26}$$

where

$$\beta' = \log S_0$$
$$K_s' = \text{salt-out constant}$$

Clearly, the value of K_s' is given by the slope of the straight lines in Fig. 13.12 for high ionic strength. The values of β' and K_s' depend on both the species of proteins and ions in the salting-out. However, Fig. 13.13 appears to show that K_s' values of hemoglobin are least sensitive to temperature and pH changes of the buffer solution.[21]

The use of organic solvents such as ethanol, methanol or acetone, etc., at lower temperatures ($<-5°C$) as referred to earlier to cause the precipitation of proteins is considered to cause a decrease of dielectric constant of the solution by the solvent addition; the decrease of the dielectric constant may result in an enhancement of the electrostatic force (Coulomb force) between protein molecules, facilitating precipitation. An alternate to this mechanism may be solvation, in which water bound to proteins is replaced by the solvent, thus decreasing the protein solubility. Though the previous salting-out procedure is less likely to denature proteins than the use of organic solvents, some advantages, that the solvents can be more easily removed from precipitated proteins, compared to the desalting associated with the former process, and in addition, that these solvents serve partly as fungicide for proteins, may lend themselves to a large-scale protein fractionation processes.

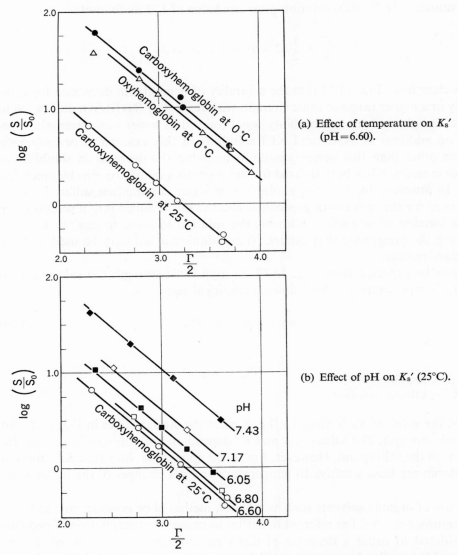

(a) Effect of temperature on K_s' (pH=6.60).

(b) Effect of pH on K_s' (25°C).

Fig. 13.13. Solubility of hemoglobin in a concentrated solution of phosphate buffer.[21]

13.5. High Resolution Techniques

13.5.1. Ultra-filtration

The process of separating components of a solution largely on the basis of molecular size, utilizing a membrane as a molecular sieve is designated as ultra-filtration; the use of this process to fractionate and purify proteins in the fermentation industry has been attempted and discussed by some workers, especially since 1966 (see Fig. 13.2 and Table 13.3 which appears later on). This development of ultra-filtration as a high resolution technique must be due to the emergence of strong and anisotropic membranes, permitting a high flux of the solution. The membrane technology originates principally from desalination projects in the U.S.A. dating back to the late 1950's. However, a sophisticated study on ultra-filtration theory, extending far beyond the introductory aspect which follows, and an extensive check of the problems which appear from case to case in practice are needed before fully fledged application of this technique in the fermentation industry is warranted.

13.5.1.1. *Theory*

Figure 13.14 shows a schematic diagram of ultra-filtration. A solution containing macromolecular substances pass along a membrane as shown in the figure; here, C_F, C_W, and C_P are designated to the solute concentrations in the feed, at the membrane surface, and of the permeate, respectively. For simplicity, the concentration profiles in the figure are assumed to remain unchanged along the direction of bulk liquid flow.

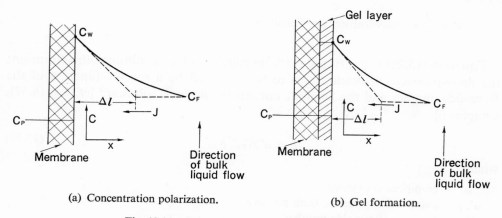

(a) Concentration polarization.　　　　　(b) Gel formation.

Fig. 13.14. Schematic diagram of ultra-filtration.[8]

For practical fractionation of solute molecules, one must necessarily assume the concentration polarization near the membrane surface as in Fig. 13.14. Solute flux through the membrane must be counterbalanced in steady state by the reverse flow of the solute due to concentration polarization; otherwise, solute transfer through the membrane may be discussed in a fashion similar to that in Section 12.3.3. (Chapter 12) with respect to oxygen flux through a polymer membrane.

Denoting the flux of solution through the membrane as J, the following equation deals with a mass balance of solute in the steady state.

$$- D\frac{dC}{dx} = J(C - C_P) \tag{13.27}$$

where

C = solute concentration
D = molecular diffusivity of solute in solution
x = coordinate, perpendicular to membrane surface

Integrating Eq. (13.27) with the boundary conditions

$$C = C_F \quad \text{at} \quad x = \Delta l$$
$$C = C_W \quad \text{at} \quad x = 0$$

provided:

Δl = boundary layer thickness (see Fig. 13.14)

$$J = \frac{D}{\Delta l} \ln \frac{C_W - C_P}{C_F - C_P}$$
$$= k \ln \frac{C_W - C_P}{C_F - C_P} \tag{13.28}$$

where

$$k = \frac{D}{\Delta l} = \text{mass-transfer coefficient}$$

Equation (13.28) is one of the working equations for membrane ultra-filtration; the mass-transfer coefficient, k, is to be expressed by a specific function of the Reynolds number and the Schmidt number of the bulk solution[20] [cf. Eq. (6.32), Chapter 6].

$$k = \alpha'' N_{Re}{}^{\alpha'} N_{Sc}{}^{\gamma} \tag{13.29}$$

where

α'' = empirical coefficient
α', γ = empirical exponents, both positive
$N_{Re} = l u \rho_m / \mu$ = Reynolds number
l = representative size of conduit for bulk flow of solution (liquid)
u = linear velocity of liquid
μ = liquid viscosity
ρ_m = liquid density
$N_{Sc} = \mu / \rho_m D$ = Schmidt number

Another working equation in membrane ultra-filtration in the steady state is:

$$J = A(\Delta P - \Delta \Pi) \tag{13.30}$$

provided:

A = membrane constant[20]

$= P_m/b$ (reference 4) [cf. Eq. (12.7), Chapter 12]

P_m = membrane permeability

   b = membrane thickness

$\varDelta P$ = hydraulic pressure difference across the membrane

$\varDelta \varPi$ = osmotic pressure difference between $\varPi_W$ at C_W and $\varPi_P$ for C_P

$\varPi_W$ = osmotic pressure at the membrane surface

$\varPi_P$ = osmotic prssure of permeate, i.e.,

$$\varDelta \varPi = \varPi_W - \varPi_P \tag{13.31}$$

If the concentration polarization can be disregarded, as in the case of pure water,

$$J = A\varDelta P \tag{13.32}$$

A sieving factor, ϕ, is introduced:[20]

$$\phi = \frac{C_P}{C_W} \tag{13.33}$$

It is assumed that the factor, ϕ, is dependent not only on the molecular size of the solute and the pore size of the membrane, but also is a function of the operating conditions, including $\varDelta P$, N_{Re}, N_{Sc}, and C_F.

Another term, R, introduced below, is designated as "rejection."

$$R = \frac{C_F - C_P}{C_F} \tag{13.34}$$

From Eqs. (13.28), (18.33) and (13.34),

$$\frac{1-R}{R} = \frac{\phi}{1-\phi} e^{J/k} \tag{13.35}$$

It is remarked from Eq. (13.30) that the increase in $\varDelta P$ entails that of the flux, J; however, the enhancement of J must be accompanied by an increase in C_W as noted from Eq. (13.28). This increase of C_W will result in the increase of $\varDelta \varPi$ in Eq. (13.30); so, the increase of $\varDelta \varPi$ tends to decrease the flux, J. It is apparent from the above reasoning that the proportionality between J and $\varDelta P$ as seen in Eq. (13.32) for pure water cannot be expected in ultra-filtration.

It may also be noted from Eq. (13.35) that the value of R depends on the working condition of a specific membrane-solute system. If a semi-logarithmic plot of J/k against $(1-R)/R$ with respect to experimental data of a given system yields a straight line, the mass-transfer coefficient, k, assessed from Eq. (13.29) is considered acceptable. Further, the use of Eq. (13.35) will be in checking whether or not another semi-logarithmic plot of $J/(N_{Re})^\alpha$ against $(1-R)/R$ yields a straight line ($\alpha = 0.7$–0.8); if the data points are aligned in a straight line, one may conclude in this case that the effect of the Reynolds number on the mass transfer coefficient, k, has been assessed properly. So far, Eq. (13.35) has been claimed to hold for solutes of lower molecular weight; for high molecular weight solutes Eq. (13.35) remains to be studied.[5]

When the concentration polarization exceeds a saturation value of solute, the solute becomes solidified at the membrane surface as shown in Fig. 13.14(b). Apparently, gel formation imposes additional resistance to solution transfer in view of the fact that the gel constant, corresponding to the membrane constant in Eq. (13.30), must be added in series to define the rate of transfer. A high shear rate of the solution to tear off the gel formation is effective for guaranteeing a higher flux of the solution. It must be mentioned that gel formation is one of the factors accounting for flux deterioration during ultra-filtration.

13.5.1.2. *Some practice*

Anisotropic ultra-filtration membranes available as of 1968 are listed in Table 13.3.[35] Though the values of water permeability and rejection are shown in the table at $\Delta P = 7$ Kg/cm², the usual operation pressure, ΔP, is less than 3.5 Kg/cm²; this is in contrast with the desalination of sea water where ΔP is of the order of 80–100 Kg/cm². This difference is principally attributable to the difference in the molecular weight of the solute; ultra-filtration here deals with macromolecules, whereas the osmotic pressure of sodium chloride in sea water far exceeds that of the macromolecules.[20]

TABLE 13.3

ANISOTROPIC ULTRA-FILTRATION MEMBRANES.[35]

No.	Membrane type	Chemical composition	Water permeability at 7Kg/cm² (ml/cm²/min)	Rejection characteristics		Max. temp. membrane can tolerate (°C)	Major application	Current manufacturer
				Compound [M.W.]	Rejection at 7Kg/cm² (%)			
1	Millipore VFWP	Cellulosic esters	0.3	Pepsin [35,000]	100	<60	Protein concentration or removal; bacterial removal	Millipore Corp.
2	Diaflo UM-10	Polyelectrolyte complex	0.5	Raffinose [594] / Dextran [10,000]	0 / 100	<60	Protein concentration, purification	Amicon Corp., Dorr-Oliver
3	Diaflo UM-2	Polyelectrolyte complex	0.2	Sucrose [342], / Raffinose [594]	50 / 90	<60	Concentration and demineralization of solutions of proteins and organic solutes	Amicon Corp., Dorr-Oliver

TABLE 13.3. (Cont'd).

4	Diaflo UM-05	The same as above, but with cation exchange capacity	0.1	NaCl [58], Sucrose [342]	5, 90	<60	Concentration and demineralization of solutions of sugars and other organic solutes	Amicon Corp., Dorr-Oliver
5	Diaflo XM-50	Substituted olefin	1.0	Dextran [10,000], Bovine serum albumin [69,000]	0, 100	<60	Macromolecular concentration and removal	Amicon Corp., Dorr-Oliver
6	Diaflo XM-100	Substituted olefin	3.0	Human serum albumin [69,000], Gamma globulin [160,000]	0, 100	<60	Macro-molecular fractionation	Amicon Corp., Dorr-Oliver
7	Diaflo PM-10	Aromatic polymer	3.0	Bacitracin [1,400], Pepsin [35,000]	70, 100	~120	Protein concentration and purification	Amicon Corp., Dorr-Oliver
8	Diaflo PM-30	Aromatic polymer	8.0	Human serum albumin [69,000], Dextran [110,000]	100, 60	~120	Macromolecular concentration and purification	Amicon Corp., Dorr-Oliver
9	Pellicon	Cellulosic	0.05	Congo red dye [697], Sucrose [342]	100, 40~60	~120	Concentration and purification of aqueous solutions	Millipore Corp.
10	HFA-100	Cellulosic	0.007	Dextran 10 [10,000]	100	<60	Macromolecular concentration and purification	Abcor Corp.
11	HFA-300	Cellulosic	1.7	Dextran 110 [110,000], Cytochrome C [13,500]	5, 20	<60	Macromolecular concentration and purification	Abcor Corp.

The dependence of ultra-filtration rate, J, on the value of ΔP is exemplified in Fig. 13.15, in which human blood plasma is subjected to ultra-filtration using a UM-10 membrane.[4] It is noted from the figure that the region where J depends on ΔP becomes extended with increase of shear rate.

A fairly linear correlation between J and ΔP when the shear rate became large (shear rate $= 1.3 \times 10^4$ sec^{-1}) in Fig. 13.15 might be ascribed to minimization of the concentration polarization due to the high shear rate.

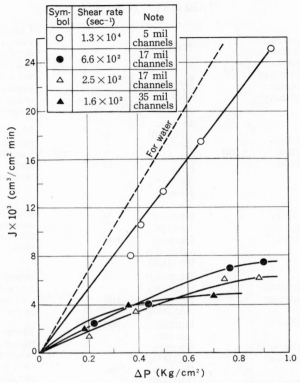

Fig. 13.15. Dependence of ultra-filtration rate on pressure in the low pressure region. Data are for human blood plasma in a recirculation system with thin channels of triangular cross-section; UM-10 membranes (*cf.* Table 13.3).[4]

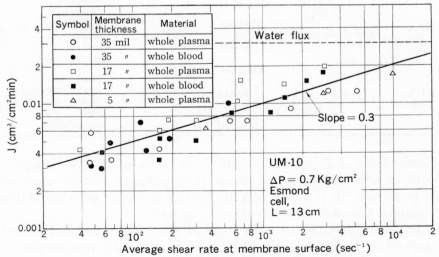

Fig. 13.16. Effect of fluid shear rate on ultra-filtration rate.[4]

A more general pattern of J as affected by shear rate is demonstrated in Fig. 13.16. In this diagram the value of flux, J, is roughly proportional to (shear rate),[0.3] and the principal contribution of the shear to an increase of J is considered to be tearing-off of gel formation as is mentioned earlier. However, ample room is left open for further experimentation and theoretical consideration on the shear rate in ultra-filtration.[5]

Experimental data showing that the flux, J, decreased with time for a casein solution under constant values of ΔP and feed rate (constant shear rate here) are shown in Fig. 13.17. Gel formation and some irreversible consolidation of the gel with time, i.e., "hardening" are claimed by Blatt *et al.*[4] to be a probable cause of the flux decay with time.

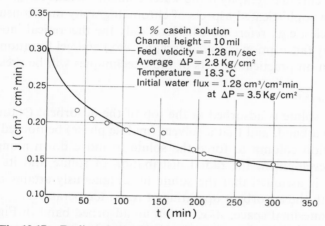

Fig. 13.17. Decline of ultra-filtration flux with time for a 1% casein solution in a turbulent-flow, thin-channel recirculating system.[4]

13.5.2. Chromatography

Chromatography refers to a general class of separations which depend on some kind of a mechanism to retard by differing amounts (and hence, achieve separation) the passage of various solute species through a column. Depending on the mechanism one may subdivide chromatography as follows:

    a. Adsorption chromatography,
    b. Liquid-liquid partition chromatography,
    c. Ion exchange chromatography, and
    d. Gel filtration chromatography.

Adsorption chromatography involves binding of solute to the solid phase (inorganic adsorbent such as alumina, silica gel, etc.) mainly by van der Waals' forces and steric interactions. Liquid-liquid partition chromatography attains separation because of the varying partition coefficient of solute between an adsorbed liquid (water) phase and the eluting solution. Ion exchange chromatography takes advantage of the different electrostatic forces between solute molecules (electrolyte) and ion exchange radicals. Gel filtration separates molecules on the basis of their size; i.e., how readily they can penetrate small pores of the column packing material.

It must be remarked at this point that a specific chromatographic separation has not necessarily been confined to any single mechanism of the above principles; e.g., ion exchange chromatography exhibits under certain conditions characteristics of adsorption chromatography and even gel filtration. On the other hand, adsorption chromatography using silica gel functions properly when the water content of the silica gel is less than 17% whereas it is claimed to work as liquid-liquid partition chromatography if the water content exceeds 32%.[37] Readers who are interested in sophisticated aspects of chromatography may consult specialized books and articles, e.g., reference 24. Here, only the theoretical interpretation of chromatography (adsorption chromatography, etc.) and gel filtration, followed by some discussion on practical aspects of these techniques will be presented.

13.5.2.1.   *Theory*[9,19]

Suppose that a solute is adsorbed at the top of the adsorbent (stationary phase), clearly forming a band, and that a solvent (mobile phase) be poured from the top of this adsorption column to force the solute to move down along the column axis. In this development the radial distribution of solute and its diffusion are disregarded; it is assumed that the solute instantaneously attains an adsorption equilibrium with the adsorbent operating at room temperatures.

Take an infinitesimal space, $A\delta x$, within an adsorbed band in Fig. 13.18. If an infinitesimal volume, δV, of a solvent (or the same solvent as that used in preparing the adsorbed band) is poured through the band, the first term of the right-hand side of Eq. (13.36) in the figure represents the amount of solute passing through the rear end of this space, $A\delta x$, while the second term equals to the amount leaving the front surface of the space. Though the concentration, C, of solute is a function both of x and V, only the partial derivative, $\partial C/\partial x$, is taken, because an addition of δV from the rear into the space is accompanied instantaneously by an emergence of the same amount of δV from the front due to the presence of the previous solvent in the void space between adsorbent particles.

Similarly, the increase of solute amount in the solvent and that of adsorbed solute, both of which are assessed at the rear surface of $A\delta x$, considering only $\partial C/\partial V$ and $\partial f(C)/\partial V$ here, respectively, are shown in Eqs. (13.37) and (13.38) in Fig. 13.17, where M = amount of adsorbent per unit volume of column, $f(C)$ = adsorption isotherm, Q = amount of adsorbed solute, $Q = Mf(C)$, and ε = void space of adsorption column ($= 1 - \alpha$).

From a material balance,

$$\Delta W_o + \Delta W_v + \Delta W_a = 0 \qquad (13.39)$$

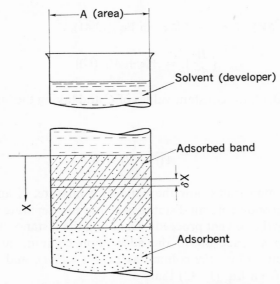

Fig. 13.18. Working principle of chromatography.

Change, ΔW, of solute content in an infinitesimal space, $A\delta x$, when an infinitesimal volume, δV, of solvent is poured into the space during a specific period of time is decomposed by:

(a) Overall

$$\Delta W_{\mathrm{o}} = C\delta V - \left\{ C + \left(\frac{\partial C}{\partial x}\right)\delta x \right\}\delta V$$

$$= -\left\{ \left(\frac{\partial C}{\partial x}\right)\delta x \right\}\delta V \qquad\qquad (13.36)$$

(b) For void-space

$$\Delta W_{\mathrm{v}} = \varepsilon A\delta x C - \varepsilon A\delta x \left\{ C + \left(\frac{\partial C}{\partial V}\right)\delta V \right\}$$

$$= -\,\varepsilon A\delta x \left(\frac{\partial C}{\partial V}\right)\delta V \qquad\qquad (13.37)$$

(c) For adsorption

$$\Delta W_{\mathrm{a}} = MA\delta x f(C) - MA\delta x \left\{ f(C) + \frac{\partial f(C)}{\partial V}\delta V \right\}$$

$$= -\,MA\delta x \frac{\partial f(C)}{\partial V}\delta V = -\,A\delta x \frac{\partial Q}{\partial V}\delta V \qquad\qquad (13.38)$$

From Eqs. (13.36) to (13.39), and rearranging,

$$\frac{\partial C}{\partial x} + A\left(\varepsilon\frac{\partial C}{\partial V} + \frac{\partial Q}{\partial V} \right) = 0 \qquad\qquad (13.40)$$

Rearranging further, Eq. (13.40) reduces to:

$$-\frac{\partial C}{\partial x} = A\{\varepsilon + Mf'(C)\}\frac{\partial C}{\partial V} \qquad\qquad (13.41)$$

Substituting

$(dV/dx)_\mathrm{c}(\partial C/\partial V)$ for $-\partial C/\partial x$ in Eq. (13.41),

$$\left(\frac{dV}{dx}\right)_\mathrm{c} = A\{\varepsilon + Mf'(C)\} \tag{13.42}$$

Integrating Eq. (13.42) for constant values of C and taking the integration constant equal to zero,

$$x = \frac{V}{A\{\varepsilon + Mf'(C)\}} \tag{13.43}$$

Figure 13.19 demonstrates how the band of a mixture, X and Y, adsorbed at the top of an adsorption column disintegrates into two separate bands of X and Y as development by the solvent proceeds. Denoting the distance of travel for solute, that of solvent in a space above the adsorption column, and the penetration distance for the solvent into the column as L_x, $L_\mathrm{s}(=V/A)$, and $L_\mathrm{c}(V/A/\varepsilon)$, respectively, it is noted from Eq. (13.43) that

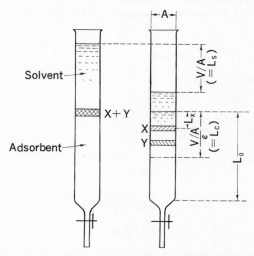

Fig. 13.19.   Working principle of chromatography[37] (cont'd).

$$\frac{L_\mathrm{x}}{L_\mathrm{s}} = R = \frac{1}{\varepsilon + MK} \tag{13.44}$$

provided:

$$K = \text{partition coefficient}$$
$$\doteqdot \frac{df(C)}{dC} = f'(C)$$

The above approximation is plausible if dilute solute is referred to, as is usually the case for chromatography in practice. The term, R, represents the ratio of distance travelled by X to that of solvent in the space above the column.

Likewise another term, R_f, which is also the ratio of travel distance for X to that for solvent in the column is introduced below.

$$R_f = \frac{L_x}{L_c} = 1 \bigg/ \left(1 + \frac{M}{\varepsilon} K\right) \tag{13.45}$$

Finally, designating the volume of solvent required to displace component, X, out of the packed column in Fig. 13.18 to V_0, a ratio, V_r, is defined by substituting L_0 and V_0/A into L_x and L_s in Eq. (13.44), respectively

$$V_r = \frac{V_0}{L_0 A} = \varepsilon + MK \tag{13.46}$$

Clearly, it is noted from Eqs. (13.44) to (13.46) that the difference in K values permits separation by this chromatography. Equation (13.43) has been derived with some assumptions to simplify the situation. In other words, Eq. (13.43) does not refer to the discontinuity of the band at all. Actually, a discrete marginal boundary for each band, minimizing the appearance of tailing is desirable in practice. For the mathematical treatment of the discontinuous boundary, readers are suggested to see reference 9.

Another theory of gel filtration deals with a particular term corresponding to K in the previous discussion. Similar to the chromatographic column discussed previously, suppose a column packed with specific materials (swollen with buffer solution and constituting a 3-dimensional network in the column; such as Sephadex, etc., see later), whose void space is designated as $V_0 (= V_e)$. If a small amount of sample solution added to the top of the bed and additional buffer is charged at a constant and very slow rate to displace the sample molecules down through the network, the liquid eluted from the column bottom, if exposed to spectrophotometric analysis, for instance, is expected to show a series of concentration spectra depending on its constitutent species, because macromolecules large enough not to penetrate into the intra-gel space pass readily through the void space, replacing buffer solution, while smaller molecules which can substitute themselves for the solution occupying the intra-gel and 3-dimensional network may require much more time (and more buffer solution to be poured through the column).

According to Gelotte, the picture mentioned above is formulated as follows:[19]

$$V_e = V_0 + K_D V_i \tag{13.47}$$

where

V_e = volume of buffer solution eluted from column bottom; this amount corresponds to a specific period of operation due to a constant feed rate of solution

V_0 = void space ($= V_\varepsilon$), implying volume of solution filling the space outside the packed gel structures in the column

V_1 = sum of intra-gel space

K_D = partition coefficient (modified definition)

If the molecules in question do not penetrate the network of intra-gel structure at all, it is apparent that $K_D = 0$, whereas for smaller molecules which can completely replace the solution in the intra-gel network, $K_D = 1$.

Rearranging Eq. (13.47),

$$V_e/V_0 = 1 + K_D(V_1/V_0) \tag{13.48}$$

Whitaker demonstrated experimentally that the ratio of elution volume to void volume, V_e/V_0, with a given species of gel remains fairly constant for each of a variety of protein molecules irrespective of the different combinations of column dimension and flow rate of buffer solution, within limits.[53] Since the ratio, V_1/V_0, is defined when the species of gel is fixed, the value of K_D characterizes the species of sample molecules least susceptible to change due to the change in gel filtration operation.

13.5.2.2. *Some practice*

Great care must be taken in preparing and operating chromatographic columns. This is particularly true with large columns where bed compaction becomes important. The column should be packed by pumped flow. The liquid head used to force solution through the column often must be restricted (sometimes to no more than 10 cm of water) to prevent compression. The solute loading volume does not matter too much. However, the mass of enzyme or protein loaded on the column depends greatly on the elution technique.

Figure 13.20 is an example of gel filtration with a mixture of sulphanilic acid, collidine and tryptophan through a column (3.5 ×35 cm) of Sephadex G-25, 50 to 100 mesh;[19] the ordinate is the optical density measured at $250-270$ mμ, whereas the abscissa represents the elution volume, V_e, eluted at a flow rate of 2 ml/min (eluant $= 0.05$ M triethylammonium carbonate at pH $= 8.0$).

If the values of $V_0 = 170$ ml, and $V_1 = 160$ ml are given in this example, the values of K_D for each material are determined from Eq. (13.48), taking V_e as each reading commensurate with a peak in optical density, i.e., $K_D = 0.8$, 1.3, and 2.0 for sulphanilic acid, collidine, and tryptophan, respectively. The fact that K_D exceeds unity, contrary to the previous discussion, might have occured due to a specific interaction between the molecules and the gel structure.

In this connection, Whitaker has suggested his preference for using the gel filtration technique as a means of measuring the M.W. of macromolecules;[53] in fact, a plot of the ratio of elution volume to void volume, V_e/V_0 against the logarithm of M.W. yielded a straight line with negative slope for a given gel filtration system and the fact that the slope was hardly sensitive to the eluant flow rate, solute concentration, sample size, etc.[53] justifies his suggestion. Generally, the height of each peak in optical density as exemplified in Fig. 13.20 is affected not only by the rate of elution, sample volume, and properties of gel particles, but also by the viscosity of the sample;[16] care must also be taken to minimize any superposition of each elution curve, as shown in the figure.

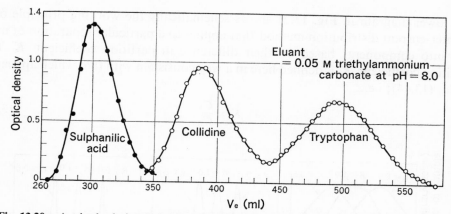

Fig. 13.20. A mixed solution (0.6 mg sulphanilic acid, 2.0 mg collidine, and 2.0 mg tryptophan) filtered through a column (3.5 × 35 cm) of Sephadex G-25, 50 – 100 mesh.[19]

Table 13.4 gives a summary of the properties of commercially available Sephadex (type = dextrans; manufacturer = Pharmacia). In addition, the following materials, such as

Biogel P (polyacrylamide gels; Bio. Rad. Labs.), Sepharose (agarose; Pharmacia), Biogel A (agarose; Bio. Rad. Labs.), Bioglas (porous silicate glass; Bio. Rad. Labs.), Biobeads (non-ionic polystyrene polymer; Bio. Rad. Labs.), etc.,

TABLE 13.4

PROPERTIES OF SEPHADEX.[15]

Sephadex type	Water regain $\left(\dfrac{\text{ml } H_2O}{\text{g dry Sephadex}}\right)$	Fractionation range for peptides and globular proteins (M.W.)
G- 10	1.0 ± 0.1	700
G- 15	1.5 ± 0.2	1,500
G- 25	2.5 ± 0.2	1,000~ 5,000
G- 50	5.0 ± 0.3	1,500~ 30,000
G- 75	7.5 ± 0.5	3,000~ 70,000
G-100	10.0 ± 1.0	4,000~150,000
G-150	15.0 ± 1.5	5,000~400,000
G-200	20.0 ± 2.0	5,000~800,000

are commercially available for gel filtration. Among them Biogel P has recently been widely employed, together with or in place of Sephadex, in gel filtration as one of the means of determining the M.W. of proteins.

13.5.3. Counter-current distribution method and other means

The lower diagram in Fig. 13.21 shows schematically the working principle of a counter-current distribution method that applies to a particular separation of mixture into components based on their difference in partition coefficient, K. The partition coefficient, K, is defined here in a way consistent with that which appeared in Eq. (13.44); i.e.,

$$K = \frac{C_A}{C_B} \tag{13.49}$$

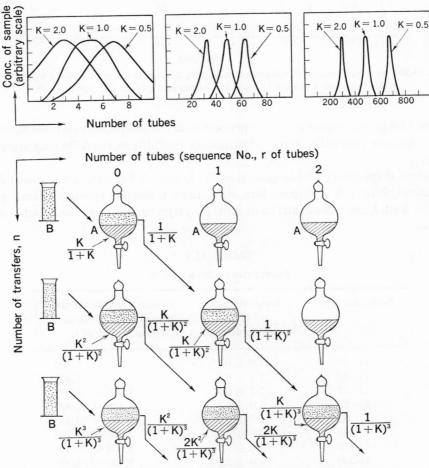

Fig. 13.21.   Schematic diagram of counter-current distribution method.[52]

where

C_A = concentration of a substance in solvent A in Fig. 13.21 (hatched area in each funnel)

C_B = concentration of the substance in solvent B in the figure (dotted portion in the funnel) in equilibrium with C_A

For simplicity, suppose that a specific material found only in solvent A in funnel No. 0 initially (C_A=unity) is exposed to the first extraction by adding another solvent, B, whose volume is assumed to be equal to that of A. After attaining an equilibrium and separating, the concentration in each solvent becomes $K/(1+K)$ and $1/(1+K)$. The effluent is transferred to funnel No. 1, while a "raffinate," A, is further added by the fresh solvent, B, in exactly the same fashion to continue the second extraction, the third and the subsequent extractions being apparent from the diagram.

The distribution of the material in each funnel (or tube) after these repetitive extractions, in other words, after n transfers, can be represented clearly by the following binomial expansion.

$$T_{n,r} = \binom{n}{r}\left(\frac{K}{1+K}\right)^n\left(\frac{1}{K}\right)^r \tag{13.50}$$

where

$T_{n,r}$ = fraction of the original substance in tube r after n transfers

TABLE 13.5

TERMS OF BINOMIAL EXPANSION.[54]

n	r			
	0	1	2	3
0	1	—	—	—
1	$\dfrac{K}{1+K}$	$\dfrac{1}{1+K}$	—	—
2	$\dfrac{K^2}{(1+K)^2}$	$\dfrac{2K}{(1+K)^2}$	$\dfrac{1}{(1+K)^2}$	—
3	$\dfrac{K^3}{(1+K)^3}$	$\dfrac{3K^2}{(1+K)^3}$	$\dfrac{3K}{(1+K)^3}$	$\dfrac{1}{(1+K)^3}$

$$K = \frac{C_A}{C_B} = \frac{\text{concentration in stationary phase}}{\text{concentration in mobile phase}}$$

To make the distribution clear, some calculation from Eq. (13.50) is illustrated in Table 13.5. It is most clear from the above reasoning that even a slight difference in K values for components in a mixture permits their separation by this method when the number of transfers is large (see the upper diagram in Fig. 13.20).

Lastly, electrophoresis will be referred to briefly. Since proteins, enzymes, amino acids, etc. are amphoteric in nature, their components in buffer solution with appropriate pH values can migrate with different velocities either to the cathode or the anode in an electric field (sometimes more than 1,000 V DC in tension). Using various materials such as filter paper, cellulosic powder, agar, starch grains, etc., as carriers, a sophisticated and high resolution of the amphoteric sample (to an extent of 1 g or so) can be conducted batchwise with large-scale equipment.

13.6. SUMMARY

1. Key equations for the mechanical separation of single cells are:
   a. For centrifugal and gravitational settling,

$$U_{c0} = \frac{gZd_p{}^2(\rho_y - \rho_m)}{18\mu} = ZU_0 \qquad \text{[Eq. (13.4)]}$$

$$F = 2U_0\Sigma \qquad \text{[Eq. (13.14)]}$$

   b. For constant pressure filtration,

$$V^2 + 2VV_0 = Kt \qquad \text{[Eq. (13.21)]}$$

2. Heat or chemical pretreatment enhances the recovery of mycelia from culture broths.

3. The mechanical methods of breaking cells include shearing with ultrasonic vibrations, grinding with or without abrasives in colloid mills, ball mills, homogenizers or shakers. In addition, cells in a liquid or frozen state can be sheared by high pressures. Only some of these equipment items (homogenizers) can be readily scaled up for industrial use.

4. Nonmechanical methods of breaking cells rely either on increasing the permeability of the cytoplasmic membrane or on interfering with the integrity of the cell wall.

5. Destruction of the cell wall gives osmotically fragile cells. Different enzymes are able to attack particular bonds in the wall of whole cells. Gram-positive bacteria are readily lyzed by enzymes but pretreatment of gram-negative cells is necessary to permit enzymatic lysis of the cell wall of these bacteria. Penicillin, which interferes with the biosynthesis of cell walls, can be added to growing cells to cause leakage of cell contents.

6. Salt solutions, acetone, perchloroacetic acid, butanol, and esters are commonly used to extract enzymes and antibiotics from fermentation media.

7. In order to prevent nucleic acids from interfering with protein fractionation

during precipitation, they must be selectively removed with a reagent such as streptomycin.

8. A useful equation for describing protein solubility in the salting-out procedure is

$$\log S = \beta' - K_s'(\Gamma/2)$$
[Eq. (13.26)]

9. Ammonium sulfate is commonly used for fractionating proteins by precipitation.

10. Membrane ultra-filtration is useful in fractionating compounds whose molecular weights range from 500 to 500,000. Ultra-filtration membranes act as molecular screens. Their pore size characterizes the particular solutes they allow to pass.

11. In ultra-filtration, concentration polarization occurs. This can cause proteins to form a gel layer next to the membrane, decreasing the flux of permeate. The effect of concentration polarization can be minimized by directing the flow parallel to the membrane.

12. Chromatography refers to a general class of separations which include:
　　a. Adsorption chromatography,
　　b. Liquid-liquid partition chromatography,
　　c. Ion exchange chromatography, and
　　d. Gel filtration chromatography.

13. Gel filtration is a particularly useful technique for fractionating enzyme mixtures. It acts by separating solute molecules on the basis of their ability to penetrate the small pores in column packing materials. Gel filtration materials are available that can fractionate solutes in the range from 700 to 800,000 in molecular weight.

14. A useful criterion for predicting the migration of solute molecules in a chromatographic column is

$$R_f = \frac{L_x}{L_c} = 1 \Big/ \left(1 + \frac{M}{\varepsilon}K\right)$$
[Eq. (13.45)]

For gel filtration, K in the above equation must be replaced by K_D which follows:

$$K_D = \frac{V_e - V_0}{V_1}$$
(*cf.* Eq. (13.47))

Unless abnormal interaction occurs between solute molecules and gel materials, the value of K_D to characterize each solute and each packing material ranges from zero to unity.

15. Factors which lead to poor resolution of solutes from a chromatographic column are:
　　a. Longitudinal diffusion,
　　b. Wall effects, and
　　c. Lack of local equilibrium.

NOMENCLATURE

A = cross-sectional area which is perpendicular to direction of motion, cm², m²; filter area, m²; membrane constant ($= P_m/b$), m/sec

b = membrane thickness, cm

C = solute concentration, wt%, g/l

C_D = drag coefficient

C_F = solute concentration in bulk liquid, wt%, g/l

C_i = molar concentration of cationic and anionic species, mole/l

C_P = solute concentration of permeate through membrane, wt%, g/l

C_W = solute concentration at membrane surface, wt%, g/l

D = molecular diffusivity of solute in liquid, cm²/sec

d_p = particle diameter, cm

F = rate of liquid flow through continuous centrifuge, m³/sec

$f(C)$ = function of C in adsorption isotherm of solute on absorbent

$f'(C)$ = derivative of $f(C)$

g = acceleration due to gravity, m/sec²

g_c = conversion factor, 9.8 kg m/Kg sec²

J = flux of solution through membrane, cm³/cm² sec

K = $(2A^2/\alpha_R\alpha''\mu)\Delta Pg_c$, m⁶/sec; partition coefficient in extraction and in chromatography

K_D = partition coefficient in gel filtration

K_s' = salt-out constant

k = mass-transfer coefficient, cm/sec

L = distance between particles, m; axial length of drum type of filter, m; conduit length in ultra-filtration, cm

L_0 = bed length of packed column, cm

L_c = distance of travel of developer (solvent) in adsorbed column, cm

L_e = effective distance of settling, m

L_s = distance of travel of developer (solvent) in a space above adsorbed column, cm

L_x = distance of travel for component, X, cm

l = length of cylindrical separator of Sharples centrifuge, m; representative size of conduit (ultra-filtration), m

Δl = boundary layer thickness, cm

M = amount of adsorbent per unit volume of column

N_{Re} = Reynolds number ($= d_p U_0 \rho_m/\mu$; $lu\rho_m/\mu$)

N_{Sc} = Schmidt number ($= \mu/\rho_m D$)

n = rotation speed of drum, sec⁻¹; number of transfers in countercurrent distribution method

n' = number of separator bowls

P_m = permeability of membrane, cm³/cm sec cmHg

ΔP = pressure drop, Kg/cm²; pressure difference across membrane, Kg/cm²

Q = amount of solute adsorbed, $Q = Mf(C)$

R = resistance, Kg; rejection, $(C_F - C_P)/C_F$; L_x/L_s

R' = resistance (in hindered settling), Kg
R_f = L_x/L_c
R_0 = radius of rotary filter drum, m
r = radial distance from rotation center, m, cm; tube No.
r_1 = radial distance between inner liquid surface and center of rotation in Sharples centrifuge, m; radius of separator bowl (De Laval centrifuge), m
r_2 = radius of outer cylinder in Sharples centrifuge, m; radius of circular center line of separator port (De Laval), m
r_c = resistance coefficient of cake, m^{-1}
r_e = effective radius of rotation in centrifuge, m
r_m = resistance coefficient of filter medium, m^{-1}
S = solubility of protein in liquid, g/l
S_0 = hypothetical solubility of protein in liquid at $\Gamma/2=0$, g/l
T = transfer (of concentration distribution in counter-current method)
t = time, sec, min, hr
U = settling velocity of a swarm of particles under gravity, m/sec
U_0 = settling velocity of single particles under gravity, m/sec
U_{c0} = settling velocity of single particles in centrifugal field, m/sec
u = linear velocity of liquid, m/sec
V = volume of filtrate, m^3; volume of liquid in centrifuge, m^3; volume of solvent (developer) poured into chromatographic column, cm^3; volume of packed column in gel filtration, cm^3
V_e = elution volume (gel filtration), cm^3
V_0 = $(r_m/\alpha_R\alpha'')A$, m^3; volume of surrounding aqueous phase in gel filtration column [$= V(1-\alpha)$], cm^3; volume of solvent required to elute one component in column chromatography, cm^3
V_i = sum of internal aqueous volume of gel substance, cm^3
V_r = $(V_0/L_0)A$
V_u = filtrate per unit time, drum type of filter, m^3/sec
v = volume of filtrate per unit area of filter, m^3/m^2
W = total solid content in original suspension, kg; amount of solute, mg
x = distance of travel, m, cm; coordinate
Z = centrifugal effect (= ratio of centrifugal acceleration to gravitational acceleration)
Z_i = cationic and anionic valences

subscripts

A, B, X, Y = immiscible liquid phases or component to be separated chromatographically
o, a, v = overall, adsorption, and void space, respectively
W, P = at membrane surface, permeate, respectively

Greek letters

α = volume fraction of cells or particles

$\alpha' = \{\beta_0 + f(\alpha)\}\beta$; empirical exponent

$\alpha'' =$ proportionality constant, kg/m³ in Eq. (13.20); empirical coefficient

$\alpha_R =$ proportionality constant (= specific resistance of cake), m/kg

$\beta =$ geometrical factor

$\beta' = \log S_0$

$\beta_0 =$ coefficient

$\beta_0' =$ empirical constant

$\Gamma/2 =$ ionic strength [$= (1/2) \Sigma C_i Z_i^2$]

$\gamma =$ empirical exponent

$\delta =$ symbol for infinitesimal scale

$\varepsilon =$ void fraction $(=1-\alpha)$, unless otherwise noted

$\mu =$ liquid (medium) viscosity or filtrate viscosity, kg/m sec, g/cm sec

$\Pi =$ osmotic pressure, Kg/cm²

$\Delta\Pi =$ osmotic pressure difference, Kg/cm²

$\rho_m =$ liquid (medium) density, g/cm³, kg/m³

$\rho_y =$ cell density, g/cm³

$\Sigma = r_e \omega^2 V / g L_e$ in Eq. (13.14), m²

$\phi =$ one-half of the apex angle of separator bowl in Fig. 13.3; sieving factor $(= C_P/C_W)$

$\phi_0 =$ radians of drum surface immersed in suspension reservoir

$\omega =$ angular velocity of rotation, radian/sec

REFERENCES

1. Aiba, S., Kitai, S., and Heima, N. (1964). "Determination of equivalent size of microbial cells from their velocities in hindered settling." *J. Gen. Appl. Microbiol.* **10**, 243.
2. Aiba, S., and Nagatani, M. (1971). "Separation of cells from culture media." *Advances in Biochemical Engineering* (Ed.) Ghose, T. K., and Fiechter, A. p. 31. Springer-Verlag, Heidelberg.
3. Ambler, C. M. (1959). "The theory of scaling up laboratory data for the sedimentation-type centrifuge." *J. Biochem. Microbiol. Tech. & Eng.* **1**, 185.
4. Blatt, W. F., Dravid, A., Michaels, A. S., and Nelsen, L. (1970). "Solute polarization and cake formation in membrane ultra-filtration: causes, consequences, and control techniques." *Membrane science and technology* (Ed.) Flinn, J. E. p. 47. Plenum, New York.
5. Brown, C. E., Tulin, M. P., and van Dyke, P. (1971). "On the gelling of high molecular weight impermeable solutes during ultra-filtration." *C. E. P. Symposium Series* **67**, 174.
6. Cohn, E. J. (1936). "Influence of the dielectric constant in biochemical systems." *Chem. Rev.* **19**, 241.
7. Creaser, E. H., Bennett, D. J., and Drysdale, R. B. (1967). "The purification and properties of histidinol dehydrogenase from *Neurospora crassa*." *Biochem. J.* **103**, 36.
8. De Filippi, R. P., and Goldsmith, R. L. (1970). "Application and theory of membrane processes for biological and other macromolecular solutions." *Membrane science and technology* (Ed.) Flinn, J. E. p. 33. Plenum, New York.
9. DeVault, D. (1943). "The theory of chromatography." *J. Am. Chem. Soc.* **65**, 532.
10. Duerre, J. A., and Ribi, E. (1963). "Enzymes released from *Escherichia coli* with the aid of a Servall cell fractionator." *Appl. Microbiol.* **11**, 467.

11. Edebo, L. (1960). "A new press for the disruption of microorganisms and other cells." *J. Biochem. Microbiol. Tech. & Eng.* **2**, 453.
12. Edebo, L. (1967). "Disintegration by freeze-pressing." *Biotech. & Bioeng.* **9**, 267.
13. Edebo, L. (1969). "Disintegration of cells." *Fermentation Advances* (Ed.) Perlman, D. p. 249. Academic Press, London.
14. Edebo, L., and Hedén, C-G. (1960). "Disruption of frozen bacteria as a consequence of changes in the crystal structure of ice." *J. Biochem. Microbiol. Tech. & Eng.* **2**, 113.
15. Ek, L. (1968). "Gel filtration—a unit operation." *Process Biochem.* **3** (9), 25.
16. Flodin, P. (1961). "Methodological aspects of gel filtration with special reference to desalting operations." *J. Chromatog.* **5**, 103.
17. Fraser, D. (1951). "Bursting bacteria by release of gas pressure." *Nature* **167**, 33.
18. Garver, J. C., and Epstein, R. L. (1959). "Method for rupturing large quantities of microorganisms." *Appl. Microbiol.* **7**, 318.
19. Gelotte, B. (1960). "Studies on gel filtration; sorption properties of the bed material Sephadex." *J. Chromatog.* **3**, 330.
20. Goldsmith, R. L. (1971). "Macromolecular ultra-filtration with microporous membranes." *Ind. Eng. Chem. Fundam.* **10**, 113.
21. Green, A. A. (1931). "Physical chemistry of proteins. VIII." *J. Biol. Chem.* **93**, 507.
22. Gunsalus, I. C. (1955). "Extraction of enzymes from microorganisms." *Methods in Enzymology* 1, (Eds.) Colowick, S. P., and Kaplan, N. O. p. 51. Academic Press, London.
23. Harold, F. M. (1970). "Anti-microbial agents and membrane function." *Adv. Microbiol. Physiol.* **4**, 46.
24. Heftmann, E. (1967). *Chromatography* 2nd Ed. Reinhold, New York.
25. Hough, J. S., and Maddox, I. S. (1970). "Yeast autolysis." *Process Biochem.* **5** (5), 50.
26. Hughes, D. E. (1951). "A press for disrupting bacteria and other microorganisms." *Brit. J. Exptl. Path.* **32**, 97.
27. Hughes, D. E. (1961). "The disintegration of bacteria and other microorganisms by the M.S.E.-Mullard ultrasonic disintegrator." *J. Biochem. Microbiol. Tech. & Eng.* **3**, 405.
28. Hughes, D. E. (1965). "Biological effect of ultrasound." *Science J.* **1965** (9), 39.
29. Hugo, W. B. (1954). "The preparation of cell-free enzymes from microorganisms." *Bact Rev.* **18**, 87.
30. Katz, D., and Rosenberger, R. F. (1970). "A mutant in *Aspergillus nidulans* producing hyphal walls which lack chitin." *Biochim. Biophys. Acta* **208**, 452.
31. Kruf, M., and Smekal, F. (1969). "Biosynthesis of bacterial lysozomes." *Process Biochem.* **4**(4), 53.
32. Kuwai, G., and Inoue, I. (1950). "Theoretical consideration about the separating characteristics of the DeLaval-type centrifuge." *Chem. Eng. (Japan)* **14**, 90.
33. Lilly M.D., and Dunnill, P. (1969). "Isolation of intracellular enzymes from microorganisms—the development of a continuous process." *Fermentation Advances* (Ed.) Perlman, D. p. 225. Academic Press, London.
34. Matile, P., and Wiemken, A. (1967). "The vacuole as the lysosome of the yeast cell." *Arch. Mikrobiol.* **56**, 148.
35. Michaels, A. S. (1968). "New separation technique for the CPI." *Chem. Eng. Progress* **64** (12), 31.
36. Milner, H. W., Lawrence, N. S., and French, C. S. (1950). "Colloidal dispersion of chloroplast material." *Science* **111**, 633.
37. Miyazaki, M., and Kotani, R. (1968). "Chromatography." *Biophysics lecture series (Japan)* (Ed.) Soc. Biophys. Japan p. 3. Yoshioka Pub. Co., Kyoto.
38. Neppiras, E. A., and Hughes, D. E. (1964). "Some experiments on the disintegration of yeast by high intensity ultrasound." *Biotech. & Bioeng.* **6**, 247.

39. Nossal, N. G., and Heppel. L. A. (1966). "The release of enzymes by osmotic shock from *Escherichia coli* in exponential phase." *J. Biol. Chem.* **241**, 3055.

40. Perrine, T. D., Ribi, E., Maki, W., Miller, B., and Oertli, E. (1962). "Production model press for the preparation of bacterial cell walls." *Appl. Microbiol.* **10**, 93.

41. Réháček, J., Beran, K., and Bíčík, V. (1969). "Disintegration of microorganisms and preparation of yeast cell walls in a new type of disintegrator." *Appl. Microbiol.* **17**, 462.

42. Ribi, E., Perrine, T. D., List, R., Brown, R., and Goode, D. (1959). "Use of a pressure cell to prepare cell walls from mycobacteria." *Proc. Soc. Expl. Biol. Med.* **100**, 647.

43. Rodgers, A., and Hughes, D. E. (1960). "The disintegration of microorganisms by shaking with glass beads." *J. Biochem. Microbiol. Tech. & Eng.* **2**, 49.

44. Rodgers, H. J. (1970). "Bacterial growth and the cell envelope." *Bact. Rev.* **34**, 194.

45. Ross, J. W. (1963). "Continuous-flow mechanical cell disintegrator." *Appl. Microbiol.* **11**, 33.

46. Ruth, B. F., Montillon, G. H., and Montonna, R. E. (1933). "Studies in filtration I. Critical analysis of filtration theory. II. Fundamental axiom of constant-pressure filtration." *Ind. Eng. Chem.* **25**, 76, 153.

47. Salton, M. J. R. (1964). *The bacterial cell wall* Elsevier, Amsterdam.

48. Shirato, S., and Esumi, S. (1963). "Filtration of the cultured broth of *Streptomyces griseus*." *J. Ferm. Tech.* (*Japan*) **41**, 87.

49. Shirato, S., Suzuki, I., and Esumi, S. (1965). "Filtration of a culture broth of *Streptomyces griseus*." *J. Ferm. Tech.* (*Japan*) **43**, 501.

50. Smoluchowski, M. S. (1912). "On the practical applicability of Stokes' law of resistance, and the modification of it required in certain cases." *Proc. 5th International Congress of Mathematics, Cambridge* **2**, 192.

51. Strominger, J. L., and Ghuysen, J-M. (1967). "Mechanisms of enzymatic bacteriolysis." *Science* **156**, 213.

52. Thesaurus in general chemistry (1960). (Ed.) Mizushima, S. p. 611. Kyoritsu Pub. Co., Tokyo.

53. Whitaker, J. R. (1963). "Determination of molecular weights of proteins by gel filtration on Sephadex." *Anal. Chem.* **35**, 1950.

54. Williamson, B., and Craig, L. C. (1947). "Identification of small amounts of organic compounds by distribution studies. V. Calculation of theoretical curves." *J. Biol. Chem.* **168**, 687.

55. Wimpenny, J. W. T. (1967). "Breakage of microorganisms." *Process Biochem.* **2** (7), 41.

56. Wiseman, A. (1969). "Enzymes for breakage of microorganisms." *Process Biochem.* **4** (5), 63.

57. Zetelaki, K. (1969). "Disruption of mycelia for enzymes." *Process Biochem.* **4** (12), 19.

Chapter 14

Immobilized Enzymes
An Alternative to Whole Cells or Enzymes

Until recently, enzymatic activities of cells have been derived from whole cell preparations and from crude or partially purified cell extracts. However, the problem of purification and removal from the end-product plus lack of large-scale continuous purification facilities has held back the full utilization of enzymes. Recently, great strides have been made in large-scale enzyme purification. This has occurred largely because of developments in ultra-filtration, gel permeation, etc.

Simultaneously, techniques have evolved for the immobilization of enzymes. Such enzyme derivatives are being used as heterogeneous catalysts in suspension in reaction mixtures, in packed beds, and in membrane systems.

Enzymes may be immobilized essentially in four ways. Each immobilization procedure will be described only briefly. For readers who are interested in the procedual details, references 6, 11, 13, and 16 of this chapter may be consulted.

14.1. Preparation of Immobilized Enzymes

14.1.1. Adsorption

There are many examples of adsorbed enzyme preparations. These include adsorption of:

    a. Ribonuclease on Dowex-50 cation exchange resin,
    b. Asparaginase, chymotrypsin and trypsin on CM-cellulose,
    c. Aminoacylase, pepsin and various proteases on DEAE-cellulose, and
    d. Invertase on charcoal.

It is interesting to note that Nelson and Griffin reported an immobilized enzyme (invertase adsorbed on charcoal) as far back as 1916.[14] Tosa *et al.* originally studied the behavior of adsorbed aminoacylase for use in the optical resolution of DL-acylated amino acids. Active preparations were made by adsorption of the enzyme on DEAE-cellulose and DEAE-Sephadex columns. These investigators were not successful in obtaining preparations from silica gel, aluminum oxide, activated carbon, and several ion exchange resins.[18]

14.1.2.   Inclusion in gel lattices and encapsulation

Active preparations of enzymes can be made by entrapping them in a gel-like structure. Many different gels have been used. Among these are starch gel and poly-acrylamide gel. Semi-permeable microcapsules made of synthetic polymers (collodion, silicone resin, etc.) are used to encapsulate enzymes. One of the early studies on gel-entrapped enzymes was that of Bernfeld et al.,[2] who obtained active preparations of trypsin, α-chymotrypsin, papain, ribonuclease, aldolase, α-, and β-amylase entrapped in a polyacrylamide gel[9] with the enzyme activity recovered ranging from 1.9 to 6.6%.

Several problems occur with gel-entrapped enzymes that must be considered in their use. Firstly, the enzymes tend to leak; secondly, they cannot be used for entrapping hydrolytic enzymes that degrade the gel, i.e., amylases cannot be entrapped in starch; thirdly, in general the hydrolytic enzymes (amylases, proteases, dextrinases, etc.) are not suitable for entrapping, because the substrates they attack are large in molecular weight and cannot penetrate the gel matrix.

Encapsulation of enzymes into collodion, nylon, polyurea, etc. may be useful particularly for medical use (see Section 14.3.1.). In addition, an interesting reactor has been developed in which concentrated cellulase reacts with particulate cellulose particles of Solka Floc and low molecular weight hydrolysates including glucose are separated from the suspension, passing through a bottom membrane of the reactor. The membrane (Amicon PM-30) used for "ultra-filtration" (cf. Section 13.5.1., Chapter 13) prevents the enzyme from passing through it, and thus the enzyme here is "immobilized."[3]

14.1.3.   Intermolecular cross-linking

Enzymes may be covalently cross-linked with appropriate bifunctional agents such as glutaraldehyde or bis-diazobenzidine disulfonic acid to give a thin enzyme membrane or a 3-dimensinal network of enzyme molecules.

Figure 14.1 illustrates the covalent cross-linking of bovine trypsin which is adsorbed first onto colloidal silica particles as a monolayer, and secondly, immobilized by intermolecular cross-linking with glutaraldehyde. This method of immobilizing bovine trypsin guaranteed a recovery efficiency of about 99% with 80% retention of esterase activity. Proteolyic activity was nearly 17% of that of native trypsin.[8]

Different from the immobilization procedures mentioned previously and that which follows next, this process is characterized by the fact that substrate molecules can achieve easier access to enzyme molecules, because the immobilized enzyme in Fig. 14.1 is in the form of an envelope around the particles and thus the enzyme is in direct contact with the surrounding substrate solution. This picture accounts for the enhanced efficiency of this preparation and in particular, the enzyme activity retained improves with respect to large substrate molecules.[8]

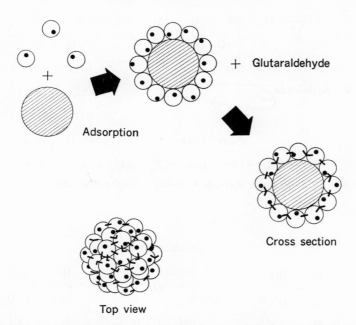

Fig. 14.1. Schematic representation of the method for preparing insoluble proteins as envelopes on colloidal particles. The shaded area represents the silica particle, the dark circles represent the active sites of a protein with biological activity and the bars represent covalent intermolecular cross-links.[8]

14.1.4.   Covalent binding[16]

Covalently coupled enzymes are by far the most important class of immobilized enzymes. The variety of carriers is extensive. It includes such materials are glass, silica, Sephadex, cellulose, nylon, polystyrene, etc. Functional groups of enzymes suitable for covalent binding are:[16]

a. α- and ϵ-amino groups,
b. α-, β- and γ-carboxyl groups,
c. -SH and -OH groups of cysteine and serine, respectively,
d. Imidazole group of histidine, and
e. Phenol ring of tyrosine.

Figure 14.2. shows covalent binding of an enzyme (trypsin) via the 1:1 copolymer of ethylene and maleic anhydride. The anhydride groups of the carrier react with protein amino groups via amide links, simultaneously yielding free carboxy groups. Cross-linking of the linear copolymer is effected by the addition of hexamethylene-diamine, as is apparent from the figure.[6]

Another example of covalent binding of an enzyme (chymotrypsin) to cellulose is shown in Fig. 14.3. One of the chlorines is sufficiently reactive with cellulose in an aqueous solution under certain conditions (20°C, pH=9–11) to form the mono-

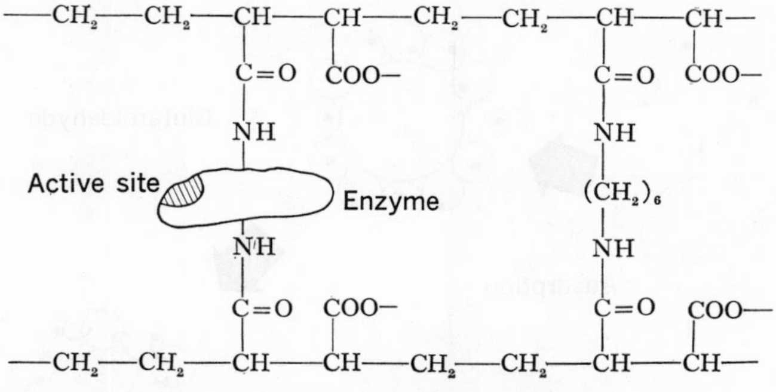

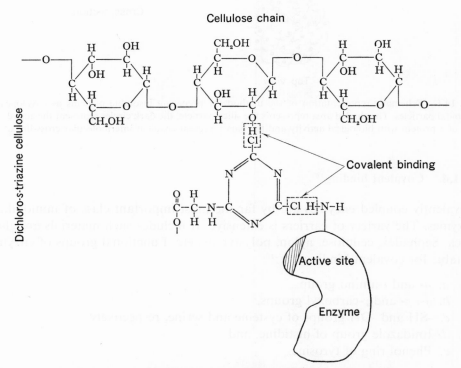

Fig. 14.2.   Example of covalent binding through copolymerization.[6]

Fig. 14.3.   Example of covalent binding of enzyme to cellulose.[12]

chloro-*s*-triazinyl-cellulose complex. NH_2 groups of enzymes will then react with the second chlorine under mild conditions (0–20°C, pH = 8.6) during a period of the order of 16 hr in this example. About 25% of the activity of native enzyme has been retained by this method. Due to the mild conditions used in this preparation, all forms of cellulose, i.e., paper and cotton cloth can be used as carriers.[12]

14.2. KINETICS OF IMMOBILIZED ENZYME SYSTEM

14.2.1. General consideration of environmental effects on kinetic behavior of immobilized enzyme system[10]

Immobilized enzymes bound to either nonporous or porous substances are shown schematically in Fig. 14.4. E, encircled in this schematic diagram, represents each enzyme molecule, whereas S and P denote the substrate and the product.

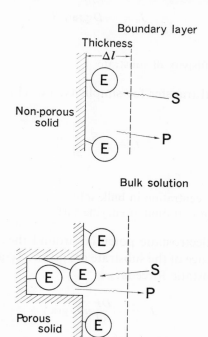

Fig. 14.4. Immobilized enzyme systems; schematic representation of nonporous and porous solids, to which enzymes are bound and immobilized. Boundary layer thickness (broken line) depends on the flow rate of bulk solution. Substrate, S, is converted to product, P, after the reaction between S and enzyme, E.

Enzymes bound to a membrane surface, in a very open and thin membrane, or to a smooth glass surface approximate to a nonporous solid system. On the other hand, enzymes bound to carboxymethylcellulose particles using chloro-s-triazines, or to p-amino benzylcellulose particles using a diazotization reaction form another category of porous solid.

It is usual to assume a boundary layer along the carrier surface (Δl = effective thickness of the boundary layer; see the broken line in Fig. 14.4). Obviously, the substrate, S must meet with diffusional resistance before it can reach the enzyme to produce the product, P. In addition, the transfer of the substrate from the bulk solution to the enzyme surface should be enhanced by electrostatic force, if both the bound enzyme and the substrate are charged. Clearly, the microenvironment illustrated schematically in Fig. 14.4 should have an appreciable effect on the kinetic behavior of the immobilized enzyme system. Here, enzymes bound to nonporous substances will be considered exclusively; in fact, kinetics for enzymes bound to porous materials are beyond the scope of the following equations, which are devoid of pore-diffusion terms.

Designating S as the substrate concentration for convenience, the flux, J_d, of substrate due to diffusion is given by:

$$J_d = -D \text{ grad } S \tag{14.1}$$

where

D = molecular diffusivity of substrate

Assuming unidirectional transfer for simplicity, Eq. (14.1) may be rearranged as follows:

$$J_d = D \frac{S_0 - S_1}{\Delta l} \tag{14.2}$$

provided:

S_0 = substrate concentration in bulk solution
S_1 = substrate concentration at enzyme surface

Supposing that the electrostatic potential around the immobilized enzyme and the electrochemical valence of the substrate are given as ϕ and n, the flux, J_e, of substrate due to the electrostatic force is:

$$J_e = -\frac{DF}{RT} n S_0 \text{ grad } \phi \tag{14.3}$$

where

F = Faraday's constant
  = Ne
N = Avogadro's number
e = electron charge
R = gas constant
T = absolute temperature

In the derivation of Eq. (14.3), no interference due to motion among substrate molecules is taken into account, i.e., the electrostatic force imposed on each molecule is equated to $en(-\text{grad } \phi)$, and further, the force is assumed to be counterbalanced by the resistance, $3\pi\mu du$, of the substrate molecule as governed by Stokes' law, provided:

μ = solution viscosity
d = diameter of substrate molecule
u = linear velocity of substrate molecule
$3\pi\mu d = kT/D$ (Nernst-Einstein equation)
$\quad\quad = RT/DN$
$k = R/N$ = Boltzmann constant
$J_e = S_0 u$

The reaction rate, v, at the enzyme surface, on the other hand, is assumed to follow a Michaelis-Menten equation, i.e.,

$$v = v_s a = -\frac{dS_1}{dt} = \frac{V_{max} S_1}{K_m + S_1} \tag{14.4}$$

where

v_s = reaction rate per unit surface area of immobilized enzyme
a = surface area per unit volume of immobilized enzyme
V_{max} = maximum reaction rate
K_m = Michaelis constant

t = time   [cf. Eq. (4.3), Chapter 4]

In a steady state,

$$v_s = J_d + J_e \tag{14.5}$$

From Eqs. (14.2) to (14.5),

$$\frac{V_{max}' S_1}{K_m + S_1} = D\frac{S_0 - S_1}{\Delta l} - \frac{DF}{RT} n S_0 \text{ grad } \phi \tag{14.6}$$

provided:

$$V_{max}' = V_{max}/a$$

Cancelling out the term S_1 from Eqs. (14.4) and (14.6), and rearranging,

$$v_s^2 - v_s \frac{D}{\Delta l}\left\{K_m + S_0 + \frac{(\Delta l)V_{max}'}{D} - \frac{n(\Delta l)F}{RT} S_0 \text{ grad } \phi\right\}$$
$$+ \frac{V_{max}'}{\Delta l} DS_0 \left\{1 - \frac{n(\Delta l)}{RT} F \text{ grad } \phi\right\} = 0 \tag{14.7}$$

Setting,

$$\left.\begin{array}{l}\alpha = \frac{D}{\Delta l}\left\{K_m + S_0 + \frac{(\Delta l)V_{max}'}{D} - \frac{n(\Delta l)}{RT} FS_0 \text{ grad } \phi\right\} \\[2mm] \beta = \frac{V_{max}'}{\Delta l} DS_0 \left\{1 - \frac{n(\Delta l)}{RT} F \text{ grad } \phi\right\}\end{array}\right\} \tag{14.8}$$

Equation (14.7) is reduced to:

$$v_s{}^2 - \alpha v_s + \beta = 0 \tag{14.9}$$

Though the approximate solution of Eq. (14.9) is $\alpha - \beta/\alpha$ or β/α, the former cannot be taken, since the value of v_s becomes infinite when Δl approaches zero. Then

$$v_s = \frac{\beta}{\alpha} = \frac{\dfrac{V_{\max}'}{\Delta l} DS_0 \left\{1 - \dfrac{n(\Delta l)}{RT} F \text{ grad } \phi\right\}}{\dfrac{D}{\Delta l}\left\{K_m + S_0 + \dfrac{(\Delta l)V_{\max}'}{D} - \dfrac{n(\Delta l)}{RT} FS_0 \text{ grad } \phi\right\}}$$

$$= \frac{V_{\max}'S_0}{S_0 + \left\{K_m + \dfrac{\Delta l}{D} V_{\max}'\right\}\left\{\dfrac{RT}{RT - n(\Delta l)F \text{ grad } \phi}\right\}}$$

$$= \frac{V_{\max}'S_0}{S_0 + K_m'} \tag{14.10}$$

where

$$K_m' = \left\{K_m + \frac{(\Delta l)V_{\max}'}{D}\right\}\left\{\frac{RT}{RT - n(\Delta l)F \text{ grad } \phi}\right\} \tag{14.11}[10]$$

It is clear from Eq. (14.11) that the kinetic constant, K_m', of the immobilized enzyme system is affected not only by the hydrodynamics of the environment, but also by the electrochemical conditions of both the carrier and the substrate. Without regard to electochemical conditions, K_m' approaches K_m when Δl can be approximated to zero. Depending on the value of the term in the second bracket on the right-hand side of Eq. (14.11), i.e., either larger or smaller than unity, K_m' becomes larger or smaller than K_m when $\Delta l \neq 0$.

14.2.1.1. K_m and K_m'

Figure 14.5 demonstrates how the electrochemical environment affects the value of K_m'.[4] The water-insoluble trypsin here is a result of the covalent binding of trypsin to a copolymer of maleic acid and ethylene (IMET), i.e., the carrier is of a polyanionic nature. The initial rate of hydrolysis of benzoyl-L-arginine amide (BAA) was measured at 25°C by changing the substrate concentration under various conditions of ionic strength extending from $\Gamma/2 = 0.04$ (pH = 7.5) to $\Gamma/2 = 0.5$. Ionic strength was adjusted by addition of sodium chloride.

It may be seen from Fig. 14.5 that the value of K_m' at low ionic strength was considerably smaller than K_m for native trypsin. When ionic strength increased ($\Gamma/2 = 0.5$) the K_m' value became nearly indentical with that of trypsin. This particular pattern of K_m' value depending on ionic strength is in agreement with Eq. (14.11) qualitatively.

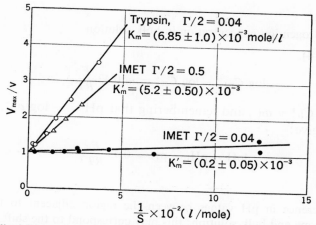

Fig. 14.5. Normalized Lineweaver-Burk plots for trypsin and IMET (water-insoluble derivative of trypsin) acting on BAA (benzoyl-L-arginine amide).[4] Substrate concentration, S, appearing as a reciprocal on the abscissa, is obviously identical with S_0 [Eq. (14.10)] in its implication.

14.2.1.2.　pH

Designating C_{H^+} as hydrogen ion concentration in a substrate solution, the concentration profile in a region near a polyionic carrier of immobilized enzymes will be discussed. Here again, no mention can be made definitely of hydrogen ion concentration within the immobilized enzyme structure. Only the concentrations near the surface of the immobilized enzyme and in the bulk solution will be considered.

If the carrier is negatively charged for instance, hydrogen ion moves to the carrier due to electrostatic force, tending to yield a higher concentration of hydrogen ion near (and presumably, within the structure of) the carrier; this polarization in concentration gives rise to another flow due to diffusion, counter to the previous flow due to electrostatic force. The following equations, assuming the steady state, may be acceptable within limits.

$$J_d + J_e = 0 \tag{14.12}$$

$$- D \operatorname{grad} C_{H1^+} - \frac{DF}{RT} C_{H1^+} \operatorname{grad} \psi = 0 \tag{14.13}$$

provided:

C_{H1^+} = concentration of hydrogen ion at the polyionic carrier surface
D = diffusivity of hydrogen ion in solution

Solving Eq. (14.13) under the condition of

$$\psi = 0 \quad \text{at} \quad C_{H1^+} = C_{H^+}$$

$$C_{H1^+} = C_{H^+} \exp\left(-\frac{F}{RT}\psi\right) \tag{14.14}$$

where

C_{H^+} = hydrogen ion concentration in bulk solution

From Eq. (14.14),

$$\log C_{Hi^+} - \log C_{H^+} = -0.43 \frac{F}{RT} \phi \qquad (14.15)$$

Assuming that $C_{H^+} \simeq a_{H^+}$, and remembering that pH $= -\log a_{H^+}$, Eq. (14.15) is rearranged to give:

$$\Delta pH = pH_i - pH = 0.43 \frac{F}{RT} \phi \qquad (14.16)[5]$$

provided:

ΔpH = difference in pH values between the region adjacent to the immobilized enzyme and bulk solution; this may correspond to the shift in pH when the enzyme is immobilized (see later)

pH_i = pH value next to immobilized enzyme

a_{H^+} = activity of hydrogen ion

Equation (14.16) points out that ΔpH becomes positive and/or negative depending on the sign of ϕ. If the polyionic carrier is negatively charged, ΔpH becomes nega-

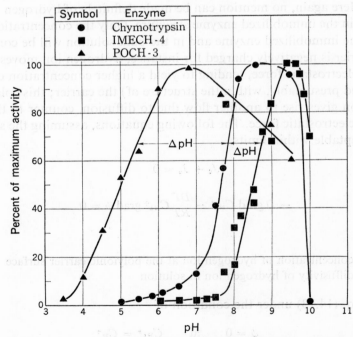

Fig. 14.6. pH-activity curves at low ionic strength ($\Gamma/2=0.001$) for chymotrypsin, a polyanionic, ethylene-maleic acid copolymer derivative of chymotrypsin (IMECH) and a polycationic, poly-L-ornithyl derivative of chymotrypsin (POCH), using acetyl-L-tyrosine ester as substrate.[5]

tive, because ϕ is negative. Concentration polarization of hydrogen ions in the steady state in this case suggests a higher concentration near the carrier structure due to the attractive force. Assuming that the value of pH_i can be represented by the pH value of native enzyme whose activity being identical with the immobilized enzyme activity under the conditions specified,[4] a negative value of ΔpH implies a shift of pH towards alkalinity; i.e., higher hydrogen ion concentration near the carrier in the above example, underlying the smaller value of pH_i in Eq. (14.16), is accompanied necessarily by a larger pH value (smaller hydrogen ion concentration in the environment outside the carrier) in the equation. The pH-activity shift of immobilized enzymes, depending on the nature of polyionic carrier, is exemplified in Fig. 14.6. The experimental data on ΔpH are supported by Eq. (14.16) and the previous discussion.

14.2.1.3. *Temperature, etc.*

In some cases, immobilization enhances the resistance of the enzyme to heat, proteases and poisons (inhibitors or organic chemicals); this characteristic, originating from a particular arrangement of enzyme and carrier in the immobilized system, is advantageous from the viewpoint of industrial application.

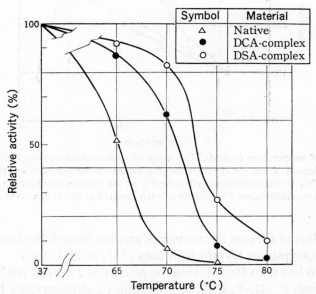

Symbol	Material
△	Native
●	DCA-complex
○	DSA-complex

Fig. 14.7. Effect of temperature on aminoacylase activity with native and immobilized enzymes–DCA complex and DSA complex, where DCA complex=DEAE-cellulose-aminoacylase complex and DSA complex=DEAE-Sephadex-aminoacylase complex. A mixture of each substance with a veronal buffer solution (pH=7.0) was incubated for 15 min at the specified temperature, followed by rapid cooling to assess the residual activity. A solution of acetyl-DL-methionine (pH=7.0) containing Co^{2+} ($\sim 1.5 \times 10^{-3}$ M) was added to the sample, incubating them at 37°C for 30 min with shaking. The amount (μmole) of L-methionine liberated during shaking gave the activity of aminoacylase via due calibration. The activity of each material at 37°C without heat treatment was taken as a basic datum.[21]

Figure 14.7 demonstrates an example where immobilized enzymes (DCA and DSA complexes) are more resistant to heat than the native form of aminoacylase. In spite of the fact that the activity of native enzyme became almost exhausted when exposed to heat at 70°C for 15 min, the residual activities of immobilized enzymes treated under the same conditions are still appreciable. The effect of temperature on the reaction rate to release L-methionine from acetyl-DL-methionine is shown in Fig. 14.8. Regarding the region of activity enhanced by heat, the activation energies for the DCA complex, the DSA complex and the native enzyme were estimated from an Arrhenius plot (not shown) as 11,100 cal/mole, 7,000 cal/mole, and 6,700 cal/mole, respectively.[21]

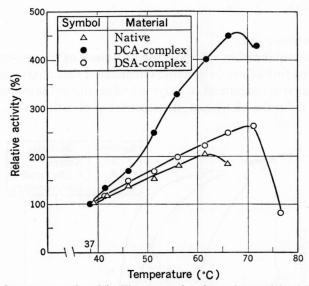

Fig. 14.8. Effect of temperature (cont'd). The assay of aminoacylase activity followed the standard procedure (see the legend of Fig. 14.7) except that the incubation temperature was changed from 37°C (standard) up to 76°C. So the temperature effect here is on the reaction rate between each material and the substrate, acetyl-DL-methionine. The activity of each material at 37 °C was taken as a basic datum.[21]

To cite the effect of urea on the activity of another immobilized enzyme, an incubation of chymotrypsin or trypsin with urea (5M) at 25°C for $t = 5$ hr decreased both activities to less than 10%, taking the activity at $t = 0$ as 100%, whereas immobilized enzymes [CMC-T, CMC-CT (trypsin or chymotrypsin bound to CM-cellulose)] retained their activities almost unchanged, except for some cases of deterioration to 80%.[17]

However, addition of Fe^{2+}, Hg^{2+} or p- chloromercuribenzoate to acetyl - DL - methionine solution ($M/15$) in place of Co^{2+} (as a control) to the extent of $10^{-3}M$, followed by incubation with DSA complex (DEAE-Sephadex-aminoacylase complex) at 37°C for 30 min (pH = 7.0) reduced the aminoacylase activity appreciably (in the case of Hg^{2+}, 7% of the control), while that of native enzyme was least susceptible to the inhibitory effects of these ions (even in the case of Hg^{2+}, the remaining activity of aminoacylase was 77% of the control).[21]

14.2.2. Performance of immobilized enzyme column

14.2.2.1. *Space velocity*

Suppose that a substrate solution passes through an enzyme column (bed length = L m) at a rate of F m³/hr. Assuming a linear correlation between S_0 in the bulk solution and S_i at the enzyme surface at the specified conditions,

$$-\frac{dS_0}{dt} = \gamma\left(-\frac{dS_i}{dt}\right) \tag{14.17}$$

provided:

γ = proportionality constant

Rearranging Eq. (14.10) with reference to Eq. (14.4),

$$v_s a = v = -\frac{dS_i}{dt} = \frac{V_{max} S_0}{K_m' + S_0} \tag{14.18}$$

From Eqs. (14.18) and (14.17),

$$-\frac{dS_0}{dt} = \frac{V_{max}\gamma S_0}{K_m' + S_0} \tag{14.19}$$

Integrating Eq. (14.19),[1]

$$(S_0)_{in} - (S_0)_{out} + K_m' \ln\frac{(S_0)_{in}}{(S_0)_{out}} = \frac{\varepsilon\gamma V_{max}}{V_s} \tag{14.20}$$

provided:

$S_0 = (S_0)_{in}$ at $t = 0$,

$S_0 = (S_0)_{out}$ at $t = \dfrac{LA\varepsilon}{F} = \dfrac{\varepsilon}{V_s}$

where

A = sectional area of packed column
V_s = space velocity
ε = void fraction of packed bed

Introducing a degree of hydrolysis, ξ, where

$$\xi = \frac{(S_0)_{in} - (S_0)_{out}}{(S_0)_{in}} \tag{14.21}$$

Eq. (14.20) is rearranged as follows:

$$\xi(S_0)_{\text{in}} = 2.303 K_{\text{m}}' \log(1 - \xi) + \frac{\varepsilon \gamma V_{\text{max}}}{V_{\text{s}}} \qquad (14.22)$$

$$\xi(S_0)_{\text{in}} = 2.303 K_{\text{m}}' \log(1 - \xi) + \frac{V_{\text{max}}''}{V_{\text{s}}} \qquad (14.22)'$$

provided:

$$V_{\text{max}}'' = \varepsilon \gamma V_{\text{max}}$$

Experimental data on the hydrolysis of acetyl-DL-methionine using a packed column of immobilized aminoacylase at 50°C are plotted referring to Eq. (14.22)' in Fig. 14.9; the ordinate is ξS_0 (this is identical with $\xi(S_0)_{\text{in}}$ in the left-hand side of Eq. (14.22)', but the subscript is omitted for simplicity). The abscissa is $\log(1-\xi)$, while the parameters are the space velocities, V_{s}, of the substrate solution through the column. The fact that the data points for each space velocity are fairly well represented by straight lines is considered to justify Eq. (14.22)'. So far as the linearity between ξS_0 and $\log(1-\xi)$ is warranted, the slope of the line is equal to 2.303 K_{m}'.

It is noted from Fig. 14.9 that the slope of each straight line tends to decrease (so, K_{m}' decreases) with increase in space velocity, V_{s}. The increase of V_{s} here im-

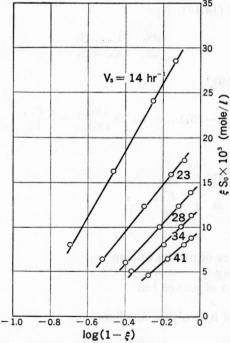

Fig. 14.9. Relationship between ξS_0 and $\log(1-\xi)$ for the hydrolysis of acetyl-DL-methionine by aminoacylase column at 50°C. Parameters are space velocities, V_{s} hr^{-1}.[22]

plies an increase in linear velocity, u, of the substrate solution over the enzyme surface. The increase of u is accompanied by a decrease of boundary layer thickness, Δl. The decrease of $K_m{}'$ corresponding to the increase of V_s (or u) in this experiment is deemed to support Eq. (14.11), if various terms appearing in the equation other than Δl are constant, and if the effect of Δl on the value of second bracket on the right-hand side of Eq. (14.11) can be disregarded.

14.2.2.2. *Pressure drop of liquid flow*

Pressure drop of liquid flow (substrate solution) through the packed bed of an enzyme column is given by the following Kozeny-Carman equation.

$$\Delta P g_c = K_c (S_0 \rho_s)^2 \frac{(1 - \varepsilon)^2}{\varepsilon^3} u \mu L$$

$$= K_c{}' u \mu L \qquad\qquad (14.23)$$

where

- ΔP = pressure drop
- g_c = conversion factor
- K_c = Kozeny-Carman constant ($\doteqdot 5.0$)
- ρ_s = true density of immobilized enzyme
- S_0 = surface area per unit mass of immobilized enzyme
- ε = void fraction of packed bed
- u = linear velocity of liquid
- μ = liquid viscosity
- L = length of packed bed
- $K_c{}' = K_c (S_0 \rho_s)^2 \dfrac{(1 - \varepsilon)^2}{\varepsilon^3}$

If the value of $K_c{}'$ is constant for a given column, the linearity betwen ΔP and L is apparent from Eq. (14.23) for constant value of u at specified temperatures. Using a packed bed of DEAE-Sephadex-aminoacylase particles (40 - 120 μ in diameter). Tosa *et al.*[22] presented the following data:

$$\Delta P = 0.265 \ \text{Kg/cm}^2 \ \text{at} \ u = 0.089 \ \text{cm/sec}, \ L = 85 \ \text{cm}$$

provided:

substrate (0.2M acetyle-DL-methionine solution) temperature = 50°C

$\Delta P = 0.161 \ \text{Kg/cm}^2$ at $u = 0.072$ cm/sec, $L = 30$ cm, when substrate temperature = 20°C,

14.2.3. Stability

In the case of enzymes adsorbed on an insoluble matrix, equilibrium is usually established between the adsorbed enzyme and free enzyme. When enzyme in this form is used in repeated batch processes where the solution is drawn off in each run, there is a constant fractional loss of enzyme per batch.

Designating M, V, and D as mass of inert carrier, volume of solution and adsorption equilibrium constant, respectively, the fractional loss of enzyme per batch can be expressed by the following equation.

$$\frac{\Delta E}{E} = \left(\frac{1}{D\dfrac{M}{V} + 1}\right) w \qquad (14.24)$$

provided:

ΔE = loss of enzyme per batch
E = total enzyme in batch
w = fraction of solution removed per batch

The equilibrium constant is defined by:

$$D \equiv \frac{y}{x} \qquad (14.25)$$

where

y = enzyme concentration in carrier matrix
x = enzyme concentration in solution

Suppose that a solution ($V = 1$ m^3) of benzyl penicillin is processed with an immobilized penicillin amidase ($M = 20$ kg for inert carrier) in batch. After the reaction the solution and the enzyme are decanted; the fraction of solution removed is taken as $w = 0.80$. Assuming that the equilibrium constant, D, be of the order of 10^3 (g enzyme/g carrier)/(g enzyme/ml solution) in Eq. (14.24),

$$\frac{\Delta E}{E} = \left(\frac{1}{10^3 \cdot \dfrac{2 \times 10^4}{10^6} + 1}\right)(0.8)$$

$$= 0.038$$

The fractional loss of enzyme per batch in this example is 3.8%.

In more general cases other than the adsorbed enzymes mentioned above, the stability in terms of maintaining the activity per batch (or in continuous operation) awaits study from case to case. Figure 14.10 presents an example of stability testing for a DEAE-Sephadex-aminoacylase column; clearly, the stability in this case depends on the carrier form, and incidentally, A-25, the OH-form was favorable.

In connection with the stability, regeneration of immobilized enzyme must be referred to briefly. For various reasons, deterioration of enzyme activity cannot be avoided. The possibility of regeneration is important from the viewpoint of industrial use on a continuous basis; an aminoacylase column deteriorated due to continuous operation could be completely regenerated by temporarily suspending op-

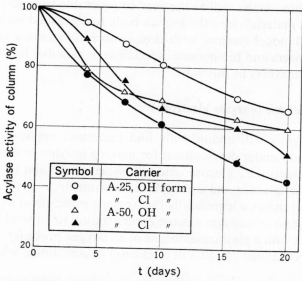

Fig. 14.10. Stability of DEAE-Sephadex-aminoacylase column. A solution of 0.2M Ac-DL-Met (acetyl-DL-methionine) (containing $5 \times 10^{-4} M$ Co^{2+}, pH=7.0) was treated with four kinds of DEAE-Sephadex-aminoacylase columns at 50°C for 20 days af flow rates of V_s=1–5 hr^{-1}.[19]

erations to wash the enzyme particles *in situ* with distilled water and then, recharging a solution of aminoacylase.[20] Generally, the regeneration of immobilized enzymes prepared by adsorption (and/or ionic binding by using DEAE-Sephadex, DEAE-cellulose, CM-cellulose, polyamino polystyrene, Amberlite XE-97, Dowex-50, etc.) is possible, whereas other preparations discussed in Sections 14.1.2. to 14.1.4. are less likely to be capable of regeneration.

14.3. APPLICATIONS OF IMMOBILIZED SYSTEMS

14.3.1. Use in medical treatment

The development of immobilized enzymes and of enzymes encased in semi-permeable microspheres offers promise in the treatment of specific diseases. Diseases exist in which the inherited abnormality of a single enzyme has profound metabolic effects, usually produced by the accumulation of a substrate for the involved enzyme (e.g., galactose phosphate in galactosemia). In other disorders, less well understood, materials accumulate and indirectly produce disease states (e.g., lipid in arteriosclerosis).

It is now feasible to think in terms of removal of these accumulated materials by means of specific enzymes of exogenous origin. Major impediments to success in this approach include the paucity of information about molecular mechanisms in most important diseases, as well as the very considerable technical problem of introducing foreign materials into the human body for extended periods. The development of animal model systems, with close cooperation among medical doctors, biochemical engineers and bio-material specialists, is essential for success in utilizing immobilized enzymes in this way.

14.3.2. Analytical use (enzyme electrode)

Immobilized enzymes will undoubtedly find important applications in analysis. Systems are already under development for making enzyme probes, but here only the following systems, i.e., glucose oxidase-oxygen probe and urease-ammonium ion probe to detect glucose and urea will be mentioned.

Figure 14.11 (a) shows schematically the working principle of a glucose oxidase-oxygen probe system to measure glucose concentration to an level of 10^{-3} mole/l.[23] Closely in contact with a plastic membrane of an oxygen probe (*cf*. Fig. 12.8, Chapter 12), a gel layer (about $25-50$ μ in thickness) of acrylamide which immobilizes

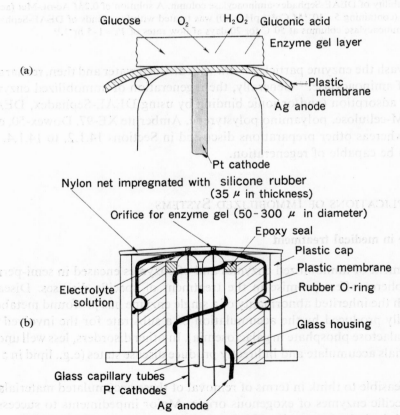

Fig. 14.11. Enzyme electrode; (a) principle and (b) example of dual construction.[23]

glucose oxidase is provided to pass glucose and oxygen through this gel layer. During their passage, the reaction of glucose and oxygen is catalyzed by glucose oxidase to produce gluconic acid and hydrogen peroxide.

$$\text{glucose} + \text{oxygen} \xrightarrow[\text{glucose oxidase}]{} \text{gluconic acid} + \text{hydrogen peroxide} \qquad (14.26)$$

When oxygen is present in excess and the concentration of glucose is considerably smaller than the K_m' value for the immobilized glucose oxidase, there is a linear relationship between glucose concentration and decrease in oxygen tension. The decrease in oxygen tension is measured by the oxygen probe.

A diagram of the dual construction of this electrode is shown in Fig. 14.11 (b). A nylon net, 35 μ in thickness, impregnated with silicone rubber, is used as a support to anchor the immobilized enzyme gel over the plastic membrane of the oxygen probe. In this example, the authors claim that the two cathodes can be used jointly as a single cathode or separately as a differential electrode.[23]

Guilbault et al. published details of their enzyme electrode to detect urea.[7] A thin film (250 – 350 μ in thickness) of urease-acrylamide gel is coated over a Beckman cationic glass electrode. When the electrode is placed in a solution containing urea, the substrate diffuses into the gel layer of immobilized enzyme. The enzyme catalyzes the decomposition of urea into ammonium ion by the following reaction.

$$\underset{\underset{O}{\|}}{NH_2-C-NH_2} + H_2O \xrightarrow{\text{urease}} 2NH_3 + CO_2 \qquad (14.27)$$

The ammonium ion produced is sensed by the monovalent cationic electrode in a manner similar to pH determination with a glass electrode. The potential of this electrode is measured after allowing sufficient time for the diffusion process to reach the steady state. For details, readers are suggested to refer to the original paper.[7]

The application of immobilized enzymes to the detection of pesticides and other trace toxic materials is a distinct possibility. Indeed, if the specificity of an enzyme can be coupled to some fluorometric systems, then one can achieve very sensitive analyses, possibly as low as 10^{-12} molar concentration.

14.3.3. Industrial use

Several immobilized enzyme processes are showing industrial promise. One such system involves penicillin amidase, which severs the side chain of benzyl penicillin, leaving a nucleus, 6-amino penicillanic acid (6-APA). 6-APA is the starting material for synthetic penicillins. In one system, purified amidase is reacted with a dichloro-s-triazine derivative of cellulose and used repeatedly in catalyzing the hydrolysis of benzyl penicillin.[15]

The use of immobilized aminoacylase to separate DL-racemic mixtures of acetylated amino acids is another good example of industrial use. The process is accomplished with a packed bed of aminoacylase adsorbed onto DEAE-Sephadex. Figure 14.12 is a flow diagram of the continuous production of L-amino acids by the use of an aminoacylase reactor. Tanabe Seiyaku Co. in Osaka, is reported to produce several deca-tons/month of amino acids using this and similar techniques.

Some interesting and possible future applications of immobilized enzymes include:

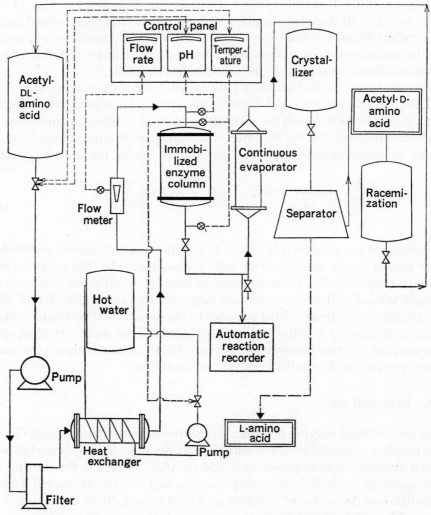

Fig. 14.12. Flow diagram (continuous system of L-amino acid production). (*Courtesy* Tanabe Seiyaku Co., Ltd., Osaka, Japan.)

Enzyme	Use
Pectinase	Clarification of fruit juices and other beverages by controlled breakdown of polygalacturonic acid
Lactase	Removal of lactose from milk
Cellulase	Conversion of cellulosic solids into water-soluble sugars
Glucose isomerase	Isomerization of glucose to fructose or invert sugar
Invertase	Conversion of sucrose to glucose and fructose
Amylase	Modification of various starches

14.3.4. Miscellaneous

Immobilized enzymes have not penetrated the household market yet. However, immobilization might prove useful for the stabilization of detergent enzymes. Likewise, some lipases might be stabilized for more effective fat and grease removal. Enzymes for dehairing creams, meat tenderization, etc. might have their shelf life enhanced through immobilization. Certainly many possibilities exist for the household use of enzymes.

Immobilized enzymes systems have at least five different configurations in which they may be used. These are:

a. Packed bed reactor,
b. Thin-membrane differential reactor,
c. Tubular wall reactor,
d. Continuous stirred tank reactor, and
e. Fluidized bed reactor.

Each system has its particular advantages. For instance, thin sheets of filter paper to which various enzymes have been bound are useful in constructing sequential reacting systems. Obviously, packed beds, membranes, and tubular wall reactors may all be used in analysis systems. The choice will depend upon the particular configuration. There is certainly a bright future for biochemical engineers to work in this area.

14.4. Summary

1. Immobilization of enzymes may extend their use in the future. Immobilization overcomes the problem of enzyme removal from the end-product. It permits repeated use of the enzyme. Immobilization may impart increased stability to the enzyme; it may allow operation at previously impractical environmental conditions by shifts in pH and temperature optima.

2. Immolized enzymes may be produced in several forms. These include,

a. Adsorption on inert carriers,
b. Inclusion in gel lattices,
    encapsulation or containment in a semi-permeable membrane systems,
c. Intermolecular cross-linking of the enzyme to itself or within a copolymer, and
d. Covalent binding of nonfunctional groups of the enzyme to inert carriers.

3. Immobilized enzymes will find application in at least four major areas. These include,

a. Medical treatment,
b. Analytical methods,
c. Industrial catalysts, and
d. Household application for cleaning and food modification.

4. Immobilized enzymes can be used in such equipment as

a. Packed bed reactor,
b. Thin-membrane differential reactor,
c. Tubular wall reactor,
d. Continuous stirred tank reactor, and
e. Fluidized bed reactor.

NOMENCLATURE

A = sectional area of packed column, m^2
a = surface area per unit volume of immobilized enzyme, m^2/m^3
a_{H^+} = activity of hydrogen ion
C_{H^+} = hydrogen ion concentration in bulk liquid
C_{Hi^+} = concentration of hydrogen ion adjacent to immobilized enzyme structure
D = molecular diffusivity of substrate or hydrogen ion in liquid, cm^2/sec; adsorption equilibrium constant, (g enzyme/g carrier)/(g enzyme/ml solution)
d = diameter of substrate molecule
E = enzyme activity
ΔE = loss of enzyme activity per batch
e = electron charge
F = Faraday's constant, 96,500 coulombs/g equivalent; flow rate, m^3/hr
g_c = conversion factor, 9.81 kg m/Kg sec^2
J_d = flux due to diffusion for substrate or hydrogen ion, mole/cm^2 sec
J_e = flux (substrate or hydrogen ion) due to electrostatic force, mole/cm^2 sec
K_c = Kozeny-Carman constant ($\doteq 5.0$)
$K_c{'}$ = $K_c (S_0 \rho_s)^2 (1-\varepsilon)^2/\varepsilon^3$
K_m = Michaelis constant, g/cm^3, mole/l
$K_m{'}$ = apparent Michaelis constant, g/cm^3, mole/l
k = Boltzmann constant, 1.38×10^{-16} erg/°K

L = bed length of packed column, m, cm

Δl = effective boundary-layer thickness, cm

M = mass of inert carrier, g

N = Avogadro's number, 6.03×10^{23}/g mole

n = electrochemical valence of substrate

P = product

ΔP = pressure drop of liquid flow, G/cm², Kg/cm²

R = gas constant, 8.31×10^7 erg/°K g mole = 1.98 cal/g mole °K

S = Substrate or substrate concentration, g/cm³, mole/cm³

S_0 = substrate concentration in bulk solution, mole/cm³; surface area per unit mass of enzyme particles, m²/kg

S_1 = substrate concentration at immobilized enzyme surface, mole/cm³

$(S_0)_{in}$ = substrate concentration in bulk solution at column inlet, mole/cm³

$(S_0)_{out}$ = substrate concentration in bulk solution at column outlet, mole/cm³

T = absolute temperature, °K

t = time, hr, day

u = linear velocity of substrate molecule or linear velocity of liquid, cm/sec, m/hr

V = solution volume, ml

V_{max} = maximum reaction rate

$V_{max}' = V_{max}/a$

$V_{max}'' = \varepsilon\gamma V_{max}$

V_s = space velocity, hr⁻¹

v = reaction rate

v_s = reaction rate per unit area of enzyme surface

w = fraction of solution removed

x = enzyme concentration in solution; coordinate

y = enzyme concentration in carrier matrix

grad = gradient

Greek letters

$$\alpha = \frac{D}{\Delta l}\left\{K_m + S_0 + \frac{(\Delta l)V_{max}'}{D} - \frac{n(\Delta l)}{RT}FS_0 \text{ grad } \phi\right\}$$

$$\beta = \frac{V_{max}'}{\Delta l}DS_0\left\{1 - \frac{n(\Delta l)}{RT}F \text{ grad } \phi\right\}$$

$\Gamma/2$ = ionic strength, $\frac{1}{2}\Sigma\{$(ionic molar concentration) (ionic valence)²$\}$

γ = proportionality constant

ε = void fraction of packed bed

μ = liquid viscosity, g/cm sec, kg/m hr

$\xi = \dfrac{(S_0)_{in} - (S_0)_{out}}{(S_0)_{in}}$, degree of hydrolysis

ρ_s = true density of immobilized enzyme, kg/m³

ϕ = electrostatic potential

REFERENCES

1. Bar-Eli, A., and Katchalski, E. (1963). "Preparation and properties of water-insoluble derivatives of trypsin. " *J. Biol. Chem.* **238**, 1690.

2. Bernfeld, P., and Wan, J. (1963). "Antigens and enzymes made insoluble by entrapping them into lattices of synthetic polymers." *Science* **142**, 678.

3. Ghose, T. K., and Kostick, J. A. (1969). "A model for continuous enzymatic saccharification of cellulose with simultaneous removal of glucose syrup." Paper No. 4, 158th National ACS Meeting, New York.

4. Goldstein, L., Levin, Y., and Katchalski, E. (1964). "A water-insoluble polyanionic derivative of trypsin. II. Effect of the polyelectrolyte carrier on the kinetic behavior of the bound trypsin." *Biochem.* **3**, 1913.

5. Goldstein, L., and Katchalski, E. (1968). "Use of water-insoluble enzyme derivatives in biochemical analysis and separation." *Z. Anal. Chem.* **243**, 375.

6. Goldstein, L. (1970). "Water-insoluble derivatives of proteolytic enzymes." *Methods in Enzymology* **19**, 935. (Eds.) Perlmann, G. E., and Lorand, L. Academic Press, New York.

7. Guilbault, G. G., and Montalvo, J. G. (1970). "An enzyme electrode for the substrate urea." *J. Am. Chem. Soc.* **92**:8 April 22, 2533.

8. Haynes, R., and Walsh, K. A. (1969). "Enzyme envelopes on colloidal particles." *Biochem. Biophys. Res. Comm.* **36**, 235.

9. Hicks, G. P., and Updike, S. J. (1966). "The preparation and characterization of lyophilized polyacrylamide enzyme gels for chemical analysis." *Anal. Chem.* **38**, 726.

10. Hornby, W. E., Lilly, M. D., and Crook, E. M. (1968). "Some changes in the reactivity of enzymes resulting from their chemical attachment to water-insoluble derivatives of cellulose." *Biochem. J.* **107**, 669.

11. Katchalski, E., Silman, I., and Goldman, R. (1971). "Effect of the microenvironment on the mode of action of immobilized enzymes." *Adv. in Enzymology* **34**, 445.

12. Kay, G., and Crook, E. M. (1967). "Coupling of enzymes to cellulose using chloro-*s*-triazines." *Nature* **216**, 514.

13. Melrose, G. J. H. (1971). "Insolubilized enzymes; biochemical application of synthetic polymers." *Rev. Pure and Appl. Chem.* **21**, 83.

14. Nelson, J. M., and Griffin, E. G. (1916). "Adsorption of invertase." *J. Am. Chem. Soc.* **38**, 1109.

15. Self, D. A., Kay, G., Lilly, M. D., and Dunnill, P. (1969). "The conversion of benzyl penicillin to 6-aminopenicillanic acid using an insoluble derivative of penicillin amidase." *Biotech. & Bioeng.* **11**, 337.

16. Silman, I. H., and Katchalski, E. (1966). "Water-insoluble derivatives of enzymes, antigens, and antibodies." *Ann. Rev. Biochem.* **35**, 873.

17. Takami, T., and Ando, T. (1968). "Studies on trypsin and chymotrypsin covalently bound to carboxymethylcellulose." *Biochem.* (*Japan*) **40**, 749.

18. Tosa, T., Mori, T., Fuse, N., and Chibata, I. (1966). "Studies on continuous enzyme reactions. I. Screening of carriers for preparation of water-insoluble aminoacylase." *Enzymologia* **31**, 214.

19. Tosa, T., Mori, T., Fuse, N., and Chibata, I. (1967). "Studies on continuous enzyme reactions. IV. Preparation of a DEAE-Sephadex-aminoacylase column and continuous optical resolution of acyl-DL-amino acids." *Biotech. & Bioeng.* **9**, 603.

20. Tosa, T., Mori, T., Fuse, N., and Chibata, I. (1969). "Studies on continuous enzyme reactions. Part V. Kinetics and industrial application of aminoacylase column for con-

tinuous optical resolution of acyl-DL-amino acids." *Agr. Biol. Chem. (Japan)* **33**, 1047.

21. Tosa, T., Mori, T., and Chibata, I. (1969). "Studies on continuous enzyme reactions. Part VI. Enzymatic properties of the DEAE-Sephadex-aminoacylase complex." *Agr. Biol. Chem. (Japan)* **33**, 1053.

22. Tosa, T., Mori, T., and Chibata, I. (1971). "Studies on continuous enzyme reactions. VIII. Kinetics and pressure drop of aminoacylase column." *J. Ferm. Tech. (Japan)* **49**, 522.

23. Updike, S. J., and Hicks, G. P. (1967). "The enzyme electrode." *Nature* **214**, 986.

AUTHOR INDEX

Numbers in italics indicate the page on which the reference numbers in parentheses here are listed without mentioning authors' names in the text.

Example: *240*(2).

Aiba, S., *141*(1,2), *142–4*(2), *145*(3,25), *146*(5,25), *147*(4, 6,7,26), *148*(4,6,7), *164*(1, 2,4), *166*(27), *167*(6,26), *170* (6), *179*(5), *180*(5), *184*(2), 186, *201*(1,2), *202*(7), 247 (1), *249*(3), *250*(3), *251*(1), *252*(1), *253*(16), *260*(2), 270 (5), 271, 272, 275, *280*(5), *283*(5), 284, *285*(3), 286 (1–4), *287*(2,6), *289*(4,6), *290*(2,4), *291*(7), 294, 296 (6,9), *320*(1), *321*(1), *328*(2, 4), *329*(4), *330*(2–4), *334*(3), *350*(1), *351*(1), *353*(2).
Aiyar, A. S., *119*(1).
Aizawa, M., *205*(11), *207*(11).
Alagaratnam, R., *306*(2).
Almolf, J. W., *283*(29).
Amaha, M., *240*(5).
Ambler, C. M., *352*(3), *353*(3).
Anderson, D. M., *279*(31).
Anderson, L. G., *112*(11).
Anderson, W. L., *283*(25).
Ando, T., *404*(17).
Appleby, M., *95*(3), *96*(3).
Asai, T., *119*(23, 24).
Ashida, K., *294*(14), *295*(14, 15), *296*(15,16).

Bar-Eli, A., *405*(1).
Barman, T. E., *102*(12).
Barnhart, E. L., *184*(17), *185* (17), 186.
Bartholomew, W. H., 204, *205* (10), *206*(10), 207.
Bartlett, M. C., *129*(8), *153*(8).
Bartley, W., 47.
Bauchop, T., 70, 71.

Beesch, S. C., 347.
Bégin-Heick, N., *166*(7).
Bennett, D. J., *366*(7).
Beran, K., *117*(2), *128*(11), *359*(41), *360*(41).
Bernfeld, P., 394.
Bícík, V., *359*(41), *360*(41).
Bigelow, W. E., 243.
Bird, R. B., *247*(10).
Blasewitz, A. G., *273*(10), *275* (10), 292.
Blatt, W. F., *373*(4), *375*(4), *376*(4), 377.
Blum, J. J., *166*(7).
Borzani, W., *128*(9).
Bowers, R. H., *108*(5), *109*(5).
Brenner, H., *260*(7), *262*(7).
Brierley, M. R., 185, 186, *187* (8).
Bright, H. J., *95*(3), *96*(3).
Brown, C. E., *374*(5), *377*(5).
Brown, C. M., 50.
Brown, R., *361*(42).
Bungay, H. R., *119*(4).
Burg, C. R., *119*(22).
Burton, H., *239*(8).

Calam, C. T., 108, *109*(5).
Caldarola, E., *324*(5), *325*(5).
Calderbank, P. H., 171, *172* (11), *177*(10), 179, 181.
Canale, R. P., *119*(6), 120, 121.
Carswell, T. S., 274.
Carver, C. E., Jr., 186.
Chain, E. B., *324*(5), *325*(5).
Chen, C. Y., 275, *280*(11), *281* (11), *282*(11), 283, 284, *294* (11).
Chian, S. K. *116*(7).

Chibata, I., *393*(18), *403*(21), *404*(21), *405*(21), *409*(19, 20).
Chikaike, T., *205*(11), *207*(11).
Citek, F. J., *273*(13).
Cohn, E. J., *368*(6).
Conners, E. W., Jr., *279*(31).
Contois, D. E., *128*(10).
Cooper, C. M., 163, 182, *183* (13), 205, 206.
Coote, S. D. J., *306*(2).
Costich, E. W., *174*(32), *175* (32).
Craig, L. C., *385*(54).
Creaser, E. H., *366*(7).
Crook, E. M., *396*(12).
Curtiss, C. F., *247*(10).

Dale, H. F., *205*(18).
Dalton, H., 50.
Daunter, B., *306*(2).
Dawes, E. A., *93*(8), *95*(8), *96*(8), *99*(8), *100*(8), *103*(8).
Decker, H. M., 273.
de Filippi, R. P., *371*(8).
Deindoerfer, F. H., 110, 111, *139*(12), *209*(5), *251*(9), 254 (9).
DeVault, D., *378*(9), *381*(9).
Dolin, M. I., 59.
Doskocil, J., *153*(32).
Doubly, J. A., *274*(12).
Dravid, A., *373*(4), *375*(4), *376*(4), *377*(4).
Driver, N., *108*(5), *109*(5).
Drysdale, R. B., *366*(7).
Duerre, J. A., 361.
Dunn, C. G., *184*(38), 186.
Dunnill, P., 359, *411*(15).

SUBJECT INDEX

A

Absolute reaction rate theory,
  in biological system, 103.
Absorption number, 182.
Acetobacter suboxydans,
  diphasic biooxidation by, 114.
N-acetylglucosamine, 102, 363–4.
N-acetylmuramic acid, 102, 363–4.
Acetylmuramyl-L-alanine amidase, 365.
Aconitase, 359.
Acridine dyes,
  as mutagen, 42,43.
Actinomycetes, 23.
Activation energy, 104, 106–7.
  for DCA (DEAE-cellulose-aminoacylase)
    complex, 404.
  for enzymes, 244.
  of isothermal death of microbes, 242–3.
Adenine, 32.
Adenosine diphosphate (ADP), 123.
Adenosine triphosphate (ATP), 56, 60, 123.
Adiabatic compressor, 273.
Aeration and agitation,
  purpose of, 163.
Aeration number,
  as a measure of decrease of power re-
    quirements in aeration, 176.
  definition of, 175.
Aerobacter aerogenes, 41.
  cell composition in continuous culture,
    48.
  continuous culture of, 152.
Aerobacter cloacae,
  simple reaction of growth for, 111.
Airborne microbes,
  frequency of distribution of, 271.
  species and number of, 271–2.
Air filter,
  purpose of, 270.
  theory of, 278.
Air sterilization,
  with corona discharge, 274.

  with electromagnetic waves, 274.
  with germicidal sprays, 274.
  with heat from air compression, 273.
Alcaligenes faecalis,
  oscillation of, 121.
Alcohol dehydrogenase, 359.
Algae, 24.
  chemical composition of, 26.
Algorithms,
  for biochemical characteristics, 342.
  for physico-chemical characteristics, 341.
  for physiological characteristics, 341.
n-alkanes,
  metabolism of, 74.
Amino acids,
  production of, 77–83.
Aminoacylase,
  temperature effect on, 403–4.
  temperature effect on immobilized, 403–4.
α-amino adipate (AAA), 85.
Aminopurine,
  as mutagen, 43.
Amoeba, 25.
Amphibolic pathways, 86–87.
Anaerobic metabolism of pyruvate,
  major products from various organisms
    in, 67.
Antibiotics,
  production of, 85.
  species of importance, 10.
Apparent rate constant,
  in complex enzyme kinetics, 96.
Apparent viscosity,
  of non-Newtonian liquid, 177.
Arrhenius equation, 104.
Ascending velocity,
  of single bubbles, 172.
  of swarms of bubbles, 173.
Ascomycete, 26, 32.
Aseptic operation,
  of pipelines and valves, 312.
  in transferring a spore suspension, 312–13.